W0231904

CELLULAR INTERACTIONS IN CARDIAC PATHOPHYSIOLOGY

Developments in Molecular and Cellular Biochemistry

Series Editor: Naranjan S. Dhalla, Ph.D., FACC

1. V.A. Najjar (ed.): *Biological Effects of Glutamic Acid and Its Derivatives.* 1981 ISBN 90-6193-841-4
2. V.A. Najjar (ed.): *Immunologically Active Peptides.* 1981 ISBN 90-6193-842-2
3. V.A. Najjar (ed.): *Enzyme Induction and Modulation.* 1983 ISBN 0-89838-583-0
4. V.A. Najjar and L. Lorand (eds.): *Transglutaminase.* 1984 ISBN 0-89838-593-8
5. G.J. van der Vusse (ed.): *Lipid Metabolism in Normoxic and Ischemic Heart.* 1989 ISBN 0-7923-0479-9
6. J.F.C. Glatz and G.J. van der Vusse (eds.): *Cellular Fatty Acid-Binding Proteins.* 1990 ISBN 0-7923-0896-4
7. H.E. Morgan (ed.): *Molecular Mechanisms of Cellular Growth.* 1991 ISBN 0-7923-1183-3
8. G.J. van der Vusse and H. Stam (eds.): *Lipid Metabolism in the Healthy and Diseased Heart.* 1992 ISBN 0-7923-1850-1
9. Y. Yazaki and S. Mochizuki (eds.): *Cellular Function and Metabolism.* 1993 ISBN 0-7923-2158-8
10. J.F.C. Glatz and G.J. van der Vusse (eds.): *Cellular Fatty-Acid-Binding Proteins, II.* 1993 ISBN 0-7923-2395-5
11. R.L. Khandelwal and J.H. Wang (eds.): *Reversible Protein Phosphorylation in Cell Regulation.* 1993 ISBN 0-7923-2637-7
12. J. Moss and P. Zahradka (eds.): *ADP-Ribosylation: Metabolic Effects and Regulatory Functions.* 1994 ISBN 0-7923-2951-1
13. V.A. Saks and R. Ventura-Clapier (eds.): *Cellular Bioenergetics: Role of Coupled Creatine Kinases.* 1994 ISBN 0-7923-2952-X
14. J. Slezák and A. Ziegelhöffer (eds.): *Cellular Interactions in Cardiac Pathophysiology.* 1995 ISBN 0-7923-3573-2

SPRINGER-SCIENCE+BUSINESS MEDIA, B.V.

Cellular Interactions in Cardiac Pathophysiology

Edited by

JÁN SLEZÁK and ATTILA ZIEGELHÖFFER

Institute for Heart Research
Slovak Academy of Sciences
Bratislava
Slovak-Republic

Reprinted from *Molecular Biochemistry*, Volume 147 (1995)

Springer-Science+Business Media, B.V.

Library of Congress Cataloging-in-Publication Data

Cellular interactions in cardiac pathophysiologing / edited by Ján Slezák and Attila Ziegelhöffer.
p. cm. – (Developments in Molecular and Cellular Biochemistry; DMCB 14)
"Reprinted from Molecular and cellular biochemistry".
ISBN 978-1-4613-5828-2 ISBN 978-1-4615-2005-4 (eBook)
DOI 10.1007/978-1-4615-2005-4
1. Heart – Pathophysiology – Congresses. 2. Heart cells – Congresses. I. Slezák, Ján. II. Ziegelhöffer, Attila. III. Molecular and cellular biochemistry. IV. Series: Developments in molecular and cellular biochemistry; v. 14.
[DNLM: 1. Heart Diseases – physiopathology – collected works.
2. Cells – physiology – collected works. W1 DE998D v. 14 1995 / WG 210 C393 1995]
RC682.9.C453 1995
618.1'207 – dc20
DNLM/DLC
for Library of Congress 95-19480
CIP

ISBN 978-1-4613-5828-2

Printed on acid-free paper

Molecular and Cellular Biochemistry:

An International Journal for Chemical Biology in Health and Disease

CONTENTS VOLUME 147, Nos. 1 & 2, 1995

CELLULAR INTERACTIONS IN CARDIAC PATHOPHYSIOLOGY
Ján Slezák and Attila Ziegelhöffer, guest editors

Preface 1

Part I: Extracellular matrix and cardiocyte interaction 5

G.L. Kukielka, K.A. Youker, L.H. Michael, A.G. Kumar, C.M. Ballantyne, C.W. Smit and M.L. Entman: Role of early reperfusion in the induction of adhesion molecules and cytokines in previously ischemic myocarium 5–12

D. Weihrauch, M. Arras, R. Zimmermann and J. Schaper: Importance of monocytes/macrophages and fibroblasts for healing of micronecroses in porcine myocardium 13–19

G. Steinhoff and A. Haverich: Cell-cell and cell-matrix adhesion molecules in human heart and lung transplants 21–27

H.N. Sabbah, V.G. Sharov, M. Lesch and S. Goldstein: Progression of heart failure: A role for interstitial fibrosis 29–31

J. Ausma, J. Cleutjens, F. Thoné, W. Flameng, F. Ramaekers, M. Borgers: Chronic hibernating myocardium: Interstitial changes 35–42

V. Pelouch, M. Milerová, B. Oštádal, B. Hučin and M. Šamánek: Differences between atrial and ventricular protein profiling in children with congenital heart disease 43–49

S. Goldstein, V.G. Sharov, J.M. Cook and H.N. Sabbah: Ventricular remodeling: insights from pharmacologic interventions with angiotensin-converting enzyme inhibitors 51–55

O. Hudlická, M.D. Brown, H. Walter†, J.B. Weiss and A. Bate: Factors involved in capillary growth in the heart 57–68

M. Gerová, O. Pecháňová, V. Stoev, M. Kittová, I. Bernátová, M. Juráni and S. Doležel: Biomechanical signals in the coronary artery triggering the metabolic processes during cardiac overload 69–73

Part II: Myocytic adaptation and myocardial injury

N. Singh, A.K. Dhalla, C. Seneviratne and P.K. Singal: Oxidative stress and heart failure 77–81

B. Oštádal, V. Pelouch, I. Oštádalová and O. Nováková: Structural and biochemical remodelling in catecholamine-induced cardiomyopathy: Comparative and ontogenetic aspects 83–88

W. Linz, G. Wiemer, J. Schaper, R. Zimmermann, K. Nagasawa, P. Gohlke, T. Unger and B.A. Schölkens: Angiotensin-converting enzyme inhibitors, left ventricular hypertrophy and fibrosis 89–92

N. Vrbjar, A. Džurba and A. Ziegelhöffer: Influence of global ischemia on the sarcolemmal ATPases in the rat heart 99–103

H.-G. Zimmer, M. Irlbeck, C. Kolbeck-Rühmkorff: Response of the rat heart to catecholamines and thyroid hormones 105–114

E.G. Krause and L. Szekeres: On the mechanism and possible therapeutic application of delayed adaptation of the heart to stress situations 115–122

T. Ravingerová, N.J. Pyne and J.R. Parratt: Ischaemic precondition in the rat heart: The role of G-proteins and adrenergic stimulation 123–128

A. Ziegelhöffer, N. Vrbjar, J. Styk, A. Breier, A. Džurba, T. Ravingerová: Adaptation of the heart to ischemia by preconditioning: Effects on energy equilibrium, properties of sarcolemmal ATPases and release of cardioprotective proteins 129–137

Q. Shao, T. Matsubara, S.K. Bhatt and N.S. Dhalla: Inhibition of cardiac sarcolemma Na^+-K^+ ATPase by oxyradical generating systems 139–144

Part III: Signal transduction

M. Böhm: Alterations of β-adrenocepter-G-protein-regulated adenylyl cyclase in heart failure 147–160

W. Schulze, W.-P. Wolf, M.L.X. Fu, R. Morwinski, I.B. Buchwalow and L. Will-Shahab: Immunocytochemical studies of the G_i Protein mediated muscarinic receptor-adenylyl cyclase system 161–168

J. Slezák, L. Okruhlicová, N. Tribulová, W. Schulze and N.S. Dhalla: Renaissance of cytochemical localization of membrane ATPases in the myocardium 169–172

M. Yasutake and M. Avkiran: Effects of selective α_{1A}-adrenoceptor antagonists on reperfusion arrhythmias in isolated rat hearts 173–180

M. Manoach, D. Varon and M. Erez: The role of catecholamines on intercellular coupling, myocardial cell synchronization and self ventricular defibrillation 181–185
A. Breier, A. Ziegelhöffer, T. Stankovičová, P. Dočolomanský, P. Gemeiner, A. Vrbanová: Inhibition of (Na/K)-ATPase by electrophilic substances: Functional implications 187–192

Index to Volume 147 193–196

Molecular and Cellular Biochemistry **147:** 1, 1995.

Preface

In spite of the multiple efforts leading to a considerable success in the treatment of the heart and vessel diseases, they still remain the major cause of mortality all over the world. Clinicians as well as researchers dealing with cardiovascular pathology believe that one of the reasons for their unsuccessful results is the lack of understanding of molecular and cellular aspects of the above processes.

To gain further insight into the problems, the International Conference on Cellular Interactions in Cardiac Pathophysiology was organized by the Institute for Heart Research of the Slovak Academy of Sciences in Bratislava – Slovakia. The main purpose of the meeting was to get together renowned scientists from all continents working in this field and to engage them into deep and fruitful discussions stimulating the exchange of the current knowledge. The Meeting was supported by many pharmaceutical companies and financial institutions. However, its main sponsor was the Council of Cardiac Metabolism of the International Society and Federation of Cardiology. An excellent venue for the conference held under the auspices of the European Section of the International Society for Heart Research was offered by the Castle Smolenice. The main topics discussed during the Meeting represented by titles of papers in this issue, served as a source of knowledge and inspiration. The participation of outstanding scientists in the field, coupled with an excellent atmosphere, created conditions for a good fellowship which altogether greatly contributed to the success of the Meeting.

The International Scientific Board of the Conference felt that many presentations brought a considerable contribution to the basic knowledge about the molecular and cellular mechanisms of cardiovascular diseases. The present focussed issue of Molecular and Cellular Biochemistry contains those contributions selected from all sessions of the Meeting.

PROF. DR. JÁN SLEZÁK & PROF. DR. ATTILA ZIEGELHÖFFER
Institute for Heart Research
Slovak Academy of Sciences
Dúbravska cesta 9
842 33 Bratislava
Slovak Republic

PART I

EXTRACELLULAR MATRIX AND CARDIOCYTE INTERACTION

Molecular and Cellular Biochemistry **147**: 5–12, 1995.

Role of early reperfusion in the induction of adhesion molecules and cytokines in previously ischemic myocardium

Gilbert L. Kukielka, Keith A. Youker, Lloyd H. Michael, Ajith G. Kumar, Christie M. Ballantyne, C. Wayne Smith and Mark L. Entman

Section of Cardiovascular Sciences, Department of Medicine, The Methodist Hospital and The DeBakey Heart Center, Speros P. Martel Section of Leukocyte Biology, Department of Pediatrics and Texas Children's Hospital, Baylor College of Medicine, Houston TX 77030, USA

Abstract

Our studies *in vitro* demonstrate that neutrophil mediated injury of isolated cardiac myocytes requires the presence of ICAM-1 on the surface of the myocyte and CD11b/CD18 activation on the neutrophil. In post-ischemic cardiac lymph, there is rapid appearance of C5a activity during the first hours of reperfusion. Interleukin-6 activity is present throughout the first 72 h of reperfusion and is sufficient to induce ICAM-1 on the surface of the cardiac myocyte. *In situ* hybridization studies suggest that ICAM-1 mRNA is found in viable myocardial cells on the edge of the myocardial infarction within 1 h of reperfusion. ICAM-1 protein expression on cardiac myocytes is seen after 6 h of reperfusion, and increases thereafter. Non-ischemic tissue demonstrates no early induction of ICAM-1 mRNA or ICAM-1 protein on myocardial cells. In our most recent experiments, we have determined that reperfusion is an absolute requirement for the early induction of myocardial ICAM-1 mRNA in previously ischemic myocardial cells. To further assess this, we have cloned and sequenced a canine interleukin-6 (IL-6) cDNA. The data suggest that early induction of IL-6 mRNA is also reperfusion dependent as it could be demonstrated in the same ischemic and reperfused segments in which ICAM-1 mRNA was found. Peak expression of IL-6 mRNA occurred much earlier than that for ICAM-1 mRNA. Similar experiments were then performed with a molecular probe for interleukin-8 (IL-8). This chemokine is a potent neutrophil stimulant and has a higher degree of specificity for neutrophils than classic chemoattractants such as C5a. The results suggest a similar pattern of induction that occurs within the first hour and is markedly increased by reperfusion. The relationship of reperfusion to ICAM-1 and cytokine induction is discussed. (Mol Cell Biochem **147**: 5–12, 1995)

Key words: myocardial reperfusion injury, myocardial infarction, interleukin-6, interleukin-8, cell adhesion molecules, neutrophils

Introduction

Acute inflammation has been postulated to play an important role in the secondary injury associated with reperfusion of the acutely ischemic myocardium [1]. The onset of reperfusion is associated with a rapid influx and accumulation of leukocytes in reperfused myocardium with the highest rates seen in the first few hours after reinstitution of blood flow [2]. Significant reductions in infarct size in models in which the myocardium was reperfused have been achieved utilizing a variety of anti-inflammatory strategies. These strategies can be grouped into three general categories: 1) agents which affect the complement cascade [3–10], 2) anti-neutrophil therapies [11, 12] and 3) interventions targeting specific adhesion molecules known to play an important part in acute inflammatory reactions, such as CD11b/CD18 and ICAM-I [2, 13–20].

We have addressed this problem by establishing a framework upon which one can examine the specific molecular processes that regulate the inflammatory reaction that occurs upon reperfusion of the previously ischemic myocardium. Our approach has utilized an awake unanesthetized canine

Address for offprints: M.L. Entman, Section of Cardiovascular Sciences, Baylor College of Medicine, One Baylor Plaza, Houston, Texas 77030–34988, USA model of ischemia and reperfusion that allows the study of inflammation under circumstances where reaction to surgical trauma has subsided. We have focused on two main areas of investigation: the understanding of the mechanisms of neutrophil localization in reperfused myocardium and the molecular basis of neutrophil-induced cellular injury.

Our initial studies provided insights into the mechanisms of complement activation as a potential signal that initiates inflammation in the context of ischemia and reperfusion. The first of these studies demonstrated localization of C1q in previously ischemic and reperfused myocardium, in the same areas where neutrophil localization was found [3]. Utilizing a chronically cannulated cardiac lymph duct model [21], we showed that a cardiolipin containing peptide from the subsarcolemmal mitochondria of the injured myocyte bound C1q and initiated the classic complement pathway [4, 22]. Subsequently, we demonstrated the presence of chemotactic activity in cardiac lymph during the first 4 h of reperfusion and its neutralization by a polyclonal antibody to C5a [7, 10]. In correlation with the chemotactic activity present in cardiac lymph, we found that post-reperfusion lymph contained neutrophils with significantly elevated levels of CD11b/CD18 on their surface not present in blood or in pre-ischemic lymph [7]. Furthermore, neutrophil accumulation in reperfused myocardium occurs during the first 4 h of reperfusion [2].

Potential mechanisms of neutrophil-induced myocyte injury were elucidated by a series of *in vitro* studies aimed at characterizing the specific molecular determinants required for neutrophils to adhere to isolated cardiac myocytes. These studies first demonstrated that neutrophils bind to cardiac myocytes only under circumstances in which neutrophil CD11b/CD18 was activated and cardiac myocytes had been incubated with cytokines to induce expression of ICAM-1 on their surface [23, 24]. Under these circumstances neutrophil-myocyte adhesion invariably led to myocyte injury. This injury occurred as a result of a compartmented transfer of reactive oxygen radicals, and was manifested morphologically by irreversible myocyte contracture [25]. To link this phenomenon to myocardial ischemia, we demonstrated that post-ischemic cardiac lymph also contained cytokine activity capable of inducing ICAM-1-dependent adhesion of neutrophils to cardiac myocytes. This activity could be completely neutralized by antibodies to IL-6 [26]. Thus, we concluded that extracellular fluid from the reperfused myocardium contained both leukotactic and cytokine activity capable of sustaining a potentially lethal neutrophil-myocyte interaction.

These observations led us to examine the induction of ICAM-1 in ischemic and reperfused myocardial cells. RNA isolated from previously ischemic (but not control) myocardial segments demonstrated striking induction of ICAM-1 mRNA occurring within the first hour of reperfusion, which continues to further increase in quantity during the first 24 h [27]. We thus hypothesized that viable myocardial cells existed in the area adjoining a reperfused myocardial infarct that expressed ICAM-1 and were vulnerable to neutrophil-induced injury. The data presented in this manuscript further characterizes not only the induction of ICAM-1 and IL-6, but also the regulation of IL-8, an important proinflammatory cytokine thought to play a major role in directing neutrophils to an area of injury or inflammation [28–31]. Evidence is provided to show that the early myocardial induction of ICAM-1, IL-6 and IL-8 is dependent upon reperfusion.

Methods

Ischemia-reperfusion protocols

Healthy mongrel dogs (15–25 kg) of either sex were surgically instrumented as previously described [27]. Anesthesia was induced intravenously with 10 mg/kg methohexital sodium (Brevital; Eli Lilly and Company, Indianapolis, IN) and maintained with the inhalational anesthetic Isoflurane (Anaquest, Madison, WI). A midline thoracotomy provided access to the heart and mediastinum. Subsequently, a hydraulically activated occluding device and a Doppler flow probe [7] were secured around the circumflex coronary artery just proximal or just distal to the first branch. Choice of location depended on the proximity and anatomical arrangement.

The animals were allowed to recover for 72 h before occlusion. Ischemia-reperfusion protocols were performed in awake animals as described [7, 21, 27]. Coronary artery occlusion was achieved by inflating the coronary cuff occluder until mean flow in the coronary vessel was zero, as determined by the Doppler flow probe. At the end of 1 h the cuff was deflated and the myocardium was reperfused. Reperfusion intervals ranged from 1–24 h. Circumflex blood flow, arterial blood pressure, heart rate, and electrocardiogram (standard limb II) were recorded continuously, Analgesia was accomplished with intravenously administered pentazocine (Talwin; Wintrop Pharmaceuticals, New York, NY)) 0.1–0.2 mg/kg.

After the reperfusion periods, hearts were stopped by the infusion of saturated potassium chloride and removed from the chest and sectioned from apex to base into four transverse rings approximately 1 cm in thickness. The posterior papillary muscle and the posterior free wall were identified. Tissue samples (0.25–1.0 g) were isolated from infarcted (I) or normally perfused (C) myocardium. Frozen tissue samples were processed in a blinded fashion for RNA isolation and analysis; duplicate samples were fixed in 10% buffered formalin for histological analysis and for blood flow determinations using radiolabeled microspheres as previously described [27, 32]. All animal protocols were reviewed by the appropriate institutional review committee and conform to institutional guide lines.

Preparation of riboprobes for ICAM-1

Digoxigenin-labeled probes were prepared *in vitro* transcription from a linearized template following the method used by Boehringer-Mannheim in the Genius RNA probe labeling kit. A 150 bp fragment of canine ICAM-1 cDNA was subcloned in both orientations into pBluescript II SK+ and PGEM-3 so that utilization of T7 polymerase would result in the generation of single-stranded antisense (3′–5′) and sense (5′–3′) RNA probes, respectively. The template (1 µg) was incubated in 20 µl of 1× NTP mixture (1 mM ATP, 1 mM GTP, 1 mM CTP, 0.35 mM digoxigenin-UTP, and 0.65 mM UTP), T7 polymerase (2 U/µl), and DEPC-treated water for 2 h at 37°C. Both RNA probes were precipitated with glycogen and sodium acetate, washed with 70% ethanol, and resuspended in DEPC-treated water. The use of two different vectors was necessary due to the availability of restriction sites for linearization to prevent any 3′ overhang that would have resulted in non-specific RNA polymerase activity. Both probes were verified by hybridization and detection on a Southern blot (both positive) and a Northern blot (antisense positive) on nylon membrane.

In situ hybridization

Samples embedded in paraffin were sectioned and deparaffinized using standard protocols and probed using riboprobes following the procedure from Boehringer-Mannheim using the Genius system. The deparaffinized sections were washed twice for 5 min each in DEPC-treated water followed by 10 min in 2× SSC buffer. Sections were then prehybridized for 1 h at room temperature in a humid chamber with prehybridization buffer containing formamide (50%), SSC (4×), Denhardt's reagent (1×), herring sperm DNA (0.5 mg/ml), yeast tRNA (0.25 mg/ml), and dextran sulfate (10%). Hybridization was performed overnight at 42°C using 500 ng/ml of either ICAM sense or antisense probe in fresh prehybridization buffer placed over the section with a small square of Parafilm[R] laid on top to prevent drying out. The slides were washed as follows: 1 h room temperature, with 2× SSC, 1 h room temperature with 1× SSC, 30 min at 37°C with 0.5× SSC, 30 min at room temperature with 0.5× SSC. Immunological detection of the hybridized probe was performed at room temperature with the following protocol: Slides washed 1 min in Buffer 1 (100 mM Tris-HCL), 150 mM NaCl, pH 7.5), 30 min in Buffer 1 containing 2% sheep serum and 0.3% Triton X-100, 5 h in a 1:500 dilution of anti-digoxigenin antibody in Buffer 1 containing 1% sheep serum and 0.3% Triton X-100, 15 min wash in Buffer 1, and a 15 min wash in Buffer 2 (100 mM Tris HCl, 100 mM NaCl, 50 mM $MgCl_2$, pH 9.5). The slides were then incubated overnight at room temperature in the dark with the color solution from the Genius detection kit containing NBT (0.33 mg/ml) and X-phosphate (0.17 mg/ml) in Buffer 2 and the reaction stopped by immersion in Buffer 3(10 mM Tris-HCl, 1 mM EDTA, pH 8.0) for 5 min. The sections were then mounted with coverslips.

Northern blots analyses

RNA was isolated from myocardial tissue segments using the acid guanidinium-thiocyanate-phenol-chloroform procedure [33]. For Northern blots, RNA (12–20 µg) was electrophoresed in 1% agarose gels containing formaldehyde, then transferred to a nylon membrane (Gene Screen Plus, New England Nuclear, Boston, MA) by standard procedures. Membranes were hybridized in QuikHyb (Stratagene, La Jolla, CA) at 68°C for 2 h with 1×10^6 dpm random hexamer ^{32}P-labeled canine cDNA probes for ICAM-1, IL-6 and IL-8 for every ml of hybridization solution. Filters were washed with 2× SSPE at 68°C for 20 min, with 1× SSPE + 1% SDS at 68°C for 15 min twice and with 1× SSPE at 21°C for 15 min with constant shaking and exposed to Hyperfilm (Amersham, Arlington Heights, IL). Canine ICAM-1 and IL-6 cDNA were prepared as previously described [27, 34].

Molecular cloning

For the preparation of a canine IL-8 cDNA by RT-PCR, peripheral mononuclear cells were isolated from citrate anticoagulated venous blood by dextran sedimentation and Ficoll-Hypaque gradient centrifugation. The banded mononuclear cells were resuspended in RPMI 1640 with 10% fetal calf serum. Cells were stimulated with LPS (lipopolysaccharide) (Sigma Chemical Company, St. Louis, MO) 100 µg/ml for 4 h at 37°C. Total RNA was isolated using a modification of the acid guanidinium-thiocyanate-phenol-chloroform extraction and resuspended in DEPC-treated (diethyl pyrocarbonate) water [33]. Reverse transcription protocols were performed with 3 µg of total RNA in each sample. After first strand synthesis primed with oligo-dt, aliquots of the reverse transcription reaction were amplified with the following primers: 5′ primer 5′-GTGTCAACATGACTTCCAAACTG-3′, 3′ primer 5′-CTTCAAAAATATCTGTACAACCTT-3′ at a final concentration of 0.25 µM. The nucleotide sequence of the primers was based on highly conserved sequences of human [35] and porcine [36] cDNA coding sequences. PCR was performed with 5 U. of *Taq* polymerase (Promega Corporation, Madison WI) for 30 cycles of 94°C, 30 sec; 55°C, 30 sec; 72°C, 60 sec. The final product was purified, cloned into pBluescript II SK-(Stratagene Cloning Systems, La Jolla CA) and both strands sequenced on an Applied Biosystems model 373A automated DNA sequencer (Applied Biosystems, Foster City, CA). This partial canine IL-8 cDNA displayed 75%

homology to human IL-8 at the amino acid level.

Results

In situ hybridization studies of ICAM-mRNA expression

In the first group of studies, we examined myocandial tissue sections from previously ischemic and control areas of the same heart after 1 h of ischemia and varying times of reperfusion. In each case, the coronary blood flow in the ischemic section was less than 20% of control blood flow. Serial sections were taken for staining with sense and anti-sense riboprobes as well as other histologic stains when appropriate.

In Figure 1, tissue samples taken after 1 h of reperfusion are shown. The hematoxylin-basic fuchsin strain specifically stains viable myocardium that has been injured [37]. Figure 1 shows the anti-sense probe specifically staining the same area of tissue that stains intensely with the hematoxylin basic-fuchsin stain. In contrast, the sense riboprobe does not stain the tissue. Sections from non-ischemic myocardium do not stain with either sense or anti-sense (data not shown).

It remained possible that tissue taken after 1 h of reperfusion might ultimately have been lethally injured but capable of expressing ICAM-1 mRNA in the first hour. To further investigate this, we examined samples obtained after 1 h of ischemia and 3 h of reperfusion. In Fig. 2, we again demonstrate staining of ICAM-1 mRNA with the anti-sense probe and no staining with the homologous sense probe. In this case, however, we have elected to stain the tissue for glycogen to make another important point. Note that the inner most layers in the subendocardium are not stained for ICAM-1 mRNA but appear to be most intensely stained with PAS stain which stains glycogen. This is because of the well described phenomenon of subendocardial sparing which results from diffusion from the endocardium [32, 38, 39]. This phenomenon allows one to demonstrate quite conclusively that there remains jeopardized myocardium with partially depleted glycogen that is viable and capable of expressing ICAM-1 mRNA even after 3 h of reperfusion. It also speaks to the exquisite specificity of this molecular induction since the inner most layers of the subendocardium which have glycogen sparing also do not stain for ICAM-1 mRNA. The remainder of the section demonstrates an area of contraction band necrosis which stains neither for glycogen nor ICAM-1 mRNA. Again, at 3 h of reperfusion, control tissue does not stain (data not shown).

Finally, we investigated the role of reperfusion in ICAM-1 mRNA induction. In Fig. 3, tissue sections were taken after 3 h of ischemia with no reperfusion. Hybridization with anti-

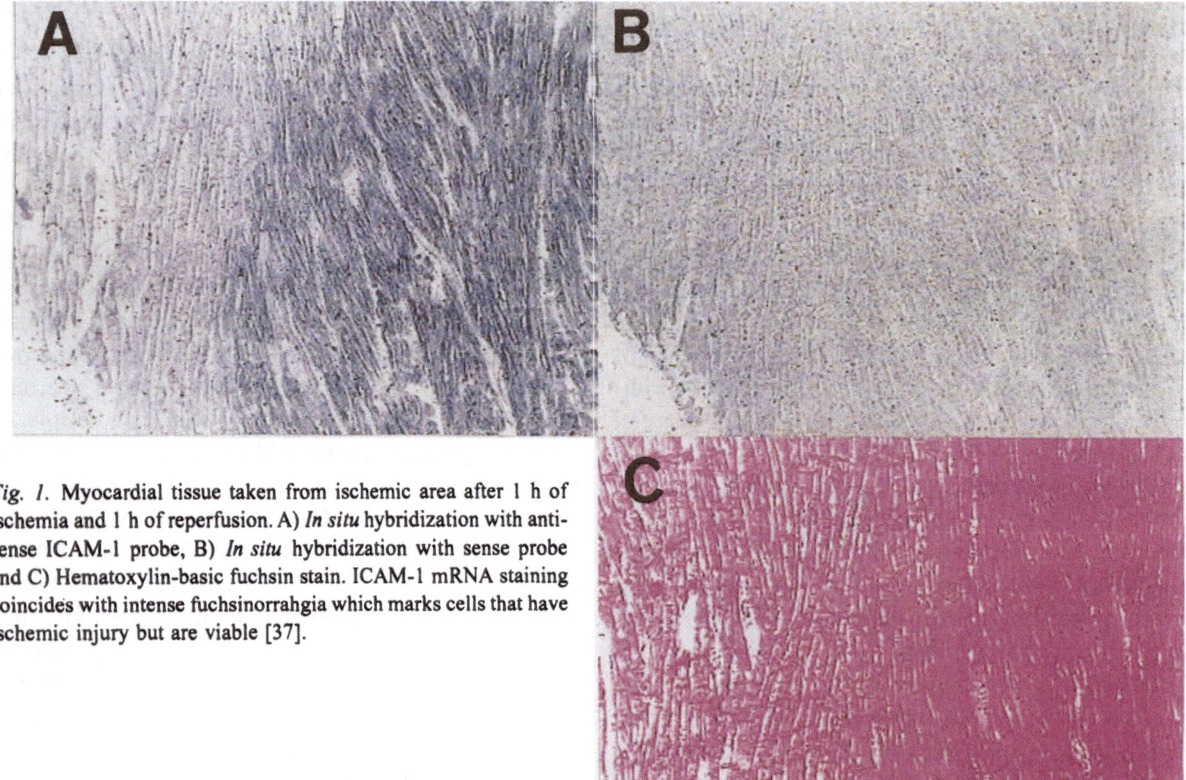

Fig. 1. Myocardial tissue taken from ischemic area after 1 h of ischemia and 1 h of reperfusion. A) *In situ* hybridization with anti-sense ICAM-1 probe, B) *In situ* hybridization with sense probe and C) Hematoxylin-basic fuchsin stain. ICAM-1 mRNA staining coincides with intense fuchsinorrahgia which marks cells that have ischemic injury but are viable [37].

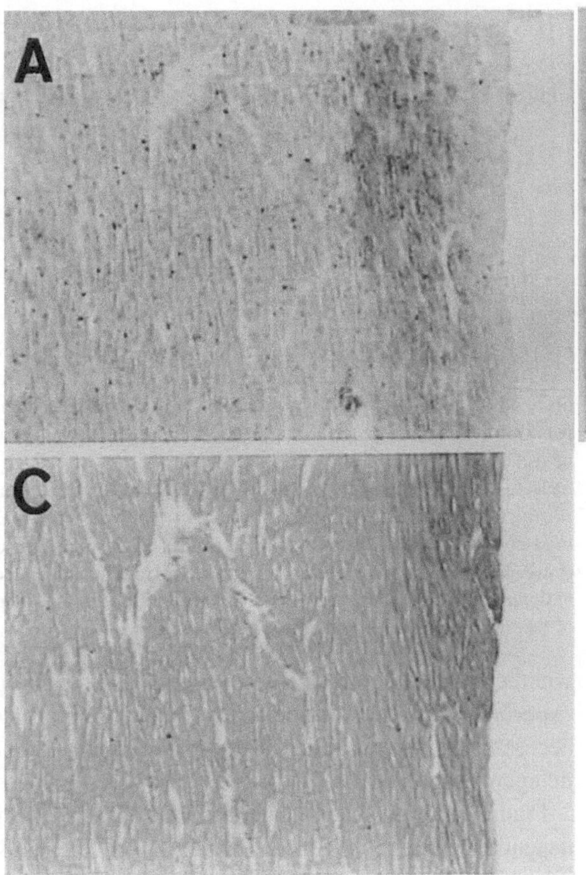

Fig. 2. Myocardial tissue taken from ischemic area after 1 h of ischemia and 3 h of reperfusion. A) *In situ* hybridization with anti-sense ICAM-1 probe, B) *In situ* hybridization with sense ICAM-1 probe and C) Periodic acid Schiff reagent for glycogen. Endocardial surface is on the right edge.

sense probes only are shown in this figure since the sense probe never stained tissue. For this experiment, tissue was taken from both ischemic and control areas of the myocardium. No significant ICAM-1 mRNA staining was seen in any of these samples suggesting that after 3 h of ischemia in the absence of reperfusion, there is no ICAM-1 mRNA induced.

Specific ICAM-1 mRNA staining is only seen in the presence of reperfusion and is easily detectable by 1 h although it continues to increase for 24 h. During the first 3 h of reperfusion, ICAM-1 mRNA staining is seen only in the viable myocardium adjoining areas of contraction band necrosis. By 24 h, intense staining is seen in both control and ischemic areas reflecting the fact that circulating cytokines [26] are inducing ICAM-1 mRNA throughout the myocardium (data not shown). However, as we have previously demonstrated [27], the protein is only expressed in the myocytes present in the ischemic areas. In the absence of reperfusion, ICAM-1 mRNA is not induced during the first 24 h in any significant amount.

Regulation of cytokines in ischemic and reperfused myocardium

Figures 4 and 5 show the deduced amino acid sequences of canine IL-6 and IL-8, and their comparison to their respective human homologues. Each of these was cloned for the dog to assure maximum specificity of our nucleic acid probes.

Figure 6 shows a Northern blot of RNA isolated from myocardial segments of two separate experimental animals. The first after 1 h of ischemia and 3 h of reperfusion and the second after 4 h of ischemia without reperfusion. In each case, the actual blood flow of the previously ischemic (or control) segment is shown along the ordinate. The Northern blot was

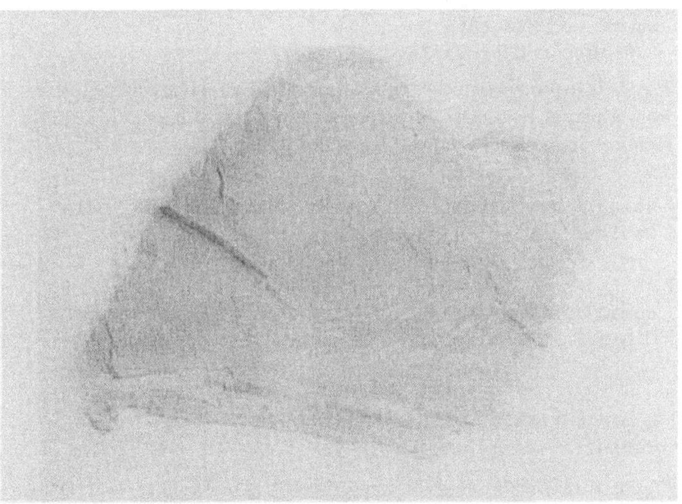

Fig. 3. In situ hybridization with anti-sense ICAM-1 probe in tissue taken after 3 h of ischemia without reperfusion. No significant staining was noted.

probed for both ICAM-1 mRNA (upper hand) and IL-6 (lower band). Following ischemia and reperfusion, both ICAM-1 and IL-6 mRNA are expressed exclusively in those segments with a significant reduction of blood flow (Fig. 6). In addition, both genes are induced in the same myocardial segments. In contrast, the heart in which no reperfusion occurred had no significant induction of either gene, despite comparable reductions in blood flow. These data are concordant with the observation in Fig. 3 with regard to ICAM-1 mRNA. They point out, however, that the cytokine which we have implicated as being the primary inducer of myocyte ICAM-1 expression [26] is likewise not induced in the absence of early reperfusion.

From the data described in the introduction, the activation, adhesion, and transmigration of neutrophils appears to play an important role in post-reperfusion injury. In our previous work, we had demonstrated that complement derived leukotactic factors were not dependent on reperfusion and appeared

```
                10         20         30         40
Human : MNSFSTSAFG PVAFSLGLLL VLPAAFPAPV PPGEDSKDVA
Canine: ---L---... ..-------- -MAT---T-G -LAG----D-

                50         60         70         80
Human : APHRQPLTSS ERIDKQIRYI LDGISALRKE TCNKSNMCES
Canine: TSNSL----A NKVEEL-K-- -GK------- M-D-F-K--D

                90        100        110        120
Human : SKEALAENNL NLPKMAEKDG CFQSGFNEET CLVKIITGLL
Canine: ---------- H---LEG--- -------Q-- --TR-T---V

               130        140        150        160
Human : EFEVYLEYLQ NRFESSEEQA RAVQMSTKVL IQFLQKKAKN
Canine: --QLH-NI-- -NY-GDK-NV KS-H----I- V-M-KS-V--

               170        180        190        200
Human : LDAITTPDPT TNASLLTKLQ AQNQWLQDMT THLILRSFKE
Canine: Q-EV------ -D---QAI-- S-DE--KHT- I------LED

               210
Human : FLQSSLRALR QM
Canine: ---F----V- I-
```

Fig. 4. IL-6 protein sequence comparison. Deduced amino acid sequences from a partial IL-6 cDNA are compared to those of human IL-6 [34]. Identical amino acids are indicated with a dash (–).

```
                10         20         30         40
Human : MTSKLAVALL AAFLISAALC EGAVLPRSAK ELRCQCIKTY
Canine: .....----- ---VL----- -A---S-VSS ---------H

                50         60         70         80
Human : SKPFHPKFIK ELRVIESGPH CANTEIIVKL SDGRELCLDP
Canine: -T-----Y-- -----D---- -E-S------ FN-N-V----

                90
Human : KENWVQRVVE KFLKRAENS
Canine: --K---.... .........
```

Fig. 5. IL-8 protein sequence comparison. A partial canine IL-8 cDNA was prepared by RT-PCR as described in Methods. The deduced amino acid sequences of the canine IL-8 cDNA, excluding the sequences derived from the original primers are compared to those of human IL-8. Identical amino acids are indicated with a dash (–).

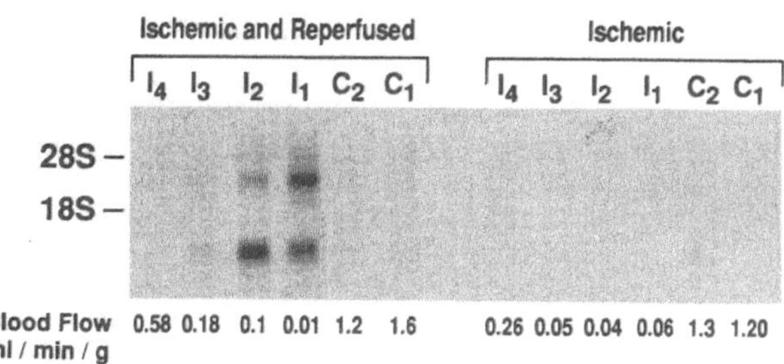

Fig. 6. Regulation of ICAM-1 and IL-6 mRNA in the myocardium by reperfusion. Experimental animals were subjected to 1 h of coronary occlusion and 3 h of reperfusion (left panel) or 4 h of ischemia (right panel). Myocardial segments are labeled C for control of I for ischemic or ischemic and reperfused depending upon the experiment. RNA was isolated from each segment and analyzed using Northern blots and blood flow determinations as described in Methods. 12 µg of RNA was loaded in each lane. Levels of ICAM-1 mRNA (upper bands) and of IL-6 mRNA (lower bands) elicited by ischemia followed by reperfusion are compared to those elicited by ischemia without reperfusion. The migration positions of the 28S and 18S rRNA bands is indicated.

whether reperfusion occurred or not [3]. The ensuing set of experiments was performed to determine if reperfusion was necessary for the induction of more cell specific chemotactic agents such as IL-8 [40].

Figure 7 suggest that the phenomenon of early reperfusion-dependence extends to this chemotactic cytokine. IL-8 is rapidly induced in the myocardium within the first hour of reperfusion and mRNA levels appear to peak between 1 and 3 h (data not shown). In similar experiments to those described in Fig. 6, IL-8 mRNA induction is seen only in the previously ischemic and reperfused myocardial segments and not in control tissue (Fig. 7). In contrast, minimal induction of the IL-8 gene is seen in the absence of reperfusion after 4 h of ischemia (Fig. 7). Thus, a cell biological event associated with reperfusion appears to be critical to the early myocardial induction of IL-8.

Discussion

In our previous experiments we demonstrated that neutrophils were capable of directly damaging cardiac myocytes under circumstances where ICAM-1 was induced on cardiac myocytes and CD11b/CD18 was activated on neutrophils [23–25]. Subsequently we provided evidence that both leukotactic factors [7, 10] and cytokine activity [26, 27] capable of inducing ICAM-1 mRNA in canine cardiac myocytes, are present within the post-ischemic cardiac lymph of experimental animals that underwent ischemia and reperfusion under circumstances in which myocardial injury occurred. In contrast, animals with sufficient collateral circulation to avoid injury resulting from coronary occlusion and reperfusion, have never demonstrated either leukotactic or

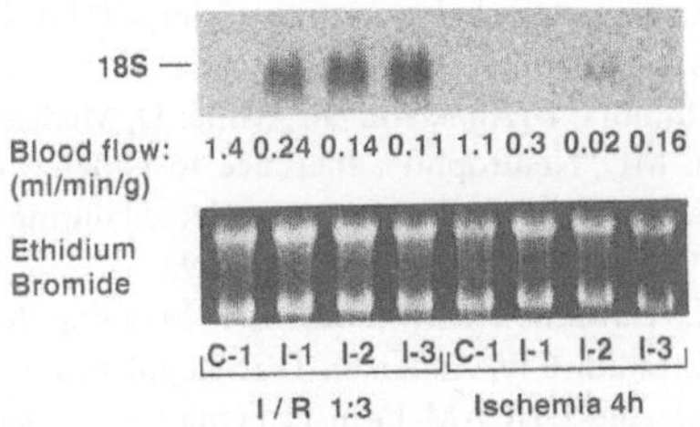

Fig. 7. Regulation of IL-8 mRNA in the myocardium by reperfusion. Representative experiments in two animals subjected to 1 h of coronary occlusion and 3 h of reperfusion (left panel) or 4 h of ischemia without reperfusion (right panel) are shown. Myocardial segments are labeled C for control or I for ischemic or ischemic and reperfused depending upon the experiment. RNA was isolated from each segment and analyzed using Northern blots and blood flow determinations as described in Methods. 20 μg of RNA was loaded in each lane as shown by the ethidium bromide staining (bottom panel). Blood flow determinations are indicated for each myocardial segment. The migration position of the 18S rRNA band is indicated.

cytokine activity in their cardiac lymph. Thus, we hypothesized that all of the components were present in the extracellular fluid of the previously ischemic myocardium to potentially allow a neutrophil-myocyte interaction resulting in myocyte injury. In the current manuscript, we have presented data which further support this concept. In addition, we report that the reperfusion event itself plays an important part in the early induction of the adhesion molecules and the cytokines necessary for neutrophil-induced myocyte injury. These observations suggest some very important principles with regard to inflammatory injury as a mechanism of secondary injury following ischemia and reperfusion.

The *in situ* hybridization demonstrated ICAM-1 mRNA in the viable tissue directly adjoining the area of contraction band necrosis seen after reperfusion. This viable area appears to persist long after reperfusion has begun suggesting that ICAM-1 mRNA is expressed in cells that have survived the ischemic insult and are now potentially jeopardized by inflammation. Since it has been frequently described that reperfusion markedly increases neutrophil influx into previously ischemic myocardium, our initial hypothesis was that reperfusion might act solely as a mechanism to markedly increase infiltrating cells. We have previously shown that neutrophil sequestration and transmigration is greatest the border between normal appearing cells and cells demonstrating contraction band necrosis [27]. This region is also the region in which ICAM-1 mRNA is induced after reperfusion as demonstrated in this and other studies [27, 32]. Likewise, cytokines responsible for inducing ICAM-1 mRNA and for activating neutrophils are also dependent on reperfusion for their early induction. These data imply that some factor or factors initiated by reperfusion are responsible for the cytokine and chemokine generation which ultimately results in ICAM-1 induction and leukocyte activation. The possibility that influx of leukocytes might in some way be responsible for expression of both the cytokines and ICAM-1 must be considered. One possibility relates to the fact that both neutrophils and mononuclear cells can express cytokines in response to stimulation by other cytokines [41, 42] and C5a stimulation [43]. Alternatively, the interaction of leukocytes with adhesion molecules on endothelial cells may well be a signal which might induce endothelial cell production of cytokines. Future experiments will be required to better elucidate this point.

However, these experiments do suggest that the inflammatory cascade occurring during reperfusion is potently influenced by events which occur very early after the onset of reperfusion and may well relate to leukocyte activation and adhesion. A greater understanding of the factors which influence this process may help identify molecular targets by which the process can be effectively interrupted.

Acknowledgements

We acknowledge the expert secretarial assistance of Lisa M. Padilla and Michelle Swarthout. This work was supported by National Institute of Health Grant HL-42550 and by grants from The Methodist Hospital Foundation and Baylor College of Medicine (GLK).

References

1. Entman ML, Kukielka GL, Ballantyne CM, Smith CW: The role of leukocytes in ischemic heart disease. In: C. Chapman (ed.). Handbook of immunopharmacology. Academic Press Limited/Harcourt Brace Hovanovich, London, 1993, pp 55–74
2. Dreyer WJ, Michael LH, West MS, Smith CW, Rothlein R, Rossen RD, Anderson DC, Entman ML: Neutrophil accumulation in ischemic canine myocardium: Insights into the time course, distribution, and mechanism of localization during early reperfusion. Circulation 84: 400–411, 1991
3. Rossen RD, Swain JL, Michael LH, Weakley S, Giannini E, Entman ML: Selective accumulation of the first component of complement and leukocytes in ischemic canine heart muscle: A possible initiator of an extra myocardial mechanism of ischemic injury. Circ Res 57: 119–130, 1985
4. Rossen RD, Michael LH, Kagiyama A, Savage HE, Hanson G, Reisberÿ JN, Moake JN, Kim SH, Weakly S, Giannini E, Entman ML: Mechanism of complement activation following coronary artery occlusion: Evidence that myocardial ischemia causes release of constituents of myocardial subcellular origin which complex with the first component of complement. Circ Res 62: 572–584, 1988
5. Pinckard RN, Olson MS, Giclas PC, Terry R, Boyer JT, O'Rourke RA: Consumption of classical complement components by heart subcellular membranes *in vitro* and in patients after acute myocardial infarction. J Clin Invest 56: 740–750, 1975
6. Hill JH, Ward PA: The phlogistic role of C3 leukotactic fragment in myocardial infarcts of rats. J Exp Med 133: 885–900, 1971
7. Dreyer WJ, Smith CW, Michael LH, Rossen RD, Hughes BJ, Entman ML, Anderson DC: Canine neutrophil activation by cardiac lymph obtained during reperfusion of ischemic myocardium. Circ Res 65: 1751–1762, 1989
8. Crawford MH, Grover FL, Kolb WP, McMahan CA, O'Rourke RA, McManus LM, Pinckard RN: Complement and neutrophil activation

in the pathogenesis of ischemic myocardial injury. Circulation 78: 1449–1458, 1988

9. Weisman HF, Barton T, Leppo MK, Marsh HC Jr, Carson GR, Concino MF, Boyle MP, Roux KH, Weisfeldt ML, Fearon DT: Soluble human complement receptor type 1: *In vivo* inhibitor of complement suppressing post-ischemic myocardial inflammation and necrosis. Science 249: 146–151, 1990
10. Hughes BJ, Hollers JC, Crockett-Torabi E, Smith CW: Recruitment of CD11b/CD18 to the neutrophil surface and adherence-dependent cell locomotion. J Clin Invest 90: 1687–1696, 1992
11. Romson JL, Hook BG, Kunkel SL, Abrams GD, Schork MA, Lucchesi BR: Reduction of the extent of ischemic myocardial injury by neutrophil depletion in the dog. Circulation 67: 1016–1023, 1983
12. Mullane KM, Read N, Salmon JA, Moncada S: Role of leukocytes in acute myocardial infarction in anesthetized dogs. Relationship to myocardial salvage by anti-inflammatory drugs. J Pharmacol Exp Ther 228: 510–522, 1984
13. Maroko PR, Carpenter CD, Chiariello M, Fishbein MC, Radvany P, Knostman JD, Hale SL: Reduction by cobra venom factor of myocardial necrosis after coronary artery occlusion. J Clin Invest 61: 661–670, 1978
14. Simpson PJ, Todd III, Fantone JC, Mickelson JK, Griffin JD, Lucchesi BR: Reduction of experimental canine myocardial reperfusion injury by a monoclonal antibody (anti-Mo1, anti-CD11b) that inhibits leukocyte adhesion. J Clin Invest 81: 624–629, 1988
15. Williams FM, Collins PD, Nourshargh S, Williams TJ: Suppression of 111In-neutrophil accumulation in rabbit myocardium by MoA ischemic injury. J Mol Cell Cardio 20: S331988
16. Seewaldt-Becker E, Rothlein R, Dammgen JW: CDw18 dependent adhesion of leukocytes to endothelium and its relevance for cardiac reperfusion. In: T.A. Springer, D.C. Anderson, A.S. Rosenthal and R. Rothlein (eds). Leukocyte Adhesion Molecules: Structure, Function and Regulation. Springer-Verlag, New York, New York, 1989, pp 138–148
17. Simpson PJ, Todd III, Mickelson JK, Fantone JC, Gallagher KP, Lee KA, Tamura Y, Cronin M, Lucchesi BR: Sustained limitation of myocardial reperfusion injury by a monoclonal antibody that alter leukocyte function. Circulation 81: 226–237, 1990
18. Tanaka M, Brooks SE, Fitzharris GP, Stoler RC, Jennings RB, Reimer KA: Effect of the IB4 anti-CD18 antibody on myocardial PMN accumulation and infarct size in dogs. FASEB J 4: A10201990
19. Ma XL, Tsao PS, Lefer AM: Antibody to CD18 exerts endothelial and cardiac protective effects in myocardial ischemia and reperfusion. J Clin Invest 88: 1237–1243, 1991
20. Lefer DJ, Suresh ML, Shandelya ML, Serrano CV, Becker LC, Kuppusamy P, Zweier JL: Cardioprotective actions of a monoclonal antibody against CD-18 in myocardial ischemia-reperfusion injury. Circulation 88: 1779–1787, 1993
21. Michael LH, Hunt JR, Weilbaecher D, Perryman MB, Roberts R, Lewis RM, Entman ML: Creatine kinase and phosphorylase in cardiac lymph: Coronary occlusion and reperfusion. Am J Physiol 248: H350–H359, 1985
22. Rossen RD: Complement activation in cardiac disease. In: C. Page, M.J. Curtis (eds). The handbook of immunopharmacology. Academic Press Limited, United Kingdom, 1993, pp 75–86
23. Entman ML, Youker KA, Shappell SB, Siegel C, Rothlein R, Dreyer WJ, Schmalstieg FC, Smith CW: Neutrophil adherence to isolated adult canine myocytes: Evidence for a CD18-dependent mechanism. J Clin Invest 85: 1497–1506, 1990
24. Smith CW, Entman ML, Lane CL, Beaudet AL, Ty TI, Youker KA, Hawkins HK, Anderson DC: Adherence of neutrophils to canine cardiac myocytes *in vitro* is dependent on intercellular adhesion molecule-1. J Clin Invest 88: 1216–1223, 1991
25. Entman ML, Youker KA, Shoji T, Kukielka GL, Shappell SB, Taylor AA, Smith CW: Neutrophil induced oxidative injury of cardiac myocytes: A compartmented system requiring CD11b/CD18-ICAM-1 adherence. J Clin Invest 90: 1335–1345, 1992
26. Youker KA, Smith CW, Anderson DC, Miller D, Michael LH, Rossen RD, Entman ML: Neutrophil adherence to isolated adult cardiac myocytes: Induction by cardiac lymph collected during ischemia and reperfusion. J Clin Invest 89: 602–609, 1992
27. Kukielka GL, Hawkins HK, Michael LH, Manning AM, Lane CL, Entman ML, Smith CW, Anderson DC: Regulation of intercellular adhesion molecule-1 (ICAM-1) in ischemic and reperfused canine myocardium. J Clin Invest 92: 1504–1516, 1993
28. Baggiolini M, Dewald B, Walz A: Interleukin-8 and related chemotactic cytokines. In: J.I. Gallin, I.M. Goldstein and R. Snyderman (eds). Inflammation: Basic principles and clinical correlates. Raven Press Limited, New York, 1992, pp 247–263
29. Baggiolini M, Walz A, Kunkel SL: Neutrophil-activating peptide-1/interleukin 8, a novel cytokine that activates neutrophils. J Clin Invest 84: 1045–1049, 1989
30. Miller MD, Krangel MS: Biology and biochemistry of the chemokines: A family of chemotactic and inflammatory cytokines. Crit Rev Immunol 12: 17–46, 1992
31. Springer TA: Traffic signals for lymphocyte recirculation and leukocyte emigration: The multistep paradigm. Cell 76: 301–314, 1994
32. Youker KA, Hawkins HK, Kukielka GL, Perrard JL, Michael LH, Ballantyne CM, Smith CW, Entman ML: Molecular evidence for induction of ICAM-1 in the viable border zone associated with ischemia-reperfusion injury of the dog heart. Circulation 1994, (in press)
33. Chomczynski P, Sacchi N: Single-step method of RNA isolation by acid guanidinium thiocyanate-phenol-chloroform extraction. Anal Biochem 162: 156–159, 1987
34. Kukielka GL, Youker KA, Hawkins HK, Perrard JL, Michael LH, Ballantyne CM, Smith CW, Entman ML: Regulation of ICAM-1 and IL-6 in myocardial ischemia: Effect of reperfusion. Ann N Y Acad Sci 1994, (in press)
35. Matsushima K, Morishita K, Yoshimura T, Lavu S, Kobayashi Y, Lew W, Apella E, Kung HF, Leonard EJ, Oppenheim JJ: Molecular cloning of a human monocyte-derived neutrophil chemotactic factor (MDNCF) and the induction of MDNCF mRNA by interleukin 1 and tumor necrosis factor. J Exp Med 167: 1883–1893, 1988
36. Goodman RB, Foster DC, Mathewes SL, Osborn SG, Kuijper JL, Forstrom JW, Martin TR: Molecular cloning of porcine alveolar macrophage-derived neutrophil chemotactic factors I and II: Identification of porcine IL-8 and another intercrine-α protein. Biochemistry 31: 10483–10490, 1992
37. Lie JT, Holley KE, Titus JL: Fuchsinorrhagia – A new histochemical indication of inapparent early myocardial ischemia. Lab Med 3: 37–40, 1972
38. Mallory GK, White PD, Salcedo-Salgar J: The speed of healing of myocardial infarction. A study of the pathologic anatomy in seventy-two cases. Am Heart J 18: 647–671, 1939
39. Fishbein MC, Maclean D, Maroko PR: The histopathologic evolution of myocardial infarction. Chest 73: 843–849, 1978
40. Miller D, Krangel MS: Biology and biochemistry of the chemokines: A family of chemotactic and inflammatory cytokines. Crit Rev Immunol 12: 17–46, 1992
41. Marucha PT, Zeff RA, Kreutzer DL: Cytokine regulation of IL-1beta gene expression in the human polymorphonuclear leukocyte. J Immunol 145: 2932–2937, 1990
42. Marucha PT, Zeff RA, Kreutzer DL: Cytokine-induced IL-1B gene expression in the human polymorphonuclear leukocyte: Transcriptional and post-transcriptional regulation by tumor necrosis factor and IL-1. J Immunol 147: 2603–2608, 1991
43. Ember JA, Sanderson SD, Hugli TE, Morgan EL: Induction of interleukin-8 synthesis from monocytes by human C5a anaphylatoxin. Am J Pathol 144: 393–403, 1994

Molecular and Cellular Biochemistry **147**: 13–19, 1995.

Importance of monocytes/macrophages and fibroblasts for healing of micronecroses in porcine myocardium

Dorothée Weihrauch, Margarete Arras, René Zimmermann and Jutta Schaper
Max Planck Institute, Department of Experimental Cardiology, Bad Nauheim, Germany

Abstract

In porcine heart, embolization of small coronary arteries with microspheres in 25 μm in diameter induces collateral capillary vessel growth by angiogenesis in and around focal necrosis. By histological analysis the inflammatory infiltrates in this porcine tissue were characterized by numerous monocytes/macrophages and fibroblasts as well as neutrophils and numerous capillaries, some in mitosis. The aim of the present study, therefore, was to clarify the role of monocytes/macrophages and fibroblasts in angiogenesis and in repair in ischemic porcine myocardium. Using a human acidic fibroblast growth factor (aFGF) cDNA probe for *in situ* hybridisation labeling for aFGF mRNA was seen in monocytes and macrophages only, beginning at day 1, with a maximum at 3 and 7 days, and minimal labeling at 4 weeks. We have also shown, with a specific antibody and fluorescence microscopy, that tumur necrosis factor alpha (TNFα) follows the same time sequence and that it is produced by monocytes/macrophages. The number of capillaries in infiltrates at 3 and 7 days as revealed by the lectin Dolichus Biflorus Agglutinin was high and declined at 4 weeks. *In situ* hybridisation using a rat cDNA probe for fibronectin showed the increased production of fibronectin mRNA in fibroblasts. To describe the expression of fibronectin and the collagens I, III, VI immunohistochemistry was used. A comparison showed that fibroblasts produced fibronectin mRNA starting at day 3, but the protein was only maximally expressed at day 7 and 4 weeks. Collagen I, III, VI expression was highest at 1–4 weeks. Conclusion: monocytes and macrophages produce the growth factors aFGF and TNFα which seem to be important for angiogenesis in the ischemic myocardium. Fibroblasts, while they produce fibronectin and collagen, exert their major function in repair and scar formation, but may take also part in angiogenesis. (Mol Cell Biochem **147:** 13–19, 1995)

Key words: monocytes/macrophages, fibrocytes, TNFα, extracellular matrix

Introduction

In previous studies we have shown the occurrence of focal necroses, followed by repair processes and angiogenesis after coronary microembolization in porcine myocardium. Acute cellular necrosis was followed by infiltration with blood cells and final scar formation as already described in many textbooks [1]. In our studies, it was obvious that the postischemic granulation tissue contained numerous monocytes/macrophages, lymphocytes, a small number of neutrophils, fibroblasts and necrotic cellular particles [2].

On the basis of these findings and the development of new morphological techniques offering the possibility to demonstrate the localisation of cytokines in tissue sections, we were interested in the possible role of monocytes/macrophages in both, repair processes and angiogenesis. We also wanted to evaluate the importance of cytokines such as fibroblast growth factor (aFGF) and tumor necrosis factor alpha (TNFα).

Material and methods

Twenty six German landrace pigs aged 3 months and weighing 21–39 kg were used in this study. The animal protocol

Address for offprints: D. Weihrauch, Max-Planck-Institute, Department of Experimental Cardiology, Benekestrasse 2, D-61231 Bad Nauheim, Germany described here was approved by the bioethical committee of the district of Darmstadt, Germany. All animals were handled in accordance with the American Physiological Society guidelines for animal welfare.

The pigs were sedated with azaperone 2 mg/kg i.m. and anesthetized with pentobarbital 30 mg/kg i.v. followed by oral intubation. In an open chest procedure after thoracotomy non-radioactive polystyrene microspheres (NEN-TRAC, DuPont, Boston, Massachusetts, USA) of 25 µm diameter were injected under sterile conditions into the left coronary circumflex artery. The chest was closed and the animals were allowed to recover. After different time intervals (24 h, 3 days, 7 days and 4 weeks) the animals were reanesthetized as described above and sacrificed with an overdose pentobarbital. The thorax was opened and the heart removed.

Tissue sampling

Tissue samples were taken from the left ventricular lateral wall, immediately fixed in liquid nitrogen and stored at –70°C for immunohistochemistry and *in situ* hybridisation. Tissue samples for electron microscopy were fixed in 3% buffered glutaraldehyde.

Electron microscopy

The samples were embedded in epon following routine procedures. Semithick and ultrathin sections were cut, stained with uranylacetate and lead citrate and viewed in a Philips CM 10 electron microscope.

Immunohistochemistry

Cryostat sections 4 µm thick were prepared, fixed in either acetone and incubated with the first antibody. After rinsing, the second antibody or the biotinylated second antibody detection system was applied. The fluorochromes Fluorescein Streptavidin or Texas Red Streptavidin were taken to visualize the antigen/antibody reaction.

First antibodies used in this study:

Antigen	Host animal	Company	Dilution	Fixation
Collagen I	rabbit	Bioscience, CH	1:100	Acetone
Collagen III	rabbit	Bioscience, CH	1:50	Acetone
Collagen VI	rabbit	Telios, USA	1:150	Acetone
Fibronectin	rabbit	ICN, USA	1:80	Acetone
TNFα	mouse	Miles, USA	1:10	Acetone
DBA		Sigma, USA	1:10	Acetone

Second antibodies used in this study:

Antigen	Host animal	Company	Dilution
Rabbit	donkey	Dianova, FRG	1:100
Mouse	donkey	Dianova, FRG	1:100

Fluorochrome used in this study:

	Company	Dilution
Fluorescein Streptavidin	Amersham, GB	1:50
Texas Red Streptavidin	Amersham, GB	1:50

The sections were viewed with a Leica Aristoplan LM microscope using the objectives NPL Fluotar 25/0.75 Oil or NPL Fluotar 50/1.00. The photographic documentation was carried out with a Leitz Orthomat E and a Leitz 35 mm camera. The film was a Kodak Ektachrome 200 for colour slides.

In situ hybridisation

Cryostat sections 4 µm thick were prepared. *In situ* hybridisation was carried out for aFGF with a cDNA probe, and for fibronectin with a RNA probe [3]. All probes were S-35 labeled. Hybridisation was carried out overnight at temperatures specific for each probe. After an extensive rinsing procedure the sections were covered with Kodak photographic emulsion and exposed for varying time intervals. After exposure the emulsion was developed, the sections were counterstained with Toluidine Blue and viewed in the microscope under transmission and darkfield conditions.

Results

Description of the model

The microspheres injected in the left coronary circumflex artery occluded arteries with a diameter smaller than 25 µm and capillaries. This vascular occlusion caused ischemia and later local necroses in the perfusion dependent area [4]. The advantage of this model is the occurrence of numerous small necrotic regions as opposed to a larger infarcted area. Thereby, functional consequences of the occurrence of necrosis are avoided. This model, therefore, is suitable to observe wound healing and angiogenesis in myocardium of pigs, or any other animal species, in dependence of time [2].

Histology

Twenty four hours after microembolization a loss of myocyte nuclei indicating cellular necrosis was observed in the

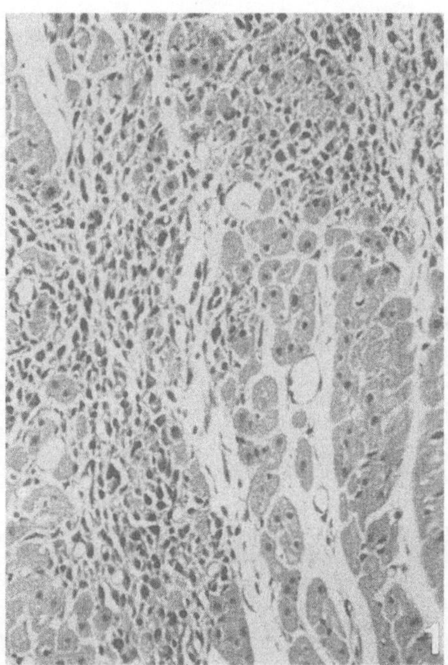

Fig. 1. H.E. staining. Three days after microembolization. Patchy necrosis with mononuclear cell infiltrate is evident.

ischemic areas by light microscopy (Fig. 1). Destruction and partial removal of necrotic myocytes were evident and mononuclear cells started to infiltrate the zone of ischemia. As early as 3 days after the ischemic insult cellular infiltrates were clearly evident. These infiltrates consisted of numerous monocytes/macrophages, lymphocytes, and a small number of neutrophils as well as fibroblasts (Fig. 2). The number of capillaries was increased after 3 days detected by staining of the tissue with the lectin DBA (Dolichus Biflorus Agglutinin) and endothelial cell mitosis, a rare event in cardiac tissue, was observed. Seven days after microembolization, the number of capillaries had increased as compared to 3 days, the removal of myocytes had progressed, and the mononuclear infiltrates were still present (Fig. 3).

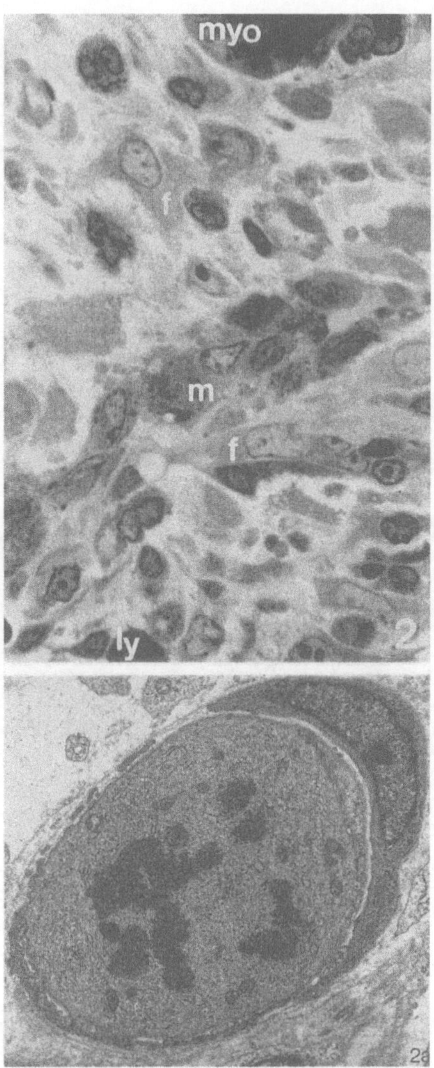

Fig. 2. Light micrograph. Remnants of irreversibly injured myocyte (myo), fibroblasts (f), lymphocytes (ly) and macrophages (m) are visible. (A) Electron microscopical picture. A capillary shows mitosis of the endothelium indicating neovascularization.

Monocytes/macrophages

Monocytes and macrophages are known to produce cytokines, enzymes and growth factors [5] and therefore we were especially interested in these cells in the experimental model presented here. Our group recently showed, using *in situ* hybridisation, that many cells within the infiltrates produce aFGF mRNA [3]. These cells were identified as monocytes/macrophages by CD68 staining which is a specific macrophage marker. The mRNA for aFGF was absent in normal myocardium, it was present at day 1 and had a maximum after 3 and 7 days (Fig. 4).

The fact that the mRNA for aFGF was observed in monocytes and macrophages is of special interest because these cells have been described to produce different angiogenic factors and to be important for the development of collateral blood vessels in the coronary circulation [6–8]. Acidic FGF is a 140 amino acid polypeptide with a 53% sequence homology to basic FGF. It was shown to be one of the most potent endothelial cell growth factors, although lacking a classical hydrophobic signal sequence [9, 10]. We would therefore like to propose that it plays an important role during neovascularization in wound healing and stimulates most probably the capillary collateral network in ischemic myocardium [11].

Additionally, many cells positively labeled for TNFα were found in the infiltrates after 3 and 7 days (Fig. 5). An immunofluorescent double staining with TNFα and CD14 – a macro-

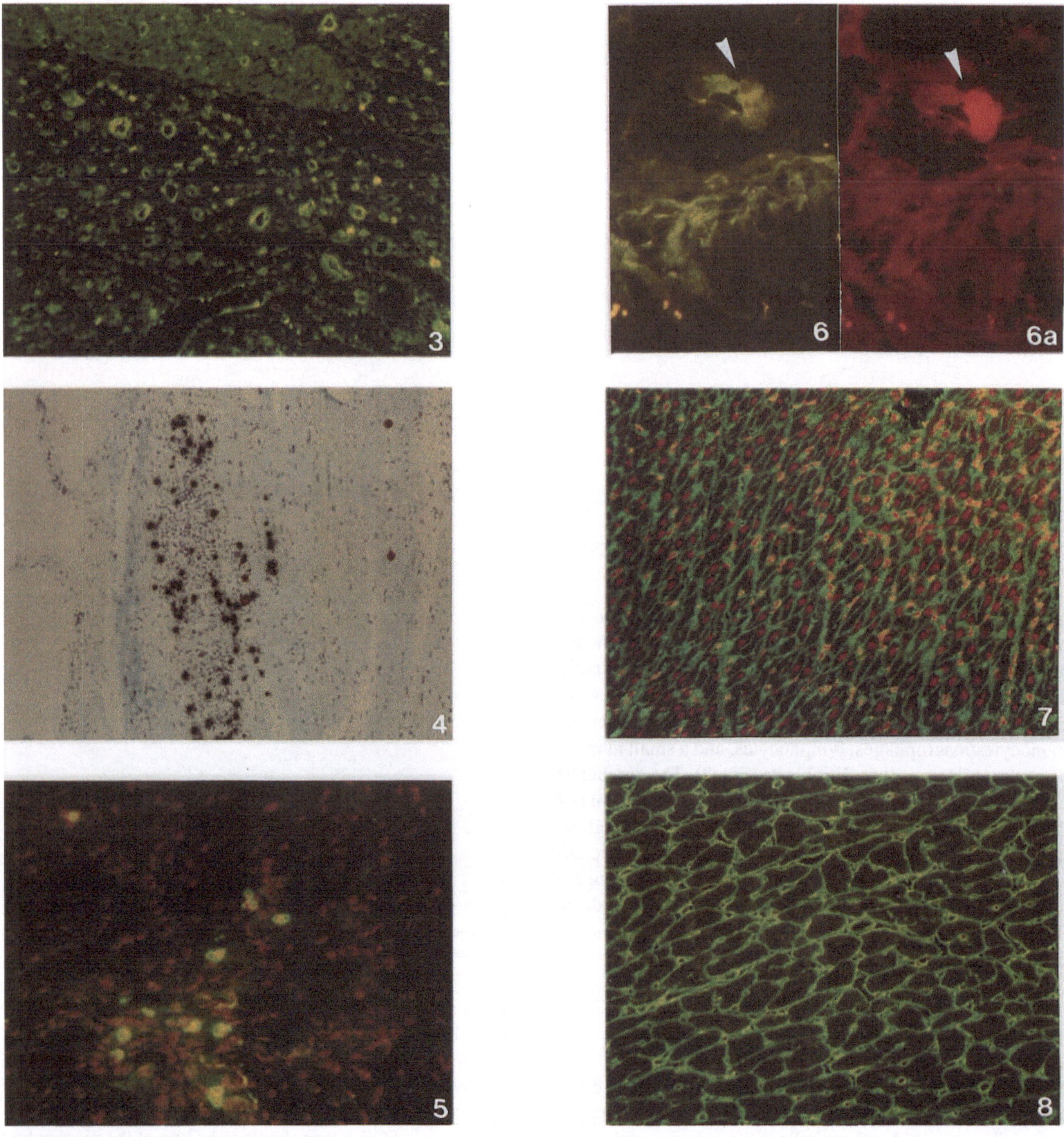

Fig. 3. DBA staining. Seven days after microembolization. Numerous microvessels are present in the necrotic area surrounding myocytes on the right side.
Fig. 4. *In situ* hybridisation for aFGF. Three days after microembolization – bright field picture showing labeling for several cells in an infiltrate.
Fig. 5. TNFα staining. Seven days after microembolization. A cluster of TNFα positive cells is situated in a mononuclear infiltrate.
Fig. 6. Double staining of macrophages with TNFα antibody and CD14 in the same section of human myocardium – a larger cell (arrow) is positive for TNFα labeled with FITC. The TNFα positive cell is also positive for CD14 (arrow) with rhodamine. (a) Macrophages are synthetizing TNFα.
Fig. 7. Distribution of fibronectin in normal myocardium.
Fig. 8. Distribution of collagen I in normal myocardium.

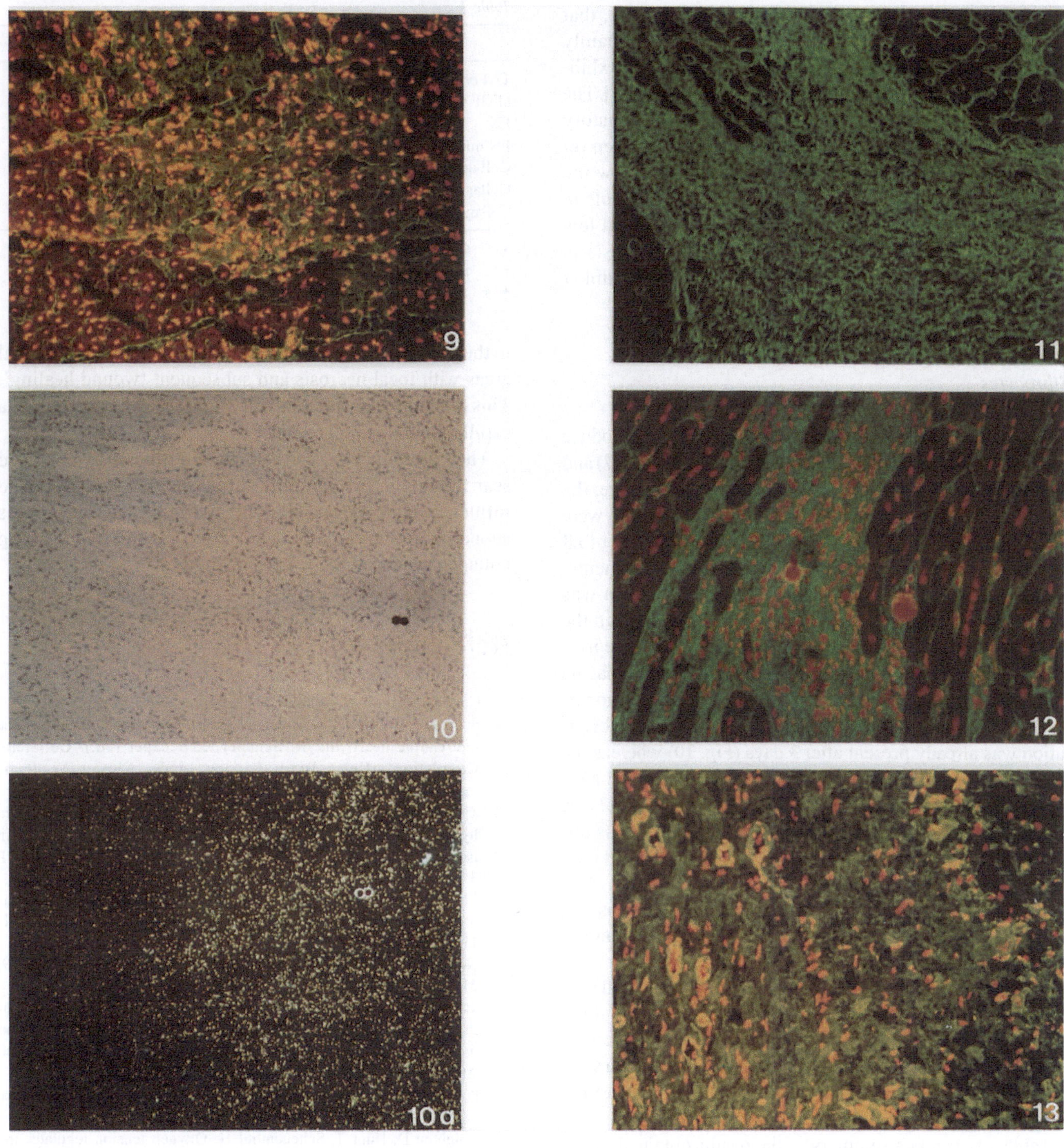

Fig. 9. Fibronectin 24 h after microembolization – localization within the myocytes is evident in the center of the picture.

Fig. 10. *In situ* hybridisation for fibronectin – 3 days after microembolization an infiltrate is present in the bright field picture. (A) The dark field picture shows many labeled cells in the infiltrate.

Fig. 11. Seven days after microembolization – fibronectin is maximally expressed in the infiltrates.

Fig. 12. Seven days after microembolization – the infiltrate shows an increased amount of collagen VI.

Fig. 13. Four weeks after microembolization – the expression of collagen VI is significantly increased.

phage marker-identified these cells as macrophages (Fig. 6). This is in agreement with reports from the literature, that TNFα, a 157 amino acid polypeptide hormone, is mainly produced by monocytes and macrophages and that it exhibits proliferating as well as angiogenic properties [12–14]. The occurrence of TNFα in increased amounts in inflammatory cells during wound healing and scar formation has been reported [15, 16]. The results of the present study allow the interpretation that TNFα most probably plays a dual role in ischemic myocardium: it may promote angiogenesis at low concentrations, and it may accelerate cellular necrosis in higher concentrations depending presumably on the number of macrophages synthetizing this cytokine.

Fibrocytes

Fibrocytes are present in normal myocardium and they produce the extracellular matrix proteins such as fibronectin (Fig. 7) and the different types of collagen (Fig. 8). In cardiac tissue, the collagen types I, III and VI are of major importance and were therefore investigated in the present study. The amount of all these proteins increased slowly with time in our experimental model. At 24 h after microembolization, fibronectin was found not only in the extracellular space but also within the myocytes (Fig. 9). Most probably, these myocytes were irreversibly injured by the ischemic insult and therefore plasma fibronectin was able to diffuse through the leaky cell membrane. The mRNA for fibronectin detected by *in situ* hybridisation was already present after 3 days (Fig. 10) whereas the protein fibronectin showed a delayed appearance and augmentation at 7 days after microembolization (Fig. 11). A slight increase of the different collagens, however, was observed at 1 week after microembolization (Fig. 12) and they showed a significant augmentation after 4 weeks (Fig. 13). The increased expression of fibronectin and the collagens was observed exclusively in the area of scar formation but not in normal myocardium.

Fibrocytes function as mediators for tissue repair and scar formation because they synthesize fibronectin [17] and the collagens. Fibronectin regulates angiogenesis by exerting growth factor activities [18]. The collagens which appear later in the time course of wound healing exert a regulating effect on neovascularization [19, 20].

Table 1 presents a summary of the results obtained in our study which shows the time course of events in wound healing and scar formation in ischemic myocardium.

Conclusion

Monocytes and macrophages produce the angiogenic factors aFGF and TNFα. These cells therefore play an important role in the process of neovascularization occurring in myocardial areas with focal necrosis and subsequent 'wound healing'. This process of neovascularization represents the origin of a capillary collateral network.

Table 1.

	12 h	24 h	72 h	1 w	4 w
TNFα	+	+++	+++	+++	+
aFGF mRNA	–	++	+++	+++	+
FN	+*	+*	++	+++	++
FN mRNA	–	–	+++	+++	+
Collagen I	–	–	–	+	+++
Collagen III	–	–	–	+	+++
Collagen VI	–	–	–	++	+++

+ slight increase
++ moderate increase
+++ significant increase
w weeks
* proteins probably taken up from plasma
FN Fibronectin

The major function of fibroblasts is attributed to repair and scar formation. In addition, fibroblasts may, however, also influence angiogenesis by synthetizing fibronectin that is supposed to exert growth factor activities and by inhibiting collagen synthesis.

References

1. Cotran R, Kumar V, Robbins S: Healing and Repair, 1989
2. Schaper J, Weihrauch D: Collateral vessel development in the porcine and canine heart. In: W. Schaper, J. Schaper (ed.). Collateral Circulation – Heart, Brain, Kidney, Limbs. Kluwer Academic Publishers, Boston/Dordrecht/London, 1993, pp 65–102
3. Zimmermann R, Schaper J, Münkel B, Mohri M, Schaper W: *In situ* hybridization studies of acidic fibroblast growth factor (aFGF) in ischemic porcine myocardium. J Mol Cell Cardiol 25 (Suppl I): IX P 3 (abstr), 1993
4. Chilian WM, Mass HJ, Williams SE: Microvascular occlusions promote coronary collateral growth. Am J Physiol 258: H 1103–H 1111, 1990
5. Nathan C: Secretory products of macrophages. J Clin Invest 79: 319–322, 1987
6. Schaper J, König R, Franz D, Schaper W: The endothelial surface of growing coronary collateral arteries. Intimal margination and diapedesis of monocytes. A combined SEM and TEM study. Virch Arch A (Pathol Anat) 370: 193–205, 1976
7. Polverini P, Cotran R, Gimbrone M: Activated macrophages induce vascular proliferation. Nature 269: 804–806, 1977
8. Knighton D, Hunt T, Scheuenthul H: Oxygen tension regulates the expression of angiogenesis factors by macrophages. Science 221: 1283–1258, 1983
9. Klagsbrun M, D'Amore PA: Regulators of angiogenesis. Annu Rev Physiol 53: 217–239, 1991
10. Folkman J, Klagsbrun M: Angiogenic factors. Sci 235: 442–447, 1987
11. Schaper W: New paradigms for collateral vessel growth. Basic Res Cardiol 88: 193–198, 1993
12. Leibovich S, Polverini P, Shepard H: Macrophage induced angiogenesis is mediated by tumor necrosis factor alpha. Nature 329: 630–632, 1987

13. Frater-Schröder M, Risau W, Hallmann R: Tumor necrosis type alpha, a potent inhibitor of endothelial cellgrowth *in vitro*, is angiogenic *in vivo*. Proc Natl Acad Sci USA 84: 5277–5281, 1987
14. Rosenbaum J, Howes E, Rubin R: Ocular inflammatory effects of intravitreally-injected tumor necrosis factor. Am J Pathol 133: 47–53, 1988
15. Fahey T, Sherry B, Tracey K: Cytokine production in a model of wound healing: the appearance of MIP-1, MIP-2, cachectin/TNF and IL-1. Cytokine 2: 92–99, 1990
16. Castagnoli C, Stella M, Berthod: TNF production and hypertrophic scarring. Cell Immunol 147: 51–63, 1993
17. Hynes RO, Yamada KM: Fibronectins: Multifunctional modular glycoproteins. J Cell Biol 95: 369–377, 1982
18. Vlodavsky I, Folkman J, Sullivan R: Endothelial cell-derived basic fibroblast growth factor: Synthesis and deposition into subendothelial extracellular matrix. Proc Natl Acad Sci USA 84: 2292–2296, 1987
19. Ingber D, Folkman J: Inhibition of angiogenesis through modulation of collagen metabolism. Lab Invest 59: 44–51, 1988
20. Nicosia RF, Belser P, Bonnano E: Regulation of angiogenesis *in vitro* by collagen metabolism. *In Vitro* Cell Dev Biol 27A: 961–966, 1991

Molecular and Cellular Biochemistry **147**: 21–27, 1995.

Cell-cell and cell-matrix adhesion molecules in human heart and lung transplants

Gustav Steinhoff and Axel Haverich
Department of Cardiovascular Surgery, Christian Albrechts University of Kiel, 24105 Kiel, Germany

Abstract

The interaction of immune cells with endothelial and target cells and extracellular matrix in human organ transplants is regulated by a number of receptor-ligand molecules. The molecules mediating intercellular adhesion and activation are classified as integrin, immunoglobulin and selectin families. In the present study the patterns of their cellular expression in human heart and lung transplants are described in normal state and during transplant rejection. The results reveal an organ specific regulation of the different adhesion molecules during transplant rejection. Specific differences were noted in the endothelial expression of vascular ligand molecules in the vascular segments of heart and lung transplants, especially in the lung capillaries. Cell type specific patterns of intercellular and cell-matrix adhesion molecules as their ligands were found in different states of graft rejection. Intravascular and interstitial differences in the expression patterns of leukocyte adhesion receptors support a concept of their stepwise function during graft infiltration. The implications for the organ specific appearance of inflammatory reactions in human heart and lung transplants as for immunosuppressive therapy are discussed. (Mol Cell Biochem **147:** 21–27, 1995)

Key words: heart transplantation, lung transplantation, adhesion molecules, integrin receptors

Introduction

In the recent years it has become clear that the MHC-directed alloantigen response as well as other inflammatory reactions by leukocytes to organ transplant endothelia are regulated by the receptor-ligand interaction of a number of cell adhesion molecules [1–6]. The presence of certain adhesion molecules on the endothelial cell surface is essential for the regulation of leukocyte-endothelium interaction (cell-cell) and the interaction with basal membrane matrix components (cell-matrix). The cellular expression of both types of molecules is influenced by cytokines. However, a stream bed specific difference of adhesion ligand molecules on arterial, venous and capillary endothelia has been described in human liver grafts [7, 8]. Similar organ specific differences are assumable to exist in heart and lung tissue. Local differences in leukocyte ligand molecule on endothelia could explain predilection sites of transplant infiltration and organ specific pecularities in the susceptibility to the rejection response.

Patients and methods

Heart transplant biopsy (n = 304) was studied in 24 patients after orthotopic heart transplantation. In all cases it was the first heart transplantation for the recipient, carried out in 1985 and 1986. Maintenance immunosuppression and rejection treatment were performed as described earlier [1]. The pathological classification of the biopsies was performed according to standard criteria. The diagnoses ranged from no rejection (n = 60), mild (n = 54), moderate (n = 65), severe (n = 2) to resolving rejection (n = 123). Time intervals after transplantation were: 1–28 days (n = 59), 29–90 days (n = 121), 91–365 days (n = 107), 1–7 years (n = 22). Normal heart tissue (n = 7) was obtained from patients undergoing open heart surgery.

Lung transplant biopsy material was studied in patients requiring retransplantation due to acute or chronic rejection [9]. The four grafts studied were resected at day 11[a],66,68, and 463 post LuTx because of acute[a] (n = 1) or chronic rejection with bronchiolitis obliterans (n = 3). Normal lung specimens from organ donors (not used for transplantation)

Address for offprints: G. Steinhoff, Klinik für Herz- und Gefäßchirurgie, Universitätsklinik Kiel, Arnold Heller Str. 7, 24105 Kiel, Germany were studied for comparison (n = 4). The postoperative maintenance immunosuppression consisted of ciclosporin A, prednisolone and azathioprine. Initial induction treatment consisted of an additional 10 day ATG (5 mg/kg/d) treatment. Acute rejection was primarily treated by steroid bolus (3 days 500 mg methylprednisolone) and secondarily upon treatment resistance with ATG or OKT3. The biopsies were all frozen immediately and stored in liquid nitrogen. The expression of intercellular immune adhesion molecules was studied on cryostat sections using standard immune peroxidase and alkaline phosphatase immunohistological techniques. The monoclonal antibodies used for detection of cell adhesion molecules are listed in Table 1.

Results

The results are given as a survey on the expression patterns of adhesion molecules found in human lung and heart transplant biopsies by standard immunohistology on cryostat sections. The monoclonal antibodies used are specified in Table 1.

Lung transplant biopsy material was studied in four patients requiring retransplantation (day 11–463 post transplantation) due to acute (n = 1) or chronic (n = 3) rejection. Heart transplant biopsies (n = 304) were studied in 24 patients after orthotopic heart transplantation with different degrees of acute and chronic rejection activity.

The results on the expression of adhesion molecules on the endothelial and parenchymal cells types are summarized in Table 2 (immunoglobulin superfamily), Table 3 (integrin matrix receptors) and Table 4 (selectin, CD44 and HECA 452). Table 5 summarizes the adhesion receptor patterns found on graft infiltrating lymphocytes in lung allografts.

Discussion

A cell-type and organ specific expression of the various adhesion molecules was found in human heart and lung transplants. Major difference was observed in the pattern of coexpressed molecules both in normal tissue as during transplant rejection. Interestingly even between the endothelia in different parts of the vascular stream bed inside the heart or the lung differences in coexpressed adhesion molecules were

Table 1. Adhesion molecules studied and monoclonal antibodies used

Adhesion molecule	Ligand structure	Monoclonal antibody	Producer/Source
IMMUNOGLOBULIN SUPERFAMILY			
CD2	LFA-3	6F10.3;8E6B3	a
LFA-3 (CD58)	CD2	G26.1;TS2/9	a; T. Springer, Boston
ICAM-1 (CD54)	CD11a/b	84H10;RR1/1;6.5B5	a; Rothlein, Boston; Haskard,
London			
ICAM-2	CD11a	CBR-IC2/1	b
VCAM-1	VLA-4	1.4C3	a
NCAM (CD56)	NCAM	T199	a
PECAM-1 (CD31)	?	5.E.6	a
INTEGRIN FAMILY			
VLA-1	laminin, collagen	TS2/7	c
VLA-2	laminin, collagen	Gi9	a
VLA-3	fibronectin, laminin, collagen	P1B5	a
VLA-4	VCAM-1, fibronectin	HP 2/1	a
VLA-5	fibronectin	SAM-1	a
VLA-6	laminin	GoH3	a
CD51	vitronectin, fibrinogen	AMF/7	a
	vWF, thrombospondin, fibronectin		
SELECTIN FAMILY			
ELAM-1	CD15, sialyl Lewis X	1.2B6; HP18	a; d
CD62	sialyl Lewis X	CBL.thromb6	a
UNCLASSIFIED			
CD44	hyaluronic acid	SBU24-32,F10-44-2	McKenzie, Melbourne;
Dalchau, London			
		Mem-85	Horeysi, Prague
HECA452	?	HECA-452	Duivestijn, Maastricht

Source of monoclonal antibodies: [a]DIANOVA, Hamburg, Germany; [b]BENDER MED/SERVA, Heidelberg, Germany; [c]BIERMANN/T-CELL SCIENCES, Bad Nauheim, Germany; [d]BECTON-DICKINSON, Heidelberg, Germany

Table 2. Expression of immunoglobulin superfamily cell-cell ligand adhesion molecules in human lung and heart transplants

	Lung transplants							Heart transplants				
	a	ao	c	v	pn	be	mac	ao	c	v	end	myo
LFA-3												
normal	++	++	+++	++	–	+	++	++	+	++	+	++
rejection	++	++	+++	++	–	+	+++	++	++	++	++	+++
ICAM-1												
normal	+	+++	+++	++	+++	–	–	++	+	++	(+)	–
rejection	++	+++	+++	+++	+++	–	++	+++	+++	++	++	+
ICAM-2												
normal	–	+++	+++	++	–	–	–	++	+	++	–	–
rejection	+++	+++	+++	+++	–	–	–	++	++	++	–	–
VCAM-1												
normal	+++	–	–	–	–	–	–	++	–	(+)	(+)	–
rejection	+++	+++	–	++	++	–	–	+++	++	++	++	–
NCAM (CD56)												
normal	–	–	–	–	–	–	–	–	–	–	–	(+)
rejection	(+)	–	–	–	(+)	–	–	–	+	–	–	++
PECAM (CD31)												
normal	+++	+++	+++	+++	–	–	++	+++	++	+++	+	–
rejection	+++	+++	+++	+++	–	–	+++	+++	+++	+++	+	–

Abbreviations used: Lung – pulmonary artery (a), arteriole (ao), capillary (c), pulmonary vein (v), pneumocyte (pn), bronchial epithelium (be), alveolar macrophage (mac); Heart – arteriole (ao), capillary (c), vein (v), endocardium (end), myocyte (myo). n.d. – not determined.

Table 3. Expression of integrin cell-matrix adhesion molecules in human lung and heart transplants

	Lung transplants							Heart transplants			
	a	ao	c	v	pn	be	mac	ao	c	v	myo
VLA-1											
normal	+++	+++	+++	+++	–	–	–	–	++	++	–
rejection	+++	+++	+++	+++	–	–	–	–	+++	+++	–
VLA-2											
normal	–	–	++	–	–	++	–	++	–	(+)	–
rejection	+	++	++	(+)	++	++	+	++	–	+	–
VLA-3											
normal	++	++	+++	+	+++	+++	–	+++	–	++	–
rejection	++	++	++	++	++	+++	–	++	+	+++	–
VLA-4											
normal	–	–	–	–	–	–	+++	–	–	(+)	–
rejection	–	–	–	–	–	–	+++	–	–	+	–
VLA-5											
normal	+++	+++	+++	+++	–	–	+++	–	–	++	–
rejection	+++	+++	+++	+++	++	+++	+++	+++	++	+++	–
VLA-6											
normal	–	++	+++	++	–	++	–	++	++	++	–
rejection	+++	+++	+++	+++	–	+	–	+++	+++	+++	–
CD51											
normal	+++	++	++	++	++	+++	++	n.d.			
rejection	+++	+++	++	++		++	++	n.d.			

Table 4. Expression of selectin and unclassified adhesion molecules in human lung and heart transplants

	Lung transplants							Heart transplants			
	a	ao	c	v	pn	be	mac	a	v	c	myo
E-selectin (ELAM-1)											
normal	(+)	++	–	–	–	–	–	(+)	–	–	–
rejection	–	–	–	–	–	–	–	++	+	++	–
P-selectin (CD62, GMP140)											
normal	++	–	–	++	–	–	–	(+)	–	–	–
rejection	+++	+++	–	++	–	–	–	++	–	+	–
CD44											
normal	+	+++	+++	+++	+++	+++	+++	–	–	–	–
rejection	+	+++	+++	+++	+++	+++	+++	–	(+)	–	–
HECA452											
normal	–	–	–	–	–	–	–	n.d.			
rejection	–	–	–	++	+++	+++	+++	n.d.			

Table 5. Intravascular and Interstitial Expression of Lymphocyte Adhesion Receptors in Human Lung Transplants

Adhesion molecule	Intravascular[1] normal	Rejection	Tissue infiltrate normal	Rejection
CD2	+	+	+	+
LFA-3	–	+	–	+
ICAM-1	–	+	+	+
ICAM-2	–	–	–	–
VCAM-1	–	–	–	–
NCAM	+	+	+	+
CD-31(PECAM)	(+)	+	–	(+)
LFA-1	+	+	+	+
CD11B	–	–	–	–
CD11C	–	–	–	–
VLA-1	–	–	–	+
VLA-3	–	–	–	–
VLA-4	+	+	+	+
VLA-5	–	–	–	+
VLA-6	–	–	–	–
CD51	–	–	–	+
LECAM/Leu8	(+)	+	–	+
ELAM-1	–	–	–	(+)
CD62	–	–	–	–
CD44	–	+	–	+
HECA452	–	(+)	–	+

Explanations: 1: 'intravascular' refers to endothelial adherent leukocytes. 'interstitial' refers to perivascular infiltrating leukocytes.

observed. It is conceivable that these differences result in a fine regulation of leukocyte reactivity inside the vascular stream bed of the organs. Furthermore, by specific ligand molecule composition a homing of leukocyte subpopulations to certain vascular compartments may be facilitated. Organ specific differences in the immunological susceptibility can be explained by the patterns of basal or induced expression of intercellular adhesion molecules on the different organ cell types. The expression of MHC-[10, 11] and other adhesion molecules including cell-matrix receptors is clearly cell-type specific. Despite major differences in the patterns of ligand molecules and their inducibility by cytokines it can be generally postulated that almost all cell types can be immunologically recognized by immune cells upon induction of the main ligand molecules: MHC (class I), LFA-3 and ICAM-1. A relative resistance against immunological recognition can exist by deficient basal expression as on cardiomyocytes or by a relative resistance against cytokine mediated induction.

The composition of such differing immunocompetent cell types in an organ transplant may determine its susceptibility to immune destruction. Generally a minor basal expression of adhesion ligand molecules or incomplete ligand pattern can be overcome by the action of local released cytokines from infiltrating leukocytes or tissue macrophages. The susceptibility to cytokine stimulation in the different cell types may also determine patterns of adhesion ligand molecules induced in a specific organ. Thus the expression of cytokine receptors on the different organ cell types can be postulated to form a regulatory link to the induction of several adhesion ligand molecules.

The ischemic damage during organ preservation and the post-reperfusion inflammatory response is a major problem affecting graft function of heart and lung in the first days after transplantation [12]. The induction of leukocyte adhesion leading to unspecific graft inflammation is a major pathomechanism of reperfusion injury in organ transplants [12]. Already intraoperatively, but also in the immediate phase after implantation immunological changes in the induction of adhesion molecules (E, P-selectin) and MHC [7, 8] have been observed. In the present study tissue of non-transplant 'control' lungs already contained E- and P-selectin positive endothelia. The inducibility of E- and P-selectin as well as VCAM-1 points to an endothelial activation even prior to transplantation that may lead to subsequent induction of leukocyte adhesion. Further steps of leukocyte infiltration and cytokine release then may be initiated in dependence on the extent of the postischemic reaction and the generation of a cascade reaction by the injured endothelia and perivascular macrophages. The extent of adhesion molecule induction may very well relate to the metabolic impairment of the transplant. It can be stated that an arteriolar or capillary leukocyte stasis induced by the endothelial reperfusion reaction may cause regional microcirculatory perfusion defects [13]. Their additional ischemic consequence may potentiate the tissue damage even in a later course after 1 or 2 days, if additional ligand molecules as ICAM-1 and MHC are induced. Secondly, the first passage sensitization of alloreactive T-lymphocytes and the induction of rejection activity may very well be influenced by the initial postischemic inflammatory reaction. This has been stated for the early induction of class I and class II MHC molecules [14] and may also relate to the coinduction of other adhesion ligand molecules. Clinical observations of an increased rejection rate in initial bad functioning grafts point in this direction [15]. In experimental research the postischemic reperfusion damage could be inhibited by anti-CD11/CD18 monoclonal antibodies [16, 17] and prostaglandin E1 [18]. This may open possibilities to modify the reaction clinically by the use of such tools. It can be assumed, however, that the selectin molecules may have central importance in the induction phase of inflammation and their manipulation to prevent reperfusion inflammation is of utmost interest.

Lung pneumocytes coexpress a variety of adhesion ligand molecules as MHC class I and II, ICAM-1, and CD44 in the normal lung. During rejection this expression was enhanced and an additional expression of VCAM-1, HECA 452 and NCAM (few cells) could be observed. It seems that the pneumocytes, either type I and type II possess a broad panel of ligand molecules enabling intensive interaction with leukocyte receptors. It is possible, that this especially is required for the intense interaction with alveolar macrophages in the defense of infectious pathogens. It is not excluded, however, that also interaction to other leukocytes is possible. This is contradicted by the fact, however, that acute lung transplant rejection finds only minor manifestation in the lung alveoli, but instead at bronchial epithelium and perivascular at arterioli. Thus a presumed high susceptibility of pneumocytes to alloantigenic lymphocyte interaction – even in the presence of a full panel of ligand molecules – seems not to occur during transplant rejection. In contrast, bronchial epithelia that express a lesser panel of ligand molecules (MHC class I, LFA-3, CD44) and are induced to express class II MHC during rejection [11] seem to be a major target of the acute and chronic immune response leeding to obliterative bronchiolitis.

The basal expression of ICAM-1, LFA-3, ICAM-2, PECAM on capillary, venous and arterial endothelia of the lung may also point to a physiological high reactivity and susceptibility of lung endothelia to circulating immune cells. In rejecting grafts this expression was increased, however main changes could be observed on the endothelia of larger arteries. These have differences in the normal state with a basal expression of VCAM-1 and P-selectin and a lesser expression of ICAM-1 (few) and ICAM-2 (negative). Only in the rejecting grafts a complete pattern was found to be induced, whereas arteriolar, capillary, and venous endothelia failed to express VCAM-1. Capillary endothelia failed to express VCAM-1, E- and P-selectin. This resembles the pattern found in liver graft sinusoidal lining cells [7, 8]. It seems that similar differences exist in the lung between the arterial, capillary and venous vessel compartments in the pattern of vascular adhesion ligand molecules both in the basal state of expression and in induced state with transplant inflammation. A true absence of inducibility for the early adhesion molecules VCAM-1, E- and P-selectin seems to exist in lung capillaries. Although it is not excluded, that especially a temporary or focal induction of E-selectin is possible [19] and not present anymore in the explanted grafts with longstanding rejection, this possibility seems to be excluded for VCAM-1 and P-selectin, as these molecules are very well induced on arteries or veins (partly). Thus a differential pattern of ligand molecule expression and differences in susceptibility to inflammatory induction clearly exist for lung vascular endothelia. The lack of inducibility of the early adhesion molecules E-, P-selectin needed for rolling of leukocytes/ thrombocytes and VCAM-1 necessary for definitive adhesion and extravasation in the capillary stream bed could explain the almost complete absence of lymphocyte infiltration in acute or chronic rejection. This mechanism could very well compensate for the high expression of other ligands as ICAM-1, -2 found on endothelia and pneumocytes and possibly facilitating leukocyte interaction in a temporary way. The lack of the initiation step, the rolling of leukocytes by selectins, for leukocyte adhesion, may effect a different mode of leukocyte-endothelial interaction and prohibit the induction of further steps leading to tissue infiltration [20].

A similar difference in adhesion receptor patterns was found for cell-matrix integrin receptors between the vascular endothelia of the heart and the lung. Differences in cell-matrix interaction or possibly in the composition of the subendothelial matrix may exist between arterial, venous, and capillary endothelia and are presumably organ or tissue specific. It is very well possible, that such differences either reflect cell differentiation or the local microenvironment composed of cell-matrix produced by perivascular cells. Such differences may have function in the special interaction of endothelial cells with the surrounding tissue cells and in their site specific anchoring to basal membrane matrix proteins as it has been demonstrated *in vitro* for fibroblasts [21]. Differences in basal membrane cell-matrix molecules may also influence leukocyte and thrombocyte adhesion by their respective integrin receptors (VLA-4 and VLA-2). Thus differences in the organ specific composition of adhesion ligand molecules may influence the manifestation of the immune response by offering distinct cell- or matrix- adhesion targets. A molecular microheterogeneity of cellular ligands as ICAM-1 isoforms or CD44 may exist in addition [22, 23] that the regulation of immune responses inside an organ may underly various special restrictions as dictated by the presence of respective activating or deactivating ligand structures.

A further indication for inflammatory changes of the resident lung transplant cells during long-standing rejection comes from an adhesion receptor change on alveolar macrophages. Although this cell type seems not to be affected by the rejection reaction and is physiologically replaced by recipients bone-marrow cells, inflammatory changes leading to lung pathology can be suspected. Especially the release of cytokines upon infections or rejection of the graft may be a major promotor of interstitial fibrotic change in the lung alveoli. A broad induction of adhesion molecules was found for ICAM-1, LFA-3, PECAM, VLA-2, VLA-4 and HECA-452 during lung transplant rejection. Both cell-cell as cell-matrix molecules are induced. This could be a general effect of cytokines released by lymphocytes or endothelial cells, but points to an activated state of the tissue macrophages. The chronic inflammatory upregulation of alveolar macrophages itself, however, could be a major threat to the integrity of the lung transplant as it could maintain interstitial inflammation even in the absence of lymphocyte infiltration. This ultimately could cause increased cytokine levels leading to T lymphocyte activation and the increased processing and presentation of alloantigenic-MHC antigens that may induce the rejection activity.

A number of new aspects arise for the immunosuppression of the rejection related immune response after solid organ transplantation in general. Experimental studies using anti-ICAM-1 [24, 25], anti-CD11/CD18 monoclonal antibodies [26, 27] and anti-VLA-4 [28, 29] support the central role of these molecular interactions for the infiltration process. For clinical purposes in modification of immunosuppression, however, not the usage of such monoclonal antibodies may be a major prospect, but the identification of anti-adhesive capacities of peptides, oligosaccharides [30] and other drug substances. Soluble forms of adhesion molecules as ICAM-1, LFA-3, E- and P-selectin and MHC are of major interest for future immunosuppressive treatment strategy. It can be estimated that these molecules have major function in the intravascular regulation of the immune response in addition to cytokines. Processes of intravascular leukocyte activation and depression may be influenced by the release and interaction of adhesion ligand peptide molecules. It is possible that the intravascular binding of soluble ICAM-1 or LFA-3 peptides to their respective receptors on leukocytes leads to a change in expression density or receptor conformation either to increase or decrease the state of intercellular affinity (leukocyte-thrombocyte, leukocyte-leukocyte, leukocyte-endothelium). It is also possible as the intravascular or interstitial blockade of interactive adhesion receptors by soluble adhesion ligands such as a downregulation to target cell binding. In this context the identification of various molecular isoforms [22, 23] may point to a differentiated use of these for the regulation of intercellular contacts. Similar effects have been postulated for the binding of soluble alloantigenic MHC to the T-cell receptor. For the alloantigenic situation of organ transplants also differential effects of the production of soluble adhesion peptides may modify the T-lymphocyte response in an organ specific way. The use of soluble adhesion molecules or special blocking peptides is likely to open new avenues for the manipulation of the immune response. This may involve not only organ transplant related questions as the induction of tolerance or modification of chronic rejection, but also may have relevance for the treatment of several non-transplant inflammatory diseases and postischemic reactions.

Acknowledgements

This study was supported by the Deutsche Forschungsgemeinschaft, grant STE 495/2-1.

References

1. Steinhoff G, Behrend M, Haverich A: Signs of endothelial inflammation in human heart allograft biopsies. Eur Heart J 12: 141–143, 1990
2. Springer TA: Adhesion receptors of the immune system. Nature 346: 425–434, 1990
3. Hogg N, Harvey J, Cabanas C, Landis RC: Control of leukocyte integrin activation. Am Rev Respir Dis 148: 55–9, 1993
4. Springer TA, Lasky TA: Sticky sugars for selectins. Nature 349: 196, 1991
5. Hynes RO: Integrins: Versatility, modulation and signaling in cell adhesion. Cell 69: 11–25, 1992
6. Steinhoff G (ed.): Cell adhesion molecules in human organ transplants. R.G. Landes Company, Austin. 1993, pp 1–110
7. Steinhoff G, Behrend M, Schrader B, Duijvestijn AM, Wonigeit K: Expression Patterns of Leucocyte Adhesion Ligand Molecules on Human Liver Endothelia: Lack of ELAM-1 and CD62 Inducibility on Sinusoidal Endothelia and Distinct Distribution of VCAM-1, ICAM-1, ICAM-2 and LFA-3. A J Path 142: 481–488, 1993
8. Steinhoff G, Behrend M, Schrader B, Pichlmayr R: Intercellular Immune Adhesion Molecules in Human Liver Transplants – Overview on Expression Patterns of Leukocyte Receptor and Ligand Molecules. Hepatology 18(2): 440–453, 1993
9. Haverich A, Hirt SW, Wahlers T, Schäfers HJ, Zink C, Borst HG: Functional results after lung retransplantation. J Heart Lung Transpl 13: 48–55, 1994
10. Steinhoff G, Wonigeit K, Schäfers HJ, Haverich A: Sequential analysis of monomorphic and polymorphic major histocompatibility complex antigen expression in human heart allograft biopsy specimens. J Heart Transpl 8: 360–370, 1989
11. Taylor PM, Rose ML, Yacoub MH: Expression of MHC antigens in normal human lung and transplanted lungs with obliterative bronchiolitis. Transplantation 48: 506–510, 1989
12. Clavien PA, Harvey PRC, Strasberg SM: Preservation and reperfusion injuries in liver allografts. Transplantation 53: 957–978, 1992
13. Marzi I, Walcher F, Menger M, Bühren V, Harbauer G, Trentz O: Microcirculatory disturbances and leucocyte adherence in transplanted livers after cold storage in Euro-Collins, UW and HTK solutions. Transplant Int 4: 45–50, 1991
14. Shackleton CR, Ettinger SL, McLoughlin MG, Scudamore CH, Miller RR, Keown PA: Effect of recovery from ischemic injury on class I and class II MHC antigen expression. Transplantation 49: 641–644, 1990
15. Howard TK, Klintmalm GB, Cofer JB, Huberg BS, Goldstein RM, Gonwa TA: The influence of preservation injury on rejection in the hepatic transplant recipient. Transplantation 49: 103–107, 1990
16. Horgan MJ, Wright SD, Malik AB: Antibody against leukocyte integrin (CD18) prevents reperfusion-induced lung vascular injury. Am J Physiol 259: 315–319, 1990
17. Simpson P, Todd III JRF, Fantone JC, Mickelson JM, Griffin JD, Lucchesi BR: Reduction of experimental canine myocardial reperfusion injury by a monoclonal antibody (anti-Mo1, anti-CD11b) that inhibits leukocyte adhesion. J Clin Invest 81: 624–629, 1988
18. Simpson PJ, Michalson J, Fatane JC, Gallagher KP, Lucchesi BR: Reduction of experimental canine myocardial infarct size with prostaglandin E1: inhibition of neutrophil migration and activation. J Pharmacol Exp Ther 244 (2): 619–624, 1988
19. Ferran C, Peuchmaur M, Desruennes M, Ghoussoub JJ, Cabrol A, Brousse N, Cabrol C, Bach JF, Chatenoud L: Implications of de novo ELAM-1 and VCAM-1 expression in human cardiac allograft rejection. Transplantation 55: 605–609, 1993
20. Lindbohm L, Xie X, Raud, Hedqvist P: Chemoattractant-induced leukocyte adhesion to vascular endothelium *in vivo* is critically dependent on initial leukocyte rolling. Acta Physiol Scand 146: 415–421, 1992
21. Dalton SL, Marcantonio EE, Assoian RK: Cell attachment controls fibronectin and a5β1 integrin levels in fibroblasts. Implications for anchorage dependent and -independent growth. J Biol Chem 267: 8186–8191, 1992
22. Seth R, Raymond FD, Makgoba MW: Circulating ICAM-1 isoforms: diagnostic prospects for inflammatory and immune disorders. Lancet 338: 83–84, 1991
23. Haynes BF, Telen MJ, Hale LP, Denning SM: CD44 – a molecule involved in leukocyte adherence and T-cell activation. Immunol Today 10: 423–428, 1989
24. Cosimi AB, Conti D, Delmonico FL, Preffer FI, Wee S-L, Rothlein R, Faanes R, Colvin R: *In vivo* effects of monoclonal antibody to ICAM-1 (CD54) in non-human primates with renal allografts. J Immunol 144: 4604–4612, 1990
25. Flavin T, Ivens K, Rothlein R, Faanes R, Clayberger C, Billingham M, Starnes VA: Monoclonal antibodies against Intercellular Adhesion Molecule 1 prolong cardiac allograft survival in cynomolgus monkeys. Transplant Proc 23: 533–534, 1991
26. Heagy W, Waltenbaugh C, Martz E: Potent ability of anti LFA-1 monoclonal antibody to prolong allograft survival. Transplantation 37: 520–523, 984
27. Kaslovsky RA, Horgan MJ, Lum H, McCandless BK, Gilboa N, Wright SD, Malik AB: Pulmonary edema induced by phagocytosing neutrophils – protective effect of monoclonal antibody against phagocyte CD18 Integrin. Circ Res 67: 795–802, 1990
28. Paul LC, Davidoff A, Paul DW, Bendiktsson H, Issekutz TB: Monoclonal antibodies against LFA-1 and VLA-4 inhibit graft vasculitis in rat cardiac allografts. Transplant Proc 25: 813, 1993
29. Orosz CG, Ohye RG, Pelletier RP, Van Buskirk AM, Huang E, Morgan C, Kincade PW, Ferguson RM: Treatment with anti-vascular cell adhesion molecule 1 monoclonal antibody induces long-term murine cardiac allograft acceptance. Transplantation 56(2): 453–60, 1993
30. Mulligan MS, Paulson JC, De Frees S, Zheng ZL, Lowe JB, Ward PA: Protective effects of oligosaccharides in P-selectin-dependent lung injury. Nature 364: 149–51, 1993

Molecular and Cellular Biochemistry **147**: 29–34, 1995.

Progression of heart failure: A role for interstitial fibrosis

Hani N. Sabbah, Victor G. Sharov, Michael Lesch and Sidney Goldstein

Department of Medicine, Division of Cardiovascular Medicine, Henry Ford Heart and Vascular Institute, Detroit, Michigan, USA

Abstract

Progressive deterioration of left ventricular (LV) function is a characteristic feature of the heart failure (HF) state. The mechanism or mechanisms responsible for this hemodynamic deterioration are not known but may be related to progressive intrinsic dysfunction, degeneration and loss of viable cardiocytes. In the present study, we tested the hypothesis that accumulation of collagen in the cardiac interstitium (reactive interstitial fibrosis, RIF), known to occur in HF, results in reduced capillary density (CD = capillary/fiber ratio) and increased oxygen diffusion distance (ODD) which can lead to hypoxia and dysfunction of the collagen encircled myocyte. Studies were performed in LV tissue obtained from 10 dogs with chronic HF (LV ejection fraction 26 ± 1%) produced by multiple sequential intracoronary microembolizations. In each dog, CD and ODD were evaluated in LV regions that manifested severe RIF (volume fraction 16 ± 2%) and in LV regions of little or no RIF (volume fraction 4 ± 1%). In regions of severe RIF, CD was significantly decreased compared to regions of no RIF (0.92 ± 0.02 vs. 1.05 ± 0.03) ($P < 0.003$). Similarly, ODD was significantly increased in regions of severe RIF compared to regions of no RIF (15.3 ± 0.4 vs. 12.2 ± 0.3 μm) ($P < 0.001$). These data suggest that in dogs with chronic HF, constituent myocytes of LV regions which manifest severe RIF may be subjected to chronic hypoxia; a condition that can adversely impact the function and viability of the collagen encircled cardiocyte. (Mol Cell Biochem **147**: 29–34, 1995)

Key words: heart failure, ventricular function, interstitial fibrosis, coronary microcirculation

Introduction

Left ventricular (LV) dysfunction, once established as a consequence of a primary myocardial injury event such as an acute myocardial infarction, can deteriorate over a period of months or years, despite the absence of clinically apparent intercurrent events [1–3]. This spontaneous and progressive deterioration of LV function often culminates in the syndrome of congestive heart failure. The mechanism or mechanisms responsible for this hemodynamic deterioration are not known but have been broadly attributed to entry into a so-called 'auto-induction phase' whereby compensatory mechanisms elicited to maintain homeostasis such as compensatory LV hypertrophy [4], dilation [5] and enhanced activity of the sympathetic nervous system and renin-angiotensin system [6, 7], themselves become factors leading to the progressive deterioration of the heart failure state. A possible working hypothesis is that activation of these compensatory mechanism lead directly or indirectly to progressive intrinsic dysfunction, degeneration and loss of residual viable cardiocytes.

At the cellular level, several structural alterations in both cardiocytes and interstitium occur in the failing heart which acting individually or in concert can adversely influence global LV contractile performance. Such alterations include 1) hypertrophy of residual myocytes [4], abnormalities of myocyte contractile structures [8], abnormalities of mitochondria [9], and progressive accumulation of collagen in the interstitial space [10]. In the present study, we tested the hypothesis that accumulation of collagen in the cardiac interstitial compartment is associated with a reduction in capillary density and with an increase in oxygen diffusion distance. If these changes are present in regions of the failing LV, they could lead to regions of chronic hypoxia, a condition which can adversely impact the function and ultimately the viability of

Address for offprints: H.N. Sabbah, Henry Ford Hospital, 2799 West Grand Boulevard, Detroit, Michigan, 48202 USA the collagen encircled myocyte. Histopathologic studies were performed in LV tissue obtained from dogs with chronic heart failure produced by multiple sequential intracoronary microembolizations.

Methods

Animal model of chronic heart failure

The dog model of chronic heart failure used in this study was previously described in detail [11]. The model manifests many of the sequelae of heart failure seen in patients, including marked and sustained depression of LV systolic and diastolic function, left ventricular hypertrophy and dilation, reduced cardiac output, increased systemic vascular resistance and enhanced activity of the sympathetic nervous system evidenced by marked elevation of plasma norepinephrine concentration [11]. In the present study, chronic LV dysfunction was produced in 10 dogs by multiple sequential intracoronary embolizations with polystyrene latex microspheres (77–102 μm in diameter). Coronary microembolizations were performed 1–3 weeks apart during cardiac catheterizations. All procedures were conducted under general anesthesia and sterile conditions. Anesthesia consisted of intravenous injections of oxymorphone hydrochloride (0.22 mg/kg), diazepam (0.17 mg/kg) and sodium pentobarbital (150–250 mg to effect). In all dogs, coronary microembolizations were discontinued when LV ejection fraction, determined angiographically, was 30–40%. Complete hemodynamic and angiographic studies were performed at an average of 3 weeks after the last embolization and were repeated at 4 months after the last embolization. At the end of the follow-up period (4 months after the last embolization), the dogs were killed and the hearts were removed and prepared for histologic examination. The study was approved by the Henry Ford Hospital Care of Experimental Animals Committee and conformed to the guiding principles of the American Physiological Society.

Hemodynamic, angiographic and neurohumoral measurements

Aortic and LV pressure were measured with catheter-tip micromanometers (Millar Instruments). Peak LV rate of change of pressure during isovolumic contraction (peak + dP/dt) and isovolumic relaxation (peak-dP/dt) were derived from the LV pressure waveform using analog differentiation. Mean right atrial pressure was measured using a Swan-Ganz catheter in conjunction with a P23 XL pressure transducer (Spectramed). Cardiac output was measured using the thermodilution method and LV stroke volume was calculated as the ratio of cardiac output to heart rate. Systemic vascular resistance was calculated as the difference between mean aortic pressure and mean right atrial pressure times 80 divided by cardiac output as described by Grossman [12]. Left ventriculograms were obtained during cardiac catheterization with the dog placed on its right side and were recorded on 35 mm cine at 30 frames/sec during the injection of 20 ml of contrast material (Hypaque meglumine 60%, Winthrop Pharmaceuticals). Correction for image magnification was made with a radiopaque calibrated grid placed at the level of the LV. LV end-systolic and end-diastolic volumes were calculated from ventricular silhouettes using the area-length method [13]. LV ejection fraction was calculated as the difference between end-diastolic and end-systolic volume divided by end-diastolic volume times 100. Venous blood samples were obtained from conscious dogs for measurement of plasma norepinephrine and angiotensin-II concentration. Plasma norepinephrine was measured using aluminium oxide absorption by high liquid chromatography. In our laboratory this technique has a day-to-day coefficient of variation of 6.8%. Plasma immunoreactive angiotensin-II was measured by radioimmunoassay after extraction by reversible absorption by phenylsilyl-silica.

Immunohistochemical methods

Four months after the last embolization, dogs were anesthetized, the chest was opened and the heart was rapidly removed and placed in ice-cold cardioplegia solution. From each heart, transmural tissue blocks were obtained from the LV free wall at the mid-ventricular level and rapidly frozen in isopentane cooled to –160°C in liquid nitrogen. Cryostat sections, 8–10 μm thick, were prepared and double stained with rabbit anti-human collagen type III polyclonal antibody (Chemicon International, Inc.) to visualize interstitial collagen and with Griffonia Simplicifolia Lectin I to visualize capillaries [14, 15]. Immunofluorescent staining was evaluated with an epifluorescent microscope. Collagen was visualized under fluorescent light and capillaries under rhodamine light.

Morphometric assessments

The same sections obtained from each dog were used for quantifying myocyte size, volume fraction of interstitial collagen, capillary density and oxygen diffusion distance. For each analysis, ten non-infarct related microscopic fields, each containing a minimum of 100 cardiocytes were selected. Five of the ten fields were selected at random from myocardial regions that manifested severe interstitial fibrosis and five from regions that manifested little or no interstitial fibrosis. For each field, the average myocyte cross sectional area (radial sections only) was calculated using computer-assisted planimetry (SigmaScan, Jandell Scientific). The volume fraction (%) of interstitial collagen, area occupied by collagen as a percent of total surface area, was quantified using computer-assisted videodensitometry (JAVA Video Analysis Software, Jandell Scientific). Capillary density was calculated using the index capillary per fiber ratio (C/F) [16]. The oxygen diffusion distance was measured as half the distance between two adjoining capillaries [16]. For each dog, morphometric data are reported as the average of each parameter calculated separately from each of the 5 selected interstitial fibrosis fields and from each of the 5 selected field which manifested little or no interstitial fibrosis.

Data analysis

To establish the presence of progressive LV dysfunction in this canine model of heart failure, comparisons of hemodynamic, angiographic and neurohumoral variable was made between measures obtained at 3 weeks after the last embolization and at 4 months after the last embolization. For these comparisons, a Student's *t*-test was used. A probability of 0.05 or less was considered significant. Comparison of morphological measures (myocyte size, collagen volume fraction, capillary density, and oxygen diffusion distance were made between LV regions that manifested severe interstitial fibrosis and regions that manifested little or no interstitial fibrosis. For these comparisons, a *t*-statistic for two means was used. A probability of 0.05 of less was considered significant. All data are reported as the mean ± standard error of the mean.

Results

Hemodynamic, angiographic and neurohumoral findings

Hemodynamic, angiographic and neurohumoral data obtained at 3 weeks and at 4 months after the last embolization are shown in Table 1. During this 17 week follow-up period, there was a significant decrease in LV ejection fraction, peak LV +dP/dt, peak LV – dP/dt, and stroke volume. Simultaneously, there was a significant increase in LV end-diastolic volume, systemic vascular resistance, plasma norepinephrine concentration and plasma angiotensin-II. These data establish the existence of spontaneous and progressive deterioration of LV function in this canine model of heart failure long after complete cessation of coronary microembolizations.

Table 1. Hemodynamic, angiographic and neurohumoral findings at 3 weeks and 4 months after the last embolization (n = 10)

	3 Weeks	4 Months	Probability
LV EF (%)	36 ± 1	26 ± 1	$p < 0.001$
EDV (ml)	54 ± 6	70 ± 7	$p < 0.001$
Peak + dP/dt	1930 ± 70	1600 ± 60	$p < 0.001$
Peak –dP/dt	1730 ± 90	1340 ± 50	$p < 0.001$
SV (ml)	34 ± 2	27 ± 2	$p < 0.001$
SVR	3090 ± 307	3601 ± 358	$p < 0.011$
PNE (pg/ml)	372 ± 33	552 ± 54	$p < 0.006$
A-II (pg/ml)	19 ± 2	32 ± 4	$p < 0.010$

LV = left ventricular; EF = ejection fraction; EDV = LV end-diastolic pressure; +/–dP/dt = LV rate of change of pressure (mmHg/sec); SV = stroke volume; SVR = systemic vascular resistance (dynes-sec-cm ^{5}); PNE = plasma norepinephrine concentration; A-II = plasma immunoreactive angiotensin-II.

Morphological findings

Reactive interstitial fibrosis was present in viable LV myocardium of all dogs studied with considerable heterogeneity in location. In general, regions of viable myocardium adjacent to old infarction and large intramural vessels tended to manifest considerably more interstitial fibrosis compared to other LV regions. The volume fraction of interstitial collagen in LV regions that manifested severe interstitial fibrosis was 16 ± 2% compared to only 4 ± 1% in LV regions that manifested little or no interstitial fibrosis ($P < 0.001$). The average myocyte cross-sectional area was modestly larger among constituent myocytes of regions of severe interstitial fibrosis compared to constituent myocytes of LV regions of little or no interstitial fibrosis (723 ± 38 vs. 642 ± 34 μm^2) but this difference was not statistically significant ($P < 0.13$). Capillary density (C/F ratio) was significantly reduced in LV regions with severe interstitial fibrosis compared to regions of little or no interstitial fibrosis (Fig. 1). Oxygen diffusion distance increased significantly in LV regions of severe interstitial fibrosis compared to regions of little or no interstitial fibrosis (Fig. 2).

Discussion

A wide gap of knowledge has existed for decades with respect to the mechanism or mechanisms that underly the progressive deterioration of LV function in patients with heart failure. This gap of knowledge, may have resulted, in part, from limitations related to detailed study of this phase of the disease in humans and from the lack of an applicable experimental animal model that can act as surrogate to the human disease. It has become abundantly evident in recent years, that preventing the transition to congestive heart failure in patients with established LV dysfunction through early pharmacologic

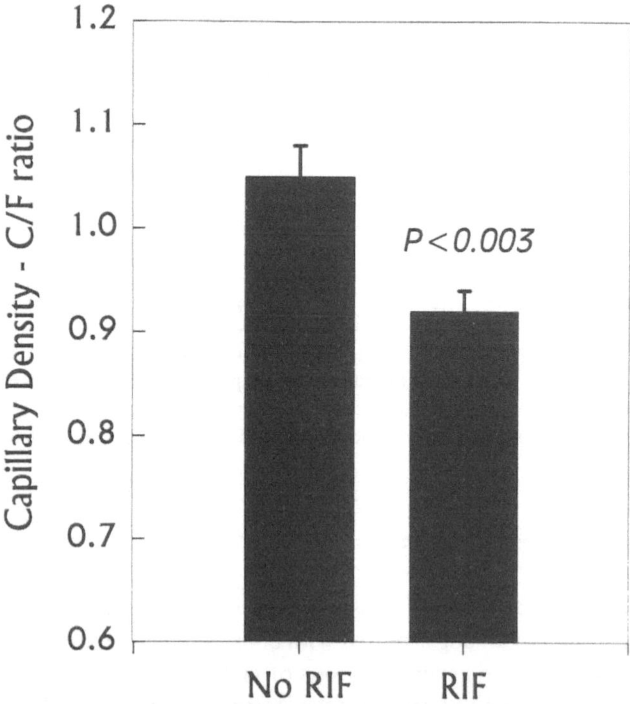

Fig. 1. Bar graph (mean ± SEM) depicting differences in capillary density (C/F Ratio) between LV regions of severe reactive interstitial fibrosis (RIF) and region of little or no interstitial fibrosis (No RIF). p-value is based on comparisons between RIF and No RIF.

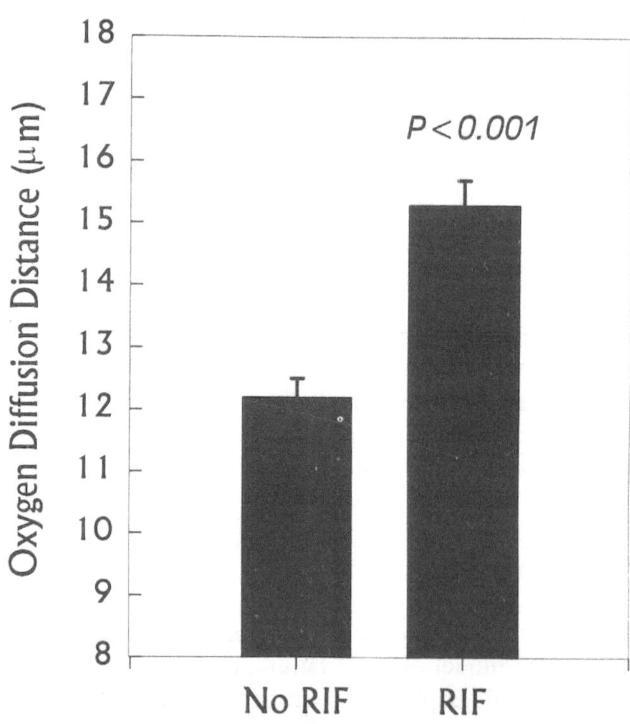

Fig. 2. Bar graph (mean ± SEM) depicting differences in oxygen diffusion distance between LV regions of severe reactive interstitial fibrosis (RIF) and region of little or no interstitial fibrosis (No RIF). p-value is based on comparisons between RIF and No RIF.

intervention, is potentially a more desirable alternative to the treatment of end-stage heart failure. Recognition of the positive merits of preventative therapy in this disease syndrome has spurred considerable interest in research aimed at expanding our understanding of the factors that promote progressive deterioration of LV function in patients who are at high risk of developing congestive heart failure. The observations made in the present study indicate that dogs with chronic heart failure produced by multiple sequential intracoronary microembolizations undergo spontaneous and progressive LV systolic and diastolic dysfunction long-after cessation of coronary microembolization. As such, this animal model simulates the continued decline in LV function seen in patients with established LV dysfunction [1–3] and provides a unique experimental tool to better discern and dissect the fundamental mechanisms responsible for this hemodynamic deterioration.

In the absence of any proven mechanisms, the following working hypothesis can be put forth to help uncover the factors responsible for the progressive deterioration of the hemodynamic state in patients with established LV dysfunction: progressive LV dysfunction results from ongoing intrinsic dysfunction, degeneration and loss of residual viable cardiocytes. It is often suggested that such cardiocyte dysfunction, degeneration and loss in the failing heart may result from sustained activation of compensatory mechanisms, elicited as a result of the initial injury, and intended to maintain homeostasis. Key among these compensatory mechanisms thought to exert a deleterious effect on the myocardium are enhanced activity of the sympathetic nervous system [6], enhanced activity of local and systemic renin-angiotensin systems [7, 17] and the development of compensatory hypertrophy of residual viable cardiocytes [4]. Activation of the sympathetic nervous system and renin-angiotensin system can have adverse hemodynamic consequences in heart failure because both systems enhance systemic vasoconstriction and promote the retention of sodium and water [18]. Recent studies, however, suggest that sustained activity of these neurohumoral systems can also have a direct effect on the heart independent of their hemodynamic action. Very high concentrations of norepinephrine and angiotensin-II have been shown to exert a direct cytotoxic effect of cardiocytes [19–21]. Although these neurohormones are undoubtedly elevated in the failing heart, as seen in the present study, it is not well established whether their concentration is sufficiently high to promote direct cardiocyte necrosis. Pathologic hypertrophy of residual cardiocytes also occurs in heart failure as a result of exposure to increased workload. Degenerative changes of residual cardiocytes have been reported in both patients and dogs with chronic heart failure [8]. According to the concept of the three degenerative stages of hypertrophy outlined by Meerson, this alteration alone can potentially lead to progressive cardiocyte degeneration and loss [22]. Direct evidence supporting this concept in the failing heart, however, is lacking.

Considerable interest has emerged in recent years with regard to alterations in the cardiac interstitium of the hypertrophied and failing heart. It is now recognized that accumulation of collagen occurs in the interstitial space of the hypertrophied and failing heart, a process termed 'reactive interstitial fibrosis' [23, 24]. The exact mechanism that promotes the accumulation of collagen in the interstitial compartment is not clear but has been attributed, in part, to enhanced activity of the renin-angiotensin-aldosterone system [23]. The fibrous tissue response of the cardiac interstitium is thought to be responsible for abnormal ventricular stiffness and has also been suggested to account for a spectrum of ventricular dysfunction that involve either the systolic or diastolic phase of the cardiac cycle or both [23]. Although intuitively it is easy to imagine how increased accumulation of collagen in the interstitial space can affect ventricular stiffness and consequently diastolic function, its ability to influence LV systolic function is not readily apparent. The observations, in the present study, of decreased capillary density and increased oxygen diffusion distance in viable LV regions that manifest severe interstitial fibrosis provide a rational explanation of the potential adverse influence that interstitial fibrosis can exert on systolic function. The observations may be used to advance the concept that interstitial fibrosis, when present, may be associated with localized chronic hypoxia; a condition which is likely to adversely influence the functional capacity and ultimately the viability of the collagen encircled cardiocyte. In the present study, oxygen diffusion distance, on average, increased by 24%. The increase in the capillary-to-myocyte oxygen diffusion distance appears to be due primarily to expansion of the interstitial space mediated by the deposition of collagen with some contribution from a modest increase in myocyte size. A modest increase in myocyte size, albeit not statistically significant, was observed in the present study in LV regions of severe interstitial fibrosis compared to regions of little or no fibrosis. Model studies by Rakusan have shown that a small increase in the oxygen diffusion distance, when the remaining oxygen determinants are normal, can result in hypoxia, while an increase to 70% of normal can decrease myocardial PO_2 to zero [16].

The concept that interstitial fibrosis can lead to hypoxia of the collagen encircled cardiocyte is supported by recent studies from this laboratory [25]. In LV tissue obtained from dogs with chronic heart failure, histological evaluations revealed a near 2-fold increase in lactate dehydrogenase activity in constituent myocytes of myocardial regions of severe interstitial fibrosis compared to myocytes of regions manifesting little or no fibrosis [25]. Recent transmission electron microscopic studies in LV tissue also obtained from dogs with chronic heart failure showed considerable degenerative changes of constituent myocytes of myocardial regions that manifested severe interstitial fibrosis compared to constituent myocytes of regions that did not manifest interstitial fibrosis [26]. These ultrastructural alterations included substantial loss of contractile elements and marked abnormalities of virtually every type of organelle including mitochondria [26]. These observations provide indirect evidence that myocytes of LV regions of severe interstitial fibrosis provide only limited contribution, if any, to overall LV contraction. Additional studies are needed to determine the degree to which interstitial fibrosis-mediated hypoxia influences the structure, function and ultimately the long-term viability of the collagen encircled cardiocyte.

In conclusion, the data from this study indicate that in dogs with progressive LV dysfunction, myocardial regions that manifest severe fibrosis are associated with reduced capillary density and increased oxygen diffusion distance. These abnormalities are likely to promote regional hypoxia which, in turn, may lead to structural and functional abnormalities of the collagen encircled cardiocytes. From this perspective, the accumulation of collagen in the cardiac interstitium can be viewed as a maladaptation that contributes the progressive deterioration of LV function in heart failure.

Acknowledgement

This study was supported by grants from the American Heart Association of Michigan and National Heart, Lung and Blood Institute HL 49090–01.

References

1. Ertl G, Kochsiek K: Development, early treatment, and prevention of heart failure. Circulation 87: IV-1-IV-2, 1993
2. Mckee PA, Castelli WP, Mcnamara PM, Kannel WB: The natural history of congestive heart failure: the Framingham study. N Engl J Med 285: 1441–1446, 1971
3. Konstam MA, Rousseau MF, Kronenberg MW, Udelson JE, Melin J, Stewart D, Dolan N, Edens TR, Ahn S, Kinan D, Howe DM, Kilcoyne L, Metherall J, Benedict C, Yusuf S, Pouleur H: Effects of the angiotensin converting enzyme inhibitor enalapril on long-term progression of left ventricular dysfunction in patients with heart failure. Circulation 86: 431–438, 1991
4. Anversa P, Olivetti G, Capasso JM: Cellular basis of ventricular remodeling after myocardial infarction. Am J Cardiol 68: 7D–16D, 1991
5. Pfeffer MA, Lamas GA, Vaughan DE, Parisi AF, Braunwald E: Effect of captopril on progressive ventricular dilatation after anterior myocardial infarction. N Engl J Med 319: 80–86, 1988
6. Levine TB, Francis GS, Goldsmith SR, Simon AB, Cohn JN: Activity of the sympathetic nervous system assessed by plasma hormone levels

and their relation to hemodynamic abnormalities in congestive heart failure. Am J Cardiol 49: 1659–1666, 1982

7. Curtiss C, Cohn JN, Vrobel T, Franciosa JA: Role of the renin-angiotensin system in the systemic vasoconstriction of chronic congestive heart failure. Circulation 58: 763–770, 1978
8. Sharov VG, Sabbah HN, Shimoyama H, Ali AS, Levine TB, Lesch M, Goldstein S: Abnormalities of contractile structures in viable myocytes of the failing heart. Intl J Cardiol 43: 287–297, 1994
9. Sabbah HN, Sharov VG, Riddle JM, Kono T, Lesch M, Goldstein S: Mitochondrial abnormalities in myocardium of dogs with chronic heart failure. J Mol Cell Cardiol 24: 1333–1347, 1992
10. Schaper J, Hein S: The structural correlate of reduced cardiac function in human dilated cardiomyopathy. Heart Failure 9: 95–111,1993
11. Sabbah HN, Stein PD, Kono T, Gheorghiade M, Levine TB, Jafri S, Hawkins ET, Goldstein S: A canine model of chronic heart failure produced by multiple sequential coronary microembolizations. Am J Physiol 260: H1379–H1384, 1991
12. Grossman W: Clinical measurements of vascular resistance and assessment of vasodilator drugs. In: W. Grossman and D.S. Baim (eds). Cardiac Catheterization, Angiography and Intervention. Lea and Febiger, Philadelphia, 1991, pp 143–151
13. Dodge HT, Sandler H, Baxley WA, Hawley RR: Usefulness and limitations of radiographic methods for determining left ventricular volume. Am J Cardiol 18: 10–24, 1966
14. Sabbah HN, Hansen-Smith F, Sharov VG, Kono T, Lesch M, Gengo PG, Steffen RP, Levine TB, Goldstein S: Decreased proportion of type I myofibers in skeletal muscle of dogs with chronic heart failure. Circulation 87: 1729–1737, 1993
15. Hansen-Smith FM, Watson L, Lu DY, Goldstein I: Griffonia simlicifolia I: Fluorescent tracer for microcirculatory vessels in nonperfused thin muscles and sectioned muscle. Microvasc Res 36: 199–215, 1988
16. Rakusan K: Oxygen in Heart Muscle. Charles C. Thomas, Springfield, 1971, pp 22–33, 66–71
17. Dzau VJ: Circulating versus local renin-angiotensin system in cardiovascular homeostasis. Circulation 77: I-4–I-13, 1988
18. Packer M: The neurohormonal hypothesis: A theory to explain the mechanism of disease progression in heart failure. J Am Coll Cardiol 20: 248–154, 1992
19. Tan LB, Jalil JE, Pick R, Janicki JS, Weber KT: Cardiac myocyte necrosis induced by angiotensin-II. Circ Res 69: 1185–1195, 1991
20. Mann DL, Kent RL, Parsons B, Cooper G: Adrenergic effects on the biology of the adult mammalian cardiocyte. Circulation 85: 790–804, 1992
21. Benjamin IJ, Jalil JE, Tan LB, Cho K, Weber KT, Clark WA: Isoproterenol-induced myocardial fibrosis in relation to myocyte necrosis. Circ Res 65: 657–670, 1989
22. Meerson FZ: The myocardium in hyperfunction, hypertrophy and heart failure. Circ Res 25: 1–163, 1963
23. Weber KT, Brilla CG: Pathological hypertrophy and cardiac interstitium. Fibrosis and renin-angiotensin-aldosterone system. Circulation 83: 1849–1865, 1991
24. Brilla CG, Weber KT: Reactive and reparative myocardial fibrosis in arterial hypertension. Cardiovasc Res 26: 671–677, 1992
25. Shimoyama H, Sabbah HN, Sharov VG, Cook J, Lesch M, Goldstein S: Accumulation of interstitial collagen in the failing left ventricular myocardium is associated with increased anaerobic metabolism among affected cardiomyocytes (Abstr). J Am Coll Cardiol, Special Issue: 98A, 1994
26. Sharov VG, Sabbah HN, Kono T, Ali AS, Shimoyama H, Lesch M, Goldstein S: Ultrastructural abnormalities of cardiomyocytes in the border zone of old infarctions: studies in dogs with chronic heart failure (Abstr). FASEB J 7: A112, 1993

Molecular and Cellular Biochemistry **147**: 35–42, 1995.

Chronic hibernating myocardium: Interstitial changes

Jannie Ausma[1], Jack Cleutjens[2], Fred Thoné[3], Willem Flameng[4], Frans Ramaekers[1] and Marcel Borgers[1,3]

[1]Department of Molecular Cell Biology and Genetics, Cardiovascular Research Institute Maastricht, University of Limburg, Maastricht, The Netherlands; [2]Department of Pathology, Cardiovascular Research Institute Maastricht, University of Limburg, Maastricht, The Netherlands; [3]Department of Morphology, Life Sciences, Janssen Research Foundation, Beerse; [4]Department of Cardiovascular Surgery, Catholic University of Leuven, Belgium

Abstract

Chronic left ventricular dysfunctional but viable myocardium of patients with chronic hibernation is characterized by structural changes, which consist of depletion of contractile elements, accumulation of glycogen, nuclear chromatin dispersion, depletion of sarcoplasmic reticulum and mitochondrial shape changes. These alterations are not reminiscent of degeneration but are interpreted as de-differentiation of the cardiomyocytes. The above mentioned changes are accompanied by a marked increase in the interstitial space. The present study describes qualitative and quantitative changes in the cellular and non-cellular compartments of the interstitial space. In chronic hibernating myocardial segments the increased extracellular matrix is filled with large amounts of type I collagen, type III collagen and fibronectin. An increase in the number of vimentin-positive cells (endothelial cells and fibroblasts) compared with normal myocardium is seen throughout the extracellular matrix.

The increase in interstitial tissue is considered as one of the main determinants responsible for the lack of immediate recovery of contractile function after restoration of the blood flow to the affected myocardial segments of patients with chronic left ventricular dysfunction. (Mol Cell Biochem **147**: 35–42, 1995)

Key words: extracellular matrix components, basement membrane, chronic hibernating myocardium, collagens

Introduction

The myocardium consists of muscle fibers and blood vessels connected and interspersed by a network of connective tissue. The scaffolding of the collagen matrix, the structural component of the connective tissue, plays an important role in maintaining the functional integrity of the myocardium [1, 2]. It has been indicated that the composition and distribution of interstitial collagen determines the stiffness of the cardiac muscle [3–7]. An increase in interstitial collagen in experimentally induced pressure overloaded cardiac hypertrophy has been described [5, 8, 9]. Also changes in the amount of interstitial collagen have been reported in non-infarcted parts of human myocardium after myocardial infarction [10]. However, as far as we are aware, nothing is known about the composition of interstitial tissue in left ventricular dysfunctional myocardium of patients with chronic hibernation. In recent studies we described the structural adaptation in cardiomyocytes from patients with chronic hibernating myocardium. The affected cardiomyocytes showed loss of sarcomeres, sarcoplasmic reticulum and T-tubules, and presented abundant plaques of glycogen, strands of rough endoplasmic reticulum, numerous mini-mitochondria, and nuclei with uniformly dispersed chromatin [11–14]. These cellular changes were not considered degenerative but were interpreted as dedifferentiation of the cells. The latter assumption was supported by 'early development' markers of the heart muscle, namely 1) the re-expression of α-smooth muscle actin; 2) the staining of titin in an embryonic-like (punctated) pattern; and 3) the disappearance of cardiotin, a late marker of heart development [15, 16]. The above mentioned changes were accompanied by a marked increase in

Address for offprints: J. Ausma, Department of Molecular Cell Biology and Genetics, University of Limburg, P.O. Box 616, 6200 MD Maastricht, The Netherlands the interstitial space, which correlated with the degree of myocardial cell change [12–14]. The interest in studying in detail extracellular matrix changes in the dysfunctioning parts of the chronic hibernating left ventricle is obviously related to the important contributary role in the degree and speed of functional recovery after revascularization.

The aim of the present study is to describe the structural changes in the interstitium of chronic hibernating myocardium. Next to the assessment of quantitative cellular and non-cellular changes in the interstitial tissue, we studied the distribution of the major extracellular matrix components collagens I, III and IV, laminin and fibronectin.

Materials and methods

Patients

The human cardiac tissue material used in this study consisted of transmural biopsies from 20 patients with severe left ventricular dysfunction. Viability of the myocardium was assessed by Positron EmissionTomography (PET). The detailed description of the patient data regarding anterior wall motion abnormalities, degree of LAD stenosis, flow-metabolic match or mismatch, and functional recovery after coronary bypass surgery has already been presented [14]. During coronary artery surgery, two transmural biopsies were taken from the anterior free wall of the left ventricle, at approximately 4 cm from the apex, between the distal LAD and the last diagonal branch. All patients gave their informed consent. The study was approved by the local Ethical Committee for Research. A first biopsy from each patient was fixed for a minimum of 2 h in 3% glutaraldehyde buffered to pH 7.4 with 90 mM KH_2PO_4, washed in the buffer and postfixed for 1 h in 2% OsO_4 buffered with 50 mM veronal acetate, dehydrated in a graded series of ethanol and embedded in epoxy resin (Epon) [17]. A second biopsy was used for studies on the immunocytochemical detection of various extracellular matrix proteins. These biopsies were directly frozen in isopentane precooled with liquid nitrogen.

Left ventricle biopsies derived from 7 donor hearts, which were either used for orthotopic transplantation or homograft prelevation, were microscopically examined and served as nonischemic controls.

Light microscopic evaluation

Morphometry of morphologic changes was performed on 2 μm thick sections of Epon-embedded biopsies, which were stained with periodic acid Schiff (PAS) and toluidine blue to quantify the loss of myofibrils and the glycogen content. The degree of cellular change was evaluated only in cells in which the nucleus was present in the plane of the section. Cells were planimetrically scored for the glycogen content [11]. To assess the area of the extracellular matrix, morphometry was carried out with a special grid with vertical and horizontal lines providing 117 intersections [17]. According to the basic principles of morphometry, counting of the number of intersections overlying a certain structure results in quantitative determination of the surface of the structure under investigation in relation to the surface of the entire tissue under the square grid. The total number of intersections was regarded as 100%, and the intersections counted in the connective tissue were expressed as the percentage of the entire tissue within the limits of the grid. Blood vessels and perivascular tissue were excluded from the analysis.

Sirius Red staining

Frozen sections of 5 μm thickness were stained with the collagen-specific dye Sirius Red (Polysciences, Warrington, PA, USA) according to the method of Junqueira *et al.* [18]. Sections were fixed overnight with 3.7% formaldehyde buffered with 0.037 M phosphate buffer (pH 7.4). After being washed for 10 min with tap water and then with distilled water (2 × 2 min), the slides were treated with 0.2% phosphomolybdic acid (5 min). Subsequently, 0.1% Sirius Red, dissolved in a saturated picric acid solution was applied for 90 min. The slides were then treated with 0.01 N HCl (2 min), dehydrated in graded series of ethanol, placed in xylol for 2 min and mounted in Entellan (Merck, Darmstadt, Germany). The collagen volume fraction was determined as the area stained with Sirius Red as a percentage of the total tissue area, by use of a Quantimet 570 morphometer (Leica, Cambridge, UK). The area occupied by blood vessels was subtracted from the total area of the interstitial tissue.

Immunohistochemical studies

The following antibodies were used in this study:

1. A rabbit polyclonal antibody against human type I collagen, AB745 (Chemicon, Tenecula, CA, USA) [19].
2. A rabbit polyclonal antibody against human type III collagen, AB747 (Chemicon, Tenecula, CA, USA) [19].
3. A mouse monoclonal antibody against human type IV collagen, 1042 [20].
4. A rabbit polyclonal antibody against human total fibronectin, A242 (DAKO A/S, Glostrup, Denmark) [21].
5. A mouse monoclonal antibody against human laminin, 4E10, which was a gift form U. Wewer [22].
6. A mouse monoclonal antibody RV202 against vimentin, which labels mesenchymal cells [23].

7. A rabbit polyclonal antibody against vimentin, pVim [24].
8. A mouse monoclonal antibody, sm-1 reacting specifically with the α-smooth muscle isoform of actin [25, 26] (Sigma, St. Louis, USA).
9. A mouse monoclonal antibody against desmin, RD301 [27, 28].

In addition, biotin-labelled lectin from psophocarpus tetragonolobus (Sigma Chemicals, St. Louis, USA), which binds specifically to endothelial cells of the human myocardium [29], was used.

Indirect immunofluorescence studies were performed on frozen sections 5 μm thick. They were air-dried before use, and dipped in methanol (5 sec) and in acetone (3 × 5 sec), both at –20°C, air-dried, and then incubated with primary antibodies for 45 min at room temperature and washed with PBS (3 × 10 min). The sections were then incubated for 45 min at room temperature with the secondary, fluorescein isothiocyanate (FITC)-conjugated goat-anti-mouse Ig (Southern Biotechnology Associates (SBA) Inc., Birmingham, AL, USA). Finally, the sections were washed in PBS (3 × 10 min) and mounted.

In the double labelling procedure, the immunostaining steps were repeated with a second primary antibody of another Ig-subclass, the sections were washed and then incubated for 45 min with the secondary, Texas Red conjugated Ig-subclass specific antibody (SBA, Birmingham, AL, USA).

After these immunohistochemical procedures the sections were placed in distilled water for 5 min, followed by post-fixation in methanol for 5 min. The sections were air-dried and mounted in Mowiol (Hoechst, Frankfurt a.M., Germany). Nuclei were routinely stained with 1:10000 diluted 4′-6-diamidine 2-o-phenylindole (DAPI: Sigma Chemicals, St. Louis, USA).

As a control, application of the primary antibody was replaced by PBS.

For the biotin-labelled lectin the same procedure was followed; in this case fluorescein isothiocyanate (FITC)-labelled avidin (Vector Laboratories, Burlingame, USA) was used as conjugate.

Results

Morphological changes

Cardiomyocytes from chronic hibernating myocardium characteristically showed depletion of sarcomeres and accumulation of glycogen. The replacement of contractile material by glycogen was limited to the vicinity of the nucleus in many cells but in others glycogen comprised the bulk of cytoplasm, leaving only a few or no sarcomeres at the periphery of the cell. Cells were considered as affected when more than 10% of the cell volume was occupied by glycogen (Table 1). The mean percentage of affected cells was 18% for this group of 20 patients. In cardiomyocytes bordering an infarcted region, a loss of contractile material was seen similar to that in non-infarcted regions.

Myocardial segments in which these cellular changes predominate, also show a marked increase in connective tissue (Fig. 1). Four of the 20 biopsies examined showed the presence of scar tissue. In the 16 non-infarcted patients the total area of connective tissue was 8%, whereas for the 4 patients with infarction it amounted to 49%. In control hearts the connective tissue area comprised approximately 2%.

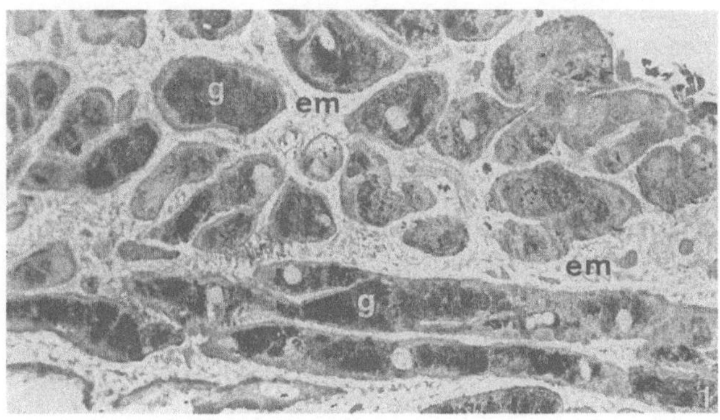

Fig. 1. Light microscopy of morphological changes in chronic hibernating myocardium. Section (2 μm thick) of an affected area, stained with PAS/toluidine blue showing that the centers of most cells are myolytic. The myolytic areas are filled with darkly stained material (glycogen, g). The extracellular matrix (em) surrounding the chronic hibernating myocardial cells is increased. Magnification: ×230.

Quantification of the collagen content

The collagen volume fraction as revealed with Sirius Red, is shown in Table 1 for the 20 patients. In chronic hibernating myocardium Sirius Red staining was visible around myocardial cells and was displayed as large bundles in the broadened interstitial spaces of the severely affected areas (Fig. 2A). In non- or less-affected zones such bundles were sparsely seen (Fig. 2B). The mean collagen volume fraction for non-infarcted chronic hibernating myocardium was about 5%, and for patients with an infarction approximately 20%. The collagen content in control myocardium was not assessed in the present samples but has been evaluated recently [10] and amounted to 2.7 ± 0.5% in a control group of 18 patients without cardiovascular disease. In some of our patients the percentage of collagen, as determined by Sirius Red morphometry, was apparently higher than the percentage of fibrosis as determined by intersection measurements. This difference may result from the fact that the two measurements were performed on different biopsies from the same patient and

Table 1. Changes in interstitial tissue and cardiomyocyte composition in chronic hibernating myocardium

Patients	Collagen volume fraction (%)	Total area of connective tissue (%)	Affected cells (%)
No infarct			
1	1.4	4	22
2	3.5	10	16
3	7.2	8	16
4	3.4	18	24
5	2.1	14	17
6	12.4	18	75
7	6.5	9	7
8	3.8	9	15
9	2.5	4	3
10	2.9	4	9
11	6.6	9	21
12	5.0	4	8
13	5.2	4	16
14	8.6	7	18
15	6.3	10	8
16	0.9	5	16
mean (n = 16)	5	8	18
Infarct			
17	14.7	23	24
18	31.0	88	7
19	18.5	28	21
20	14.2	56	12
mean (n = 4)	20	49	16
Controls			
mean (n = 7)	not done	2 (range 1–5)	0 (range 0–0)

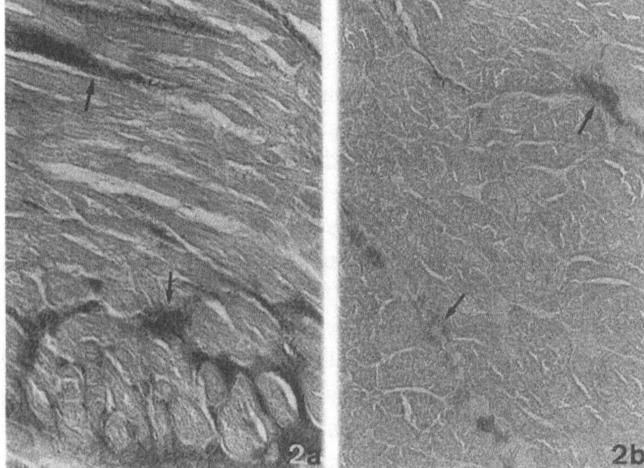

Fig. 2. Sirius Red staining on frozen sections of the left ventricle. (A) Chronic hibernating myocardium surrounded by extensive collagen deposits (arrows). (B) Chronic hibernating myocardium with a low amount of Sirius Red-positive staining (arrows) and correspondingly a small increment of the extracellular matrix. Magnification: ×256.

small differences in fibrosis do occur in adjacent areas of chronic hibernating myocardium.

Immunofluorescence studies

Collagen types I and III

As studied with polyclonal antibodies, type I and III collagens were present throughout the extracellular matrix of the myocardium, having a fibrillar appearance. In normal myocardium, bundles of interstitial collagen were concentrated mainly around blood vessels, while cardiomyocytes were found interspersed by small amounts of fine fibers of type I and III collagen (Fig. 3A). In chronic hibernating myocardial segments increased amounts of type I and III collagen were seen throughout the enlarged interstitial space (Fig. 3B). The cardiomyocytes at the border of an infarcted zone were surrounded by these two collagen subtypes (Fig. 3C).

Fibronectin

In normal myocardium the immunoreactivity to fibronectin was present in the interstitial space between the cardiomyocytes and at the sarcolemma of myocardial cells (Fig. 4A). In chronic hibernating myocardial cells fibronectin was

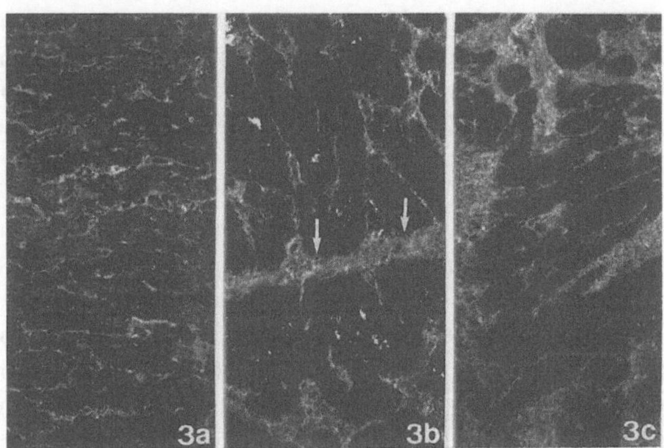

Fig. 3. Immunofluorescence micrographs of frozen sections of myocardium incubated with anti-collagen type I. (A) Staining in control myocardium. Collagen I is sparsely distributed throughout the extracellular matrix. (B) Staining in chronic hibernating myocardium. Marked increase in the amount of collagen is seen in the enlarged interstitial space (arrows). (C) The border of an infarct (upper part) shows extensive staining of the markedly enlarged connective tissue compartment. Magnification: ×192.

found in high amounts throughout the increased interstitial space, with a distribution pattern similar to that of collagens types I and III (Fig. 4B). The chronic hibernating myocardial cells were never positive for fibronectin in their cytoplasm. As for type I and III collagens, the cardiomyocytes bordering an infarcted zone were surrounded by large amounts of fibronectin.

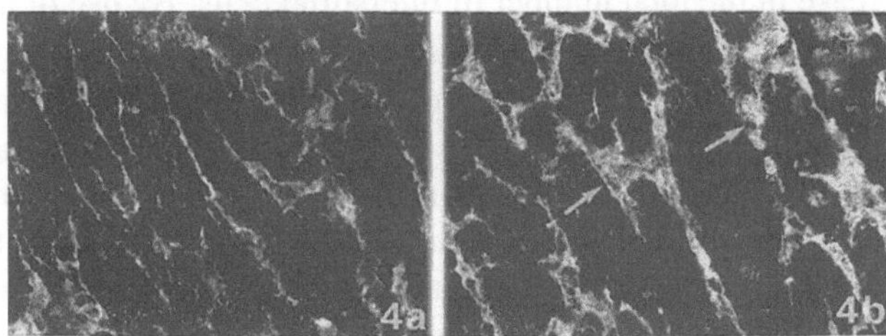

Fig. 4. Immunofluorescence micrographs of frozen biopsies from chronic hibernating myocardium, showing staining of fibronectin in an area with a nearly normal interstitial space (A), and in a zone with a markedly increased interstitial space (B) Note the increment in the amount of fibronectin in the latter (arrows). Magnification: ×192.

Collagen IV and laminin

Like fibronectin, type IV collagen and laminin are components of the basement membrane of the cardiomyocytes. In normal myocardium, type IV collagen and laminin were colocalized at the sarcolemma of cardiomyocytes. In chronic hibernating myocardium the basal lamina of cardiomyocytes stained in a normal way for type IV collagen and laminin (Fig. 5). In the interstial space, type IV collagen and laminin were both seen surrounding the endothelial cells of capillaries and small interstitial cells.

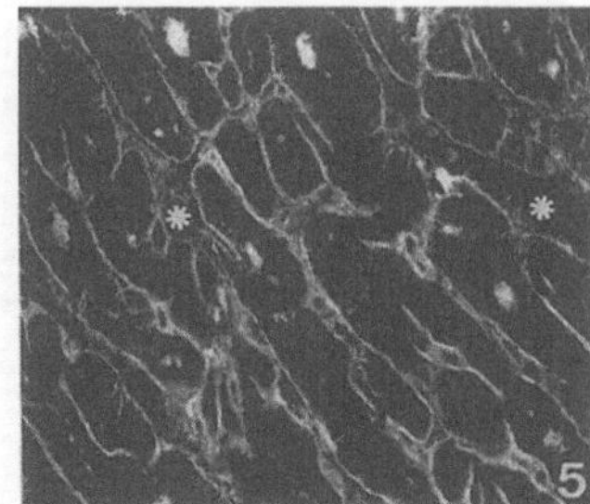

Fig. 5. Immunofluorescence micrograph of a section from chronic hibernating myocardium showing immunohistochemical staining of the basement membrane with anti-collagen type IV. There are no abnormalities in the staining pattern compared with control myocardium. The extended extracellular spaces are indicated by asterisks. Magnification: ×192.

Vimentin

Vimentin normally stains mesenchymal cells. In normal myocardium vimentin-positive staining was detected within fibroblasts, endothelial cells and smooth muscle cells (Fig. 6A). In chronic hibernating myocardium the whole interstitial space was filled with numerous vimentin-positive cells (Fig. 6B). Most of the vimentin-positive cells were recognized as endothelial cells or fibroblast-like cells. This is also clearly evidenced by the large number of small DAPI-positive nuclei (Fig. 6B). The very intense staining for vimentin in the broadened interstitial spaces was due to the presence of a higher number of vimentin-containing cells. To make a clear distinction between cells of endothelial, fibroblast, myofibroblast and smooth muscle origin, specific markers for these cell types, respectively lectin, desmin and α-smooth muscle actin were used. The vimentin-positive cells were not reactive for desmin (compare Fig. 6B and C). Some of the vimentin-positive cells were identified as endothelial cells by their lectin-binding capacity (compare Fig. 6D and E). The reactivity to α-smooth muscle actin was restricted to smooth muscle cells of blood vessels (compare Fig. 6F and G).

Discussion

Cardiomyocytes of chronic hibernating myocardium undergo typical ultrastructural changes of which the replacement of sarcomeres by glycogen is the hallmark [11–14]. Events of a purely degenerative nature, such as acute necrosis of myocytes, abnormal storage of lipids or multilamellar bodies, gross intracellular edema, or the presence of inflammatory cells, are only rarely seen. Cells with this hibernating phenotype are found in severely stenosed myocardial segments derived from patients with or without previous infarction. In the former case, such cells frequently border the infarcted zone. The changes in the cardiomyocytes are accompanied by a marked increase in the amount of connective tissue material [11]. This increase is most pronounced

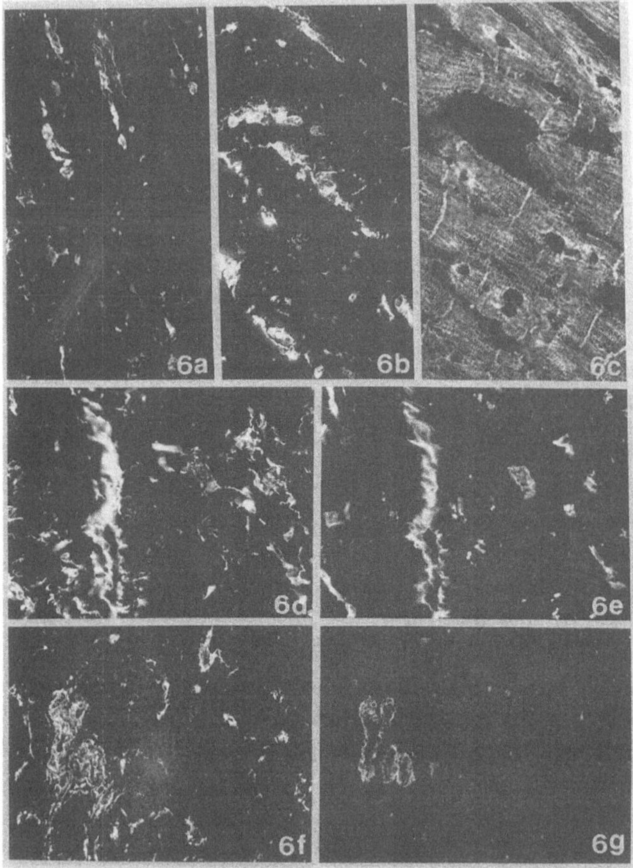

Fig. 6. Immunofluorescence micrographs of control (A) and chronic hibernating myocardium (B–G) incubated with antibodies to vimentin (A, B, D, F), desmin (C) α-smooth muscle actin (G), or the endothelial marker lectin (E). Figures B and C, D and E, F and G are double-label pictures of exactly the same area. For detailed explanation see text. Magnification: ×192.

in areas where the structurally affected cardiomyocytes prevail [12, 14].

Quantification of the increase of extracellular matrix

Quantification of the collagen content in chronic hibernating myocardium by Sirius Red morphometry emphasizes an increased volume fraction of collagen throughout the extracellular matrix. The average collagen concentration in chronic hibernating myocardium is about two times higher than in normal hearts [10]. Increases in the concentration of collagen similar to that of chronic myocardial segments were also found in non-infarcted segments of the rat myocardium after infarction [30].

In the present study a markedly increased immunofluorescence of types I and III collagen and fibronectin could be demonstrated in the extracellular matrix of chronic hibernating myocardium. Fibronectin was never seen in the cytoplasm of cardiomyocytes, as previously described for necrotic cardiomyocytes after myocardial infarction [31]. Necrosis of cardiomyocytes was seldom observed in our cases. In pressure-overload hypertrophy, the accumulation of collagen (called reactive fibrosis) occurred in the absence of cardiomyocyte necrosis [5, 32]. This process also occurs in chronic hibernating myocardium.

The basement membrane

In addition to its presence in the extracellular matrix, fibronectin is also part of the basement membrane of cardiomyocytes. There is no difference in its location between normal and chronic hibernating cardiomyocytes. Other basement membrane components, such as type IV collagen and laminin, also show an identical distribution pattern between affected and non-affected tissues, suggesting that the basement membrane of chronic hibernating myocardial cells is intact. This is in strong contrast to what has been observed in necrotic cardiomyocytes during an acute infarction [33] or in dilated cardiomyopathy [34], where an irregular staining pattern for these constituents is displayed.

The nature of the interstitial cells

The question arises whether the increase in extracellular matrix is caused by an increased production of extracellular matrix components by existing interstitial cells or by an increase in the total number of interstitial cells. By use of the combination of DAPI staining of nuclei and an immunohistochemical approach for the determination of the phenotype of interstitial cells, it was demonstrated that in hibernating myocardium the number of cells of endothelial and fibroblast origin throughout the interstitial space is dramatically increased. Since these interstitial cells do not contain desmin or α-smooth muscle actin, they can be clearly differentiated from myofibroblasts. This suggests that no transformation of fibroblasts into myofibroblasts has occurred in these chronic hibernating segments, as is for instance the case in healing wounds [35–37] or in scar tissue of the heart during pressure-overload of the right ventricle [38]. It is therefore likely that the increased amounts of types I and III collagens and fibronectin in chronic hibernating myocardium are a result of increased fibroblast proliferation.

Effect of the increased extracellular matrix

The presence of an increased collagen content in the extracellular matrix may contribute to increased stiffness of the left ventricle, especially because the rigidity of type I collagen.

This has been observed in patients with coronary artery disease and dilated cardiomyopathy [39]. Increased collagen concentrations leading to greater stiffness of the myocardium have also been described in conditions of pressure-overload hypertrophy [5–7, 9, 32, 38, 40, 41]. In the rat hypertrophy model, fibroblast proliferation and collagen synthesis increased within several days after abdominal aorta constriction and resulted in an abnormal stiffness of the myocardium [5].

Furthermore, the excess collagen may impair contractile function of the cardiomyocytes, as a consequence of the disruption of force transmission between the contracting cells [41]. After restoration of the blood flow by coronary artery bypass surgery, recovery of contractile function is often delayed in cases of chronic hibernating myocardium [12, 43]. Although such a recovery to normal function is, at least in part, dependent on the building of a normal amount of sarcomeres, the presence of an extensive extracellular matrix compartment may certainly be one of the major causes of the delay in recovery. Full recovery conceivably might only occur when the interstitial tissue is reduced to normal proportions.

References

1. Eghbali M, Eghballi M, Robinson TF, Seifter S, Blumenfeld OO: Collagen accumulation in heart ventricles as a function of growth and aging. Cardiovasc Res 23: 723–729, 1989
2. Caufield JB, Borg TK: The collagen network of the heart. Lab Invest 40: 364–372, 1979
3. Weber KT, Janicki JS, Shroff SG, Pick R, Chen RM, Bashey KI: Collagen remodelling of the pressure overloaded, hypertrophied nonhuman primate myocardium. Circ Res 62: 757–765, 1988
4. Weber KT: Cardiac interstitium in health and disease: the fibrillar collagen network. J Am Coll Cardiol 13: 1637–1652, 1989
5. Doering CW, Jalil JE, Janicki JS, Pick R, Aghili S, Abrahams C, Weber KT: Collagen network remodelling and diastolic stiffness of rat left ventricle with pressure overload hypertrophy. Cardiovasc Res 22: 686–695, 1988
6. Jalil JE, Doering W, Janicki JS, Pick R, Clark WA, Abrahams C, Weber KT: Structural vs. contractile protein remodelling and myocardial stiffness in hypertrophied rat left ventricle. J Mol Cell Cardiol 20: 1179–1187, 1988
7. Jalil JE, Doering W, Janicki JS, Pick R, Shroff SG, Weber KT: Fibrillar collagen and myocardial stiffness in intact hypertrophied rat left ventricle. Circ Res 64: 1041–1050, 1989
8. Pick R, Janicki JS, Weber KT: Myocardial fibrosis in nonhuman primate with pressure overload hypertrophy. Am J Pathol 135: 771–781, 1989
9. Contard F, Koteliansky V, Marotte F, Dubus I, Rappaport I, Samuel JL: Specific alterations in the distribution of extracellular matrix components within rat myocardium during the development of pressure overload. Lab Invest 64: 65–75, 1991
10. Volders PGA, Willems IEMG, Cleutjens JPM, Arends J-W, Havenith MG, Daemen MJAP: Interstitial collagen is increased in the noninfarcted human myocardium after myocardial infarction. J Mol Cell Cardiol 25: 1317–1323, 1993
11. Borgers M, Thoné F, Wouters L, Ausma J, Shivalkar B, Flameng W: Structural correlates of regional myocardial dysfunction in patients with critically coronary artery stenosis: chronic hibernation? Cardiovasc Pathol 2: 237–245, 1993
12. Vanoverschelde J-L, Wijns W, Depré C, Essamri B, Heyndricks GR, Borgers M, Bol A, Melin J: Mechanisms of chronic regional postischemic dysfunction in humans: new insights from the study of non-infarcted collateral dependent myocardium. Circulation 87: 1513–1523, 1993
13. Ausma J, Ramaekers F, Shivalker B, Thoné F, Flameng W, Borgers M: Cellular adaptation in hibernating myocardium in the human. In: M. Hori, Y. Maruyama, RS Reneman, (eds). Cardiac adaptation and failure. Springer-Verlag, Tokyo, 85–99, 1994
14. Maes A, Shivalkar B, Flameng W, Nuyts J, Borgers M, Ausma J, Bormans G, Schiepers C, De Roo M, Mortelmans L: Histological alterations in chronically hypoperfused myocardium: correlation with PET findings. Circulation 90: 735–745, 1994
15. Schaart G, Vander Ven PMF, Ramaekers FCS: Characterization of cardiotin, a structural component in the myocard. Eur J Cell Biol 62: 34–48, 1993
16. Ausma J, Schaart G, Thoné F, Shivalker B, Flameng W, Depré Ch, Vanoverschelde J-L, Ramaekers F, Borgers M: Chronic ischemic viable myocardium in man: Aspects of dedifferentation. Cardiovasc Pathol 4: 29–37, 1995
17. Flameng W, Wouters L, Sergeant P, Lewi P, Borgers M, Thoné F, Suy R: Multivariate analysis of angiographic histologic and electrocardiographic data in patients with coronary heart disease. Circulation 70: 7–17, 1984
18. Junqueira LCU, Bignolas G, Bretani RR: Picrosirius staining plus polarization microscopy: a specific method for collagen detection in tissue sections. Histochem J 11: 447–455, 1979
19. Bedossa P, Bacci J, Lemaigre G, Martin E: Effects of fixation and procession on the immunohistochemical visualisation of type-I, -III and -IV collagen in paraffin-embedded liver tissue. Histochemistry 88: 85–89, 1987
20. Havenith MG, Cleutjens JPM, Beek C, vd Linden E, De Goey AFPM, Bosman FT: Human specific anti-type IV collagen monoclonal antibodies, characterization and immunochemical application. Histochemistry 87: 123–128, 1987
21. Van Helden WCH, Kok-Verspuy A, Harff GA, van Kamp GJ: Ratenephelometric determination of fibronectin in plasma. Clin Chem 31: 1182–1184, 1985
22. Wewer U, Albrechtsen R, Manthorpe M, Varon S, Engvall E, Ruoslathi E: Human laminin isolated in a nearly intact biologically active form from placenta by limited proteolysis. J Biol Chem 20: 12654–12660, 1983
23. Ramaekers FCS, Huijsmans A, Schaart G, Moesker O, Vooijs GP: Tissue distribution of keratin 7 as monitored by a monoclonal antibody. Expl Cell Res 170: 235–249, 1987
24. Ramaekers FCS, Puts JJG, Moesker O, Kant A, Huysmans A, Haag D, Jap PHK, Herman CJ, Vooijs GP: Antibodies to intermediate filament proteins in immunohistochemical identification of human tumors: an overview. Histochem J 15: 691–713, 1983
25. Woodcock-Mitchell J, Mitchell JJ, Low RB, Kieny M, Sengel P, Rubbia L, Skalli O, Jackson B, Gabbiani G: α-Smooth muscle actin is transiently expressed in embryonic rat cardiac and skeletal muscles. Differentiation 39: 161–166, 1988
26. Skalli O, Ropraz P, Trzeciak A, Benzonana G, Gillesen D, Gabbiani G: A monoclonal antibody against α-smooth muscle actin: a new probe for smooth muscle differentiation. J Cell Biol 103: 2787–2796, 1986
27. Raats FR, Henderik JB, Verdijk M, van Oort FLG, Gerards WLM, Ramaekers FCS: Bloemendaal H: Assembly of the carboxy-terminally deleted desmin in vimentin free cells. Eur J Cell Biol 56: 84–103, 1991

28. Schaart G, Viebahn C, Langmann W, Ramaekers FCS: Desmin and titin expression in early postimplantation mouse embryos. Development 107: 585–596, 1989
29. Laitinen L, Hormia M, Virtanen I: *Psophocarpus tetragonolobus* agglutinin reveals N-acetyl galactosaminyl residues confined to endothelial cells and some epithelial cells in human tissues. J Histochem Cytochem 38: 875–884, 1990
30. van Krimpen C, Smits JFM, Cleutjens JPM, Debels JJ, Schoenmaker RG, Struyker-Boudier HAJ, Bosma FT, Daemen MJAP: DNA synthesis in the non-infarcted cardiac interstitium after left coronary artery ligation in the rat: effects of captopril. J Mol Cell Cardiol 23: 1245–1253, 1991
31. Cassells W, Kimura H, Sanchez JA, Yu Z-X, Ferrans VJ: Immunohistochemical study of fibronectin in experimental myocardial infarction. Am J Pathol 137: 801–810, 1990
32. Weber KT, Jalil JE, Janicki JS, Pick R: Myocardial collagen remodelling in pressure overload hypertrophy. A case for interstitial heart disease. Am Heart J 2: 931–940, 1989
33. Vracko R, Cunninghasm D, Frederickson G, Thorning D: Basal lamina of rat myocardium. Its fate after death of cardiac myocytes. Lab Invest 58: 77–87, 1988
34. Wolff PG, Kühl U, Schultheiss H-P: Laminin distribution and autoantibodies to laminin in dilated cardiomyopathy and myocarditis. Am Heart J 117: 1303–1309, 1989
35. Gabbiani G, Le lous M, Bailey AJ, Bazin S, Delaunary A: Collagen and myofibroblasts of granulation tissue. Virchows Arch B Cell Path 21: 133–145, 1976
36. Darby I, Skalli I, Gabbiani G: α-Smooth muscle actin is transiently expressed by myofibroblasts during experimental wound healing. Lab Invest 63: 21–29, 1990
37. Skalli O, Schürch, Seemayer T, Lagacé R, Montandon D, Pittet B, Gabbiani G: Myofibroblasts from diverse pathologic settings are hterogeneous in their content of actin isoforms and intermediate filament proteins. Lab Invest 60: 275–285, 1989
38. Lesie KD, Taatjes DJ, Schwarz J, von Turkovich M, Low RB: Cardiac myofibroblasts express alpha smooth muscle actin during right ventricular pressure overload in rabbit. Am J Pathol 139: 207–216, 1991
39. Bishop JE, Greenbaum R, Gibson DG, Yacoub M, Laurent GJ: Enhanced deposition of predominantly type I collagen in myocardial disease. J Mol Cell Cardiol 22: 1157–1165, 1990
40. Weber KT, Brilla CG: Pathological hypertrophy and cardiac interstitium. Circulation 83: 1849–1865, 1991
41. Villarreal FJ, Dillmann WH: Cardiac hypertrophy changes in mRNA levels for TGF-β1, fibronectin and collagen. Am J Physiol 262: H1861–H1866, 1992
42. Weber KT, Brilla CG, Janicki JS: Myocardial fibrosis: functional significance and regulatory factors. Cardiovascular Research 27: 341–348, 1993
43. Tubau JF, Rahimtoola SH: Hibernating myocardium: a historical perspective. Cardiovasc Drugs Ther 6: 267–271, 1992

Molecular and Cellular Biochemistry **147**: 43–49, 1995.

Differences between atrial and ventricular protein profiling in children with congenital heart disease

Václav Pelouch, Marie Milerová[1], Bohuslav Oštádal, Bohumil Hučín[1] and Milan Šamánek[1]

Institute of Physiology, Academy of Sciences of the Czech Republic and [1]Centre of Pediatric Cardiology and Cardiovascular Surgery, University Hospital Motol, Prague, Czech Republic

Abstract

The purpose of the present study was to compare protein profiling of atria and ventricles in children operated for congenital heart disease. Tissue samples were obtained during surgery from patients with normoxemic (ventricular and atrial septal defects) and hypoxemic (tetralogy of Fallot) diseases. Protein fractions were isolated by stepwise extraction from both right ventricular and atrial musculature. The concentration of total atrial protein in the normoxemic patients exceeded the ventricular value (110 ± 2.1 vs 99.9 ± 4.0 $mg.g^{-1}$ wet weight, respectively); in the hypoxemic group this atrio-ventricular difference disappeared. The concentration of contractile proteins in all cardiac samples was significantly higher in the ventricles as compared with atria, while the concentration of collagenous proteins was significantly higher in the atria (due to a higher amount of the insoluble collagenous fraction). The concentration of sarcoplasmic proteins (containing predominantly enzyme systems for aerobic and anaerobic substrate utilization), however did not differ between ventricles and atria. Furthermore, ventricular contractile fractions obtained from both normoxemic and hypoxemic patients were contaminated with the myosin light chain of atrial origin. Soluble collagenous fractions (containing newly synthesized collagenous proteins, predominantly collagen I and III), derived from all ventricular samples, were contaminated by low molecular weight fragments (mol. weight 29–35 kDa). The proportion of the soluble collagenous fraction was significantly higher in atrial but not in ventricular myocardium of hypoxemic children as compared with the normoxemic group. It seems, therefore, that lower oxygen saturation affects the synthesis of collagen preferentially in atrial tissue. (Mol Cell Biochem **147: 43 –49, 1995**)

Key words: protein profiling of cardiac muscle, collagenous proteins, contractile proteins, myosin light chains, metabolic proteins-congenital heart disease, tetralogy of Fallot, ventricular septal defect, atrial septal defect

Introduction

Cardiac mammalian muscle is a heterogeneous tissue compared of distinct muscle cell populations. Ultrastructural studies have revealed the existence of significant differences between ventricular and atrial fibers. The large ventricular myocytes are organized into layers in each of which they follow a uniform direction. On the other hand, atrial cells are more slender; they tend to form bundles of varying sizes that criss-cross over each other frequently [1–3]. Gross biochemical analysis of both myocardial compartments showed no differences in dry weight and glycogen concentration, but atria were found to have a higher lipid content [4]. There are differences on the level of myofibrillar proteins: atrial myosin has a higher Ca ATPase activity and contains an electrophoretically and immunologically distinct structure of both heavy and light chains [1, 5–7]. However, very few quantitative data are available on the fine structure of cardiac extracellular space. This compartment contains a variety of collagenous proteins, charged proteglycans and glycoproteins [8–12]. Geometrically, the space varied in size; the above elements form a complex weave of fibrillar matrix. The network has been described in anatomic terms; all myocytes are interconnected to adjacent myocytes by short collagen struts measuring about 150 nm in rats and somewhat thicker in humans [3, 13]. A number of studies have appeared which are con-

Address for offprints: V. Pelouch, Institute of Physiology, Academy of Sciences of the Czech Republic, Vídeňská 1083, 142 00 Prague 4, CZ Czech Republic cerned with ventricular collagenous structures [8, 9, 14–16].

The precise characterization of cellular composition of the heart has been hampered by the lack of well-defined molecular markers for different types of cardiac muscle cells. However, the protein profiling together with the discovery of multiple myosin isoenzymes or different collagenous types are powerful tools for distinguishing of myofibrillar and extracellular compartments of cardiac muscle. Our previous studies showed that there are atrio-ventricular differences both in the activity of energy-supplying enzymes [17] and the proportion of collagenous and noncollagenous protein fractions isolated from human myocardium of patients with congenital heart disease; the concentration of contractile proteins is higher in the ventricles, the concentration of collagenous proteins is significantly higher in the atrium, the concentration of sarcoplasmic proteins is not different [16, 18]. Many of congenital heart defects have, however, a hypertrophic myocardium as a consequence: in hypoxemic defects this is occasioned by the increased pressure load, in the normoxemic diseases the volume load is prime factor [19, 54]. In the present paper, the main attention was paid to the biochemical pattern of cardiac proteins isolated from patients with normoxemic and hypoxemic congenital heart disease. The work aimed at determining whether: a) the protein profiling of atria and ventricle is dependent on arterial oxygen saturation, b) chronic hypoxaemia affects the qualitative composition of collagenous and myofibrillar proteins.

Subject of study

The study was performed on 51 samples of human cardiac tissue obtained during surgery of children with congenital heart disease (age: 6 months to 16 years). The samples were taken from right atria (n = 25) and right ventricles (n = 8) of patients with *normoxemic* (ventricular septal defect: n = 18, arterial oxygen saturation – 94.6%; atrial septal defect: n = 11, arterial oxygen saturation – 93.6%) and from right atria (n = 9) and right ventricles (n = 9) of patients with *hypoxemic* congenital heart disease (tetralogy of Fallot: n = 17, arterial oxygen saturation – 61.5%). A preliminary observation did not reveal significant age-related differences in individual protein fractions; the samples from children of different ages were, therefore, pooled [16, 18]. The atrial tissue was excised before commencement of the cardiopulmonary bypass; the ventricular samples were taken 2–80 min after the heart beat had been arrested: no *in vitro* aging of collagenous and noncollagenous proteins was observed during this time [16].

Quantitative analysis of protein profile

a) Isolation of protein fractions

Samples were rapidly weighed and transferred into precooled homogenization test tubes with 200 ul of 50 mmol.l^{-1} sodium-potassium-phosphate buffer, pH 7.4, containing 10 mmol .l^{-1} EDTA and 1% Triton X-100. Subsequently they were frozen to –50°C and kept at this temperature until the next isolation stage. Tissue samples were later thawed, the volume made up to 20-fold original with buffer, homogenized and centrifuged at 15 000 g. This step was repeated once more and pooled supernatants (a 40-fold multiple of the original amount) were used for the determination of sarcoplasmic proteins (this fraction contains predominantly metabolic proteins [17]). In subsequent steps the pellet was resuspended and fractions of contractile and collagenous proteins were obtained in a stepwise manner by extracting contractile proteins into a supernatant with phosphate buffer (100 mmol.l^{-1}, pH 7.4, containing 1.1 mol.l^{-1} KCl); the pellet was shortly washed with 0.5 mol acetic acid and then extracted with 0.5 mol.l^{-1} CH_3COOH-pepsin (pH 1.45); pepsin concentration was kept in the range 1:50–100. After 24 h at 4°C the extracts were centrifuged. The supernatant contained the fraction of soluble collagenous proteins. The pellet was further suspended in 1.1 mol.l^{-1} NaOH and left for 45 min at 105°C. This fraction contained insoluble collagenous proteins [for details see 11, 12, 17, 20, 21].

b) Determination of proteins and hydroxyproline

Protein concentration in individual fractions was determined according to Lowry *et al.* [22]. Total protein was the sum of concentrations of total noncollagenous (sum of sarcoplasmic and contractile proteins) and both fractions of collagenous proteins (pepsin-soluble and pepsin-insoluble fractions). The results are expressed as mg.g^{-1} wet weight. Hydroxyproline concentration in both collagenous fractions was determined [23]. Furthermore, individual myocardial samples were digested in 6 M HCl for 16 h at 105°C, the resulting solution was decolorized and supernatant was analyzed for 4-hydroxyproline [21]. Total collagen was calculated from the concentrations of the amino acid hydroxyproline multiplied by a factor of 7.46 assuming that it constitutes an average of 13.4% of the collagen molecule [21].

c) Effect of hypoxaemia on protein composition of human atrial and ventricular myocardium

The concentration of total atrial protein in normoxemic patients exceeded the ventricular value (110.4 ± 2.1 vs 99.9 ± 4.0 mg.g^{-1} wet weight); in hypoxemic group this atrio – ventricular difference disappeared (97.8 ± 3.6, n = 9 vs 96.2 ± 2.2 mg.g^{-1} wet weight, n = 9 for ventricular and atrial samples, resp.). There was no statistical atrio-ventricular difference in the concentration of sarcoplasmic proteins, recovered

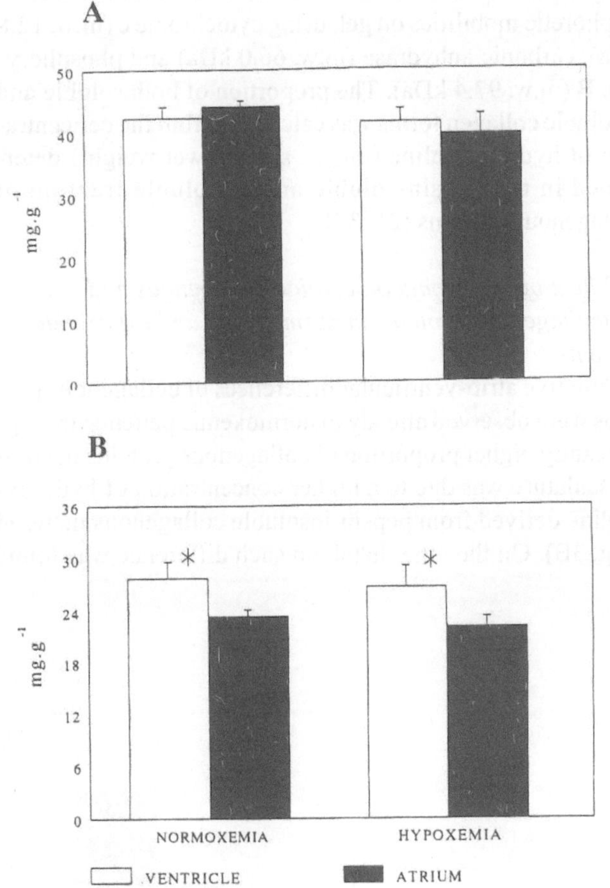

Fig. 1. Protein profiling of noncollagenous fractions. Concentration of sarcoplasmic proteins (A) and contractile proteins (B) in the human right ventricle (white bars) and right atrium (black bars) of normoxemic and hypoxemic patients. Values are means ± S.E.M. (expressed as mg per g of wet weight). Significance: *p < 0.05 or less vs atrium. Data from [16].

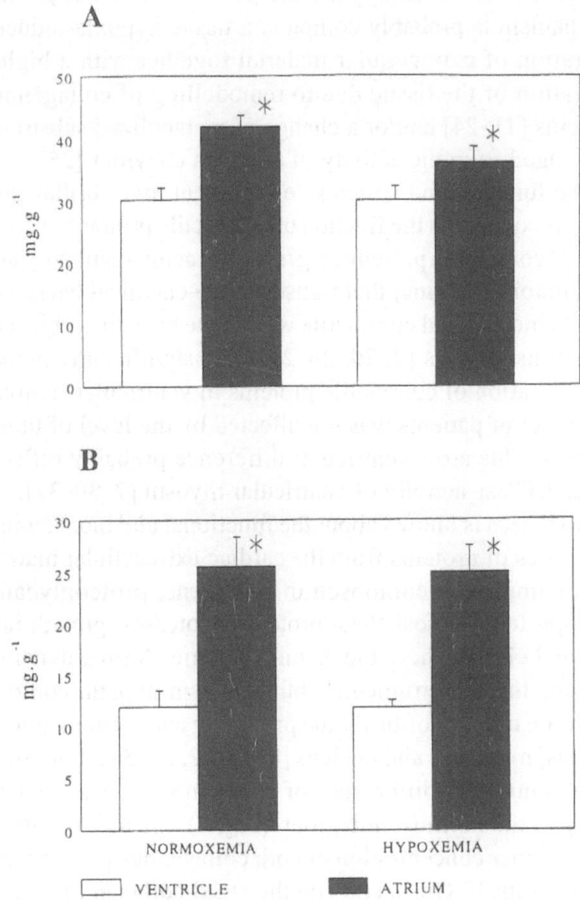

Fig. 2. Protein profiling of collagenous fractions. Concentration of total collagenous proteins (was obtained by pooling the concentration of proteins in soluble and insoluble collagenous structures) (A) and total collagen (hydroxyproline in the fraction of soluble and insoluble collagenous structures multiplied by factor 7.46) (B) in human right ventricle (white bars) and right atrium (black bars) in normoxemic and hypoxemic patients. Values are means ± S.E.M. (expressed as mg per g of wet weight). Significance: *p < 0.05 or less vs atrium. Data from [16].

in group the part of the normoxemic human myocardia (Fig. 1A), the concentration of sarcoplasmic proteins in hypoxemic atrial samples was significantly lower (44.10 ± 1.01 vs 39.64 ± 3.21, $p < 0.05$, for normoxemic and hypoxemic atrial musculature resp., see also Fig. 1A). The concentration of contractile proteins was significantly lower in atrial samples in both normoxemic and hypoxemic groups (Fig. 1B). The higher concentration of the total atrial collagenous proteins (Fig. 2A) as compared with the ventricular values corresponded to a higher concentration of collagen (Fig. 2B). These atrio-ventricular differences of total collagenous proteins were not affected by hypoxaemia; the concentration of total atrial collagenous proteins was, however, significantly decreased (42.01 ± 1.76 vs 36.28 ± 2.20 mg/g, $p < 0.05$ for normoxemic and hypoxemic groups resp. – see Fig. 2A). In contrast, the concentration of total collagen in hypoxemic atrial musculature was not affected (Fig. 2B).

These results imply that the proportions of individual protein fractions as well as atrio-ventricular differences in human myocardium were affected by blood oxygenation; atrial musculature at normoxemic saturation contained a higher concentration of total proteins as compared with hypoxemic one. On the other hand, the total concentration of proteins in ventricular musculature was not dependent on arterial oxygen saturation (the same amount of proteins was found in both normoxemic and hypoxemic ventricular as well as in hypoxemic atrial musculature). These atrio-ventricular differences reflected the amount of collagenous proteins. However, atrial musculature of hypoxemic patients exhibited a reduced concentration of both collagenous and sarcoplasmic (metabolic) proteins.

Analysis of the protein profile did not provide an ambiguous explanation of why the total concentration of atrial musculature is lower in hypoxemic patients. The underlying mechanism is probably complex: a tissue hypoxia-induced alteration of extracellular material together with a higher hydration of the tissue due to remodelling of collagenous proteins [11, 24] and/or a change of metabolized substrates (as judged from the activity of different enzymes [25].

The function and structure of different myofibrillar proteins, recovered in the fraction of contractile proteins (a complex of contractile proteins, e.g. myosin, actin, regulatory and modulatory proteins) that transduce the chemical energy of ATP to mechanical contractile work have been the subject of numerous reviews [7, 20, 26–29]. The significantly higher concentration of contractile proteins in ventricular samples in our set of patients was not affected by the level of blood oxygen. This atrio-ventricular difference probably reflects lower ATPase activity of ventricular myosin [7, 30–32].

Much less is known about the functional and biochemical properties of proteins from the cardiac extracellular matrix. This complex is composed of collagens, proteoglycans, glycoproteins, cytoskeletal proteins, proteases, growth factors and elastin; these biochemical entities form a dynamic network that is instrumental both in transmitting the contractile force from myofibrills and providing connections among vessels, myocytes and nucleus [8, 10–12, 33, 34]. Important atrio-ventricular differences of collagen were observed already in the normoxemic group (Fig. 2); the atrial samples had a higher concentration of both collagenous proteins and collagen itself. Our results on the atrial collagen of human myocardium are in agreement with previously reported values for other species [15, 35]. We assume that this may correlate with the higher amount of basic growth factors detected as a rule in the atrial part of the myocardium [36]; it thus reflects a different activation of fibroblasts. The finding hypoxaemia that did not elevate the concentration of cardiac collagenous proteins is seemingly surprising: however, a higher concentration was induced in newborn rat myocardium only when the animals were kept under severe hypoxia (7.5% of oxygen) for 5 weeks [37, 38]. In a more moderate hypoxia (10–15% of oxygen) the growth of myocardial collagenous stroma was proportional to the growth of noncollagenous proteins [39, 40].

Qualitative analysis of protein profile

a) Separation of cardiac collagenous and noncollagenous proteins

Different cardiac protein fractions (see above) were characterized by SDS-electrophoresis. Discontinuous SDS-PAGE [41] was performed on 1.5 mm thick and 14 cm long slab gels. For separation of collagenous and non-collagenous fractions the concentration of the stacking gel was 4% T with 2.6% C; separating gel was 5–15% gradient slab gel. The molecular weight of individual bands were determined from the electrophoretic mobilities on gel, using cytochrome c (m.w. 12.8 kDa), carbonic anhydrase (m.w. 66.0 kDa) and phosphorylase B (m.w. 97.4 kDa). The proportion of both soluble and insoluble collagen forms was calculated from the concentration of hydroxyproline ($mg.g^{-1}$ sample wet weight) determined in the pepsin-soluble and -insoluble fraction of collagenous proteins [21, 37].

b) Effect of hypoxemia on cardiac collagenous and noncollagenous proteins in normoxemic and hypoxemic patients

Qualitative atrio-ventricular differences of collagenous proteins were observed already in normoxemic patients: the significantly higher proportion of collagenous proteins in atrial musculature was due to a higher concentration of hydroxyproline derived from pepsin-insoluble collagenous material (Fig. 3B). On the other hand, no such difference was found

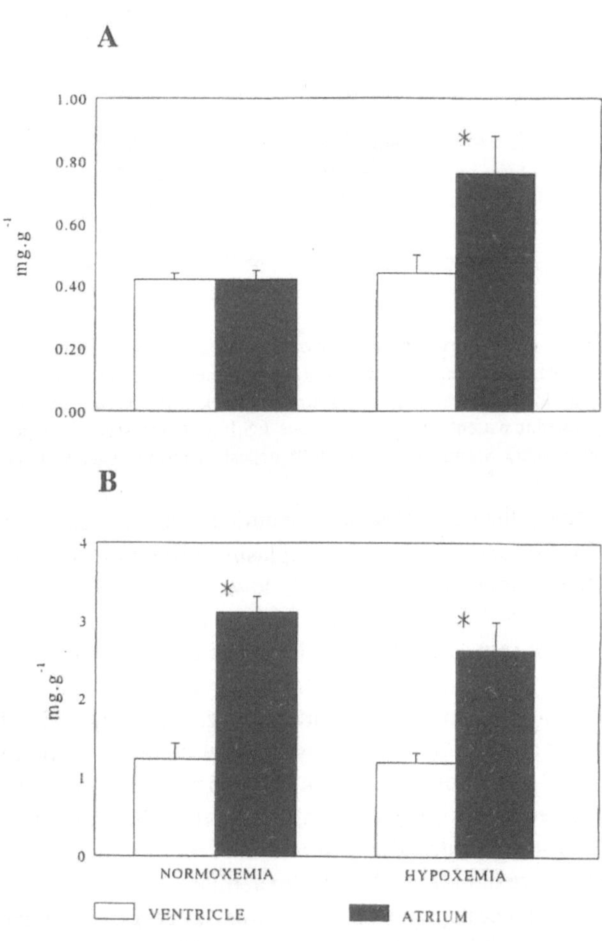

Fig. 3. Hydroxyproline profiling of collagenous fraction. Concentration of pepsin soluble (A) and pepsin insoluble (B) collagenous forms in human right ventricle (white bars) and right atrium (black bars) in normoxemic and hypoxemic patients. Values are means ± S.E.M. (expressed as mg per g of wet weight). Significance: *$p < 0.05$ or less vs atrium. Data from [16].

in the fraction of pepsin-soluble collagenous proteins (Fig. 3A). The amount of soluble collagen (as judged from the hydroxyproline measurement) was 12% and 26% for atrial and ventricular musculature of normoxemic patients, resp. Hypoxaemia elevated only the concentration of hydroxyproline in atrial soluble collagenous proteins (Fig. 3A), while the concentration of hydroxyproline in insoluble collagenous proteins of hypoxemic atrial musculature was significantly lower (Fig. 3B). The proportion of soluble collagen in atrial hypoxemic musculature therefore reached 22% of total collagen. The elevated concentration of hydroxyproline in the fraction of soluble collagenous proteins in the atria of hypoxemic patients attests to a faster collagen synthesis.

At present little is known about collagen synthesis and degradation of collagenous structures even in normal myocardium [8, 9, 42–46]. Lower arterial oxygen saturation probably affects the synthesis of collagen preferentially in the atrial compartment; the physiological meaning of this phenomenon is still far from being clear.

A typical electrophoretic pattern of the soluble collagenous protein fraction is shown in Fig. 4. Both atrial and ventricular collagen is composed from a mixture of collagens I, III and V. There are at least two different types of cardiac collagens: *fibrillar* (e.g. collagen I, III, V) and *nonfibrillar* forms (collagen types IV and VI). Both types have been detected in the cardiac muscle [8, 10, 11]. The most abundant subtypes, collagen I and III, serve to direct the contractile force generated by cardiac myocytes and contribute to the passive stretch. However, the data on atrio-ventricular differences in collagen typing are still lacking. The samples of soluble collagenous proteins from ventricular musculature (Fig. 4) contained a different amount of low molecular weight fragments (mol. weight 29–35 kDa); the composition was not dependent on the oxygen saturation. This electrophoretic pattern did not allow us to determine whether these proteins originate from the degradation of collagen molecules or from both the intermediate filament network and the myocyte cytoskeleton material [24].

There were significant atrio-ventricular differences in the profile of myofibrillar proteins (Fig. 5). Both ventricular and atrial samples of contractile proteins contained heavy chains of myosin (HC), actin (A) and tropomyosin (TM). On the other hand, the electrophoretic pattern of the myosin light chains (LC) was not identical: samples of ventricular musculature, beside LC_{1V} and LC_{2V}, were contaminated with atrial myosin light chains (LC_{1A}). The amount of contamination in normoxemic and hypoxemic patients was not different (Fig. 5).

Two types of myosin subunits play a role in cardiac contraction: HC and LC chains; atrial and ventricular isoforms of both types have been described [6, 7]. In embryonic and the very early postnatal period only one, atrial, type occurs in both ventricular and atrial human myocardium. During further cardiac development, distinct types are synthesized

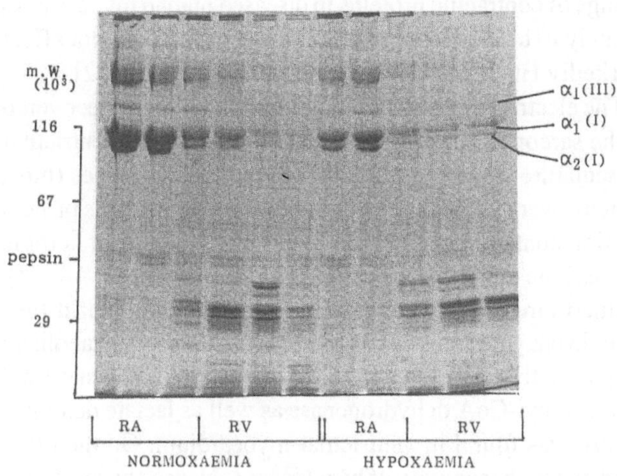

Fig. 4. SDS-polyacrylamide gel electrophoresis of pepsin-soluble fraction of collagen. Reference grade collagen type I [α_1 (I) and α_2 (I)] and collagen type III [α_1 (III)] standards and collagen extracted from human myocardial tissue were run. Samples were derived from human right atrium (RA) and right ventricle (RV) of normoxemic and hypoxemic patients.

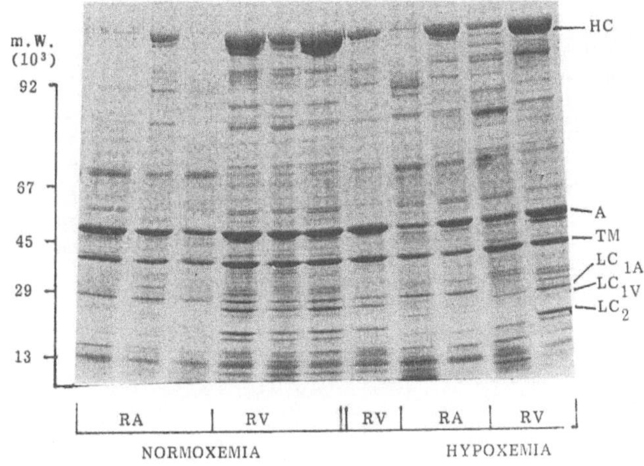

Fig. 5. SDS-polyacrylamide gel electrophoresis of contractile proteins of human cardiac muscle. Molecular weight standards and contractile proteins extracted from human myocardial tissue were run. Samples were derived from human right atrium (RA) and right ventricle (RV) of normoxemic and hypoxemic patients. Positions of the main contractile proteins are identified: HC – myosin heavy chains, A – actin, TM – tropomyosin, LC_1, LC_2 of atrial (A) or ventricular (V) origin.

in the ventricle [47, 48]. The reappearance of atrial types in ventricles and vice versa is typical for the protein profile of hypertrophied adult human myocardium [30, 49, 50]. Recently, Tahair *et al.* [51] hypothesized that the reexpression of myosin LC of the atrial type in human ventricle is a marker of diseased myocardium. It was therefore not surprising that all ventricular samples of our setup were contaminated by atrial LC. On the other hand no ventricular LC pattern was seen in atrial samples. It seems therefore that the primary change of contractile proteins in diseased human myocardium is likely to be a different pattern of LC which does not affect markedly HC or myosin ATPase activity [16, 32, 52].

The electrophoretic pattern of metabolic proteins recovered in the sarcoplasmic fraction from both atrial and ventricular musculature exhibited no atrio-ventricular differences (most proteins were in the range of 20–67 kDa (Fig. 6). This protein fraction contains predominantly enzyme systems for aerobic and aerobic substrate utilization [for details see 17, 18, 53]. Furthermore, there were significant atrio-ventricular differences in enzyme activities connected with aerobic metabolism: a higher activity of citrate synthase, malate dehydrogenase and hydroxyacyl-CoA dehydrogenase as well as lactate dehydrogenase was found in ventricular myocardium. On the other hand, the higher activity of hexokinase (glucose phosorylating enzyme) was detected in the atrial tissue. Hypoxaemia lowered the capacity of oxidative enzymes in both parts; the changes were more pronounced in atrial musculature [17, 25, 53]. However, atrio-ventricular differences persisted and it may be assumed that they do exist also in healthy individuals.

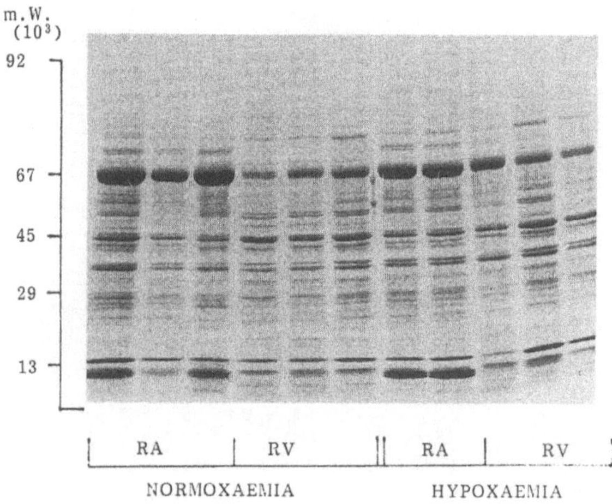

Fig. 6. SDS-polyacrylamide gel electrophoresis of metabolic proteins (isolated as sarcoplasmic fraction) of human cardiac muscle. Molecular weight standards and sarcoplasmic proteins extracted from human myocardial tissue were run. Samples were derived from human right atrium (RA) and right ventricular (RV) of normoxemic and hypoxemic patients.

In conclusion, there are atrio-ventricular differences in the protein profile of human myocardium of patients both normoxemic and hypoxemic with congenital heart disease; atrial musculature exhibits a higher amount of collagenous proteins with the predominance of insoluble collagenous forms, and a significantly lower amount of contractile proteins. The myofibrillar samples of ventricular musculature contained LC of atrial type. Hypoxia did not elevate the concentration of total collagen, but atrial musculature contains a higher proportion of soluble collagenous forms.

References

1. Gorza L, Sartore S, Schiaffino S: Myosin types and fiber types in cardiac muscle. II-Atrial myocardium. J Cell Biol 95: 838–845, 1982
2. Sartore S, Gorza L, Bormioli SP, Dalla-Libera L, Sciaffino S: Myosin types and fiber types in cardiac muscle. I-Ventricular myocardium. J Cell Biol 88: 226–233, 1981
3. Sommer JR, Jennings RB: Ultrastructure of cardiac muscle. In: Fozzard *et al.* (eds). The heart and cardiovascular system. Raven Press Ltd., New York, 1992, pp 3–50
4. Arminger LC, Seelye RN, Morrison MA, Holliss DG: Comparative biochemistry and fine structure of atrial and ventricular myocardium during autolysis *in vitro*. Basic Res Cardiol 79: 218–229, 1984
5. Dalla-Libera L, Carraro U, Pauletto P: Light and heavy chains of myosin from atrial and ventricular myocardium of turkey and rat. Basic Res Cardiol 78: 671–678, 1983
6. Dechesne C, Legher J, Bouvagnet P, Claviez M, Leger JJ: Fractionation and characterization of two molecular variants of myosin from adult human atrium. J Mol Cardiol 17: 753–767, 1985
7. Swynghedauw B: Developmental and functional adaptation of contractile proteins in cardiac and skeletal muscles. Physiol Rev 66: 710–711, 1986
8. Weber K: Cardiac interstitium: Extracellular space of the myocardium. In: H.A. Fozzard *et al.* (eds). The heart and Cardiovascular system Raven press Ltd., New York, 1992, pp 1465–1480
9. Borg TK, Terracio L: Interaction of the extracellular matrix with cardiac myocytes during development and disease. In: T.F. Robinson, R.K.H. Kinne, (eds). Cardiac myocyte-connective tissue interactions, in health and disease Basel: 13, 1990, pp 113–129
10. Borg TK, Burgess ML: Holding it all together: Organization and function of the extracellular matrix in the heart. Heart Failure 8: 230–238, 1992
11. Pelouch V, Jirmar R: Biochemical characterization of cardiac collagen and its role in ventricular remodelling following infarction. Physiol Res 42: 283–292, 1993
12. Pelouch V, Dixon IMC, Golfronan L, Beamish RE, Dhalla NS: Role of extracellular matrix proteins in heart function. Mol Cell Biochem 129: 101–120, 1994
13. Caufield JB, Norton P, Weaver D: Cardiac dilatation associated with collagen. Mol Cell Biochem 118: 171–179, 1992
14. Weber K: Cardiac interstitium in health and disease. The fibrilar collagen network. J Am Coll Cardiol 13: 1637–1652, 1989
15. Imataka K, Naito S, Seko Y, Fujii J: Hydroxyproline in all part of the rabbit heart in hypertension and in its reversal. J Mol Cell Cardiol 21: (Suppl. V) 133–139, 1989
16. Pelouch V, Milerova M, Ostadal B, Samanek M, Hucin B: Protein profiling of human atrial and ventricular musculature: the effect of normoxemia and hypoxemia in congenital heart diseases. Physiol Res 42: 235–242, 1993
17. Bass A, Šamánek M, Ošt'ádal B, Hucín B, Stejskalová M, Pelouch V: Differences between atrial and ventricular energy supplying enzymes in children. J App Cardiol 3: 397–405, 1988
18. Pelouch V, Milerová M, Šamánek M, Ošt'ádal B, Hučín B: Protein composition of the atria and ventricles in children with congenital heart disease. Physiol bohemoslov 37: 530–531, 1988
19. Yazaki Y, Tsuchimochi H, Kurabayashi M, Komuro I: Molecular adaptation to pressure overload, in human and rat hearts. J Mol Cell Cardiol 21: 91–101, 1989
20. Pelouch V, Oštádal B, Kolář F, Milerová M, Grünermel J: Chronic hypoxia-induced age-dependent changes of collagenous and non-collagenous cardiac protein fractions. In: B. Oštádal, N.S. Dhalla (eds). Heart Function in Health and Disease. Kluwer Academic

Publishers, Boston/Dordrecht/London, 1992, pp 209–218

21. Pelouch V, Dixon IMC, Sethi R, Dhalla NS: Alteration of collagenous protein profile in congestive heart failure secondary to myocardial infarction. Mol Cell Biochem 129: 121–131, 1994
22. Lowry OH, Rosenbrough HJ, Farr AL, Randall RJ: Protein measurement with Folin phenol reagent. J Biol Chem 193: 265–275, 1951
23. Huszar G: Monitoring of collagen and collagen fragments in chromatography of protein mixture. Anal Biochem 105: 424–429, 1980
24. Lockard VG, Bloom S: Trans-cellular desmin-laminin B intermediate filament network in cardiac myocytes. J Mol Cell Cardiol 25: 303–309, 1993
25. Samánek M, Bass A, Ošt'ádal B, Hucín B, Stejskalová M: Effect of hypoxaemia on enzymes supplying myocardial energy in children with congenital heart disease. Int J Cardiol 25: 265–270, 1989
26. Humprey JE, Cummins P: Regulatory proteins of the myocardium. Atrial and ventricular tropomyosin and troponin in the developing and adult bovine and human heart. J Mol Cell Cardiol 16: 647–657, 1984
27. Nakanishi T, Okuda H, Kamata K, Abe K, Sekiguchi M, Takao A: Development of myocardial contractile system in fetal rabbit. Pediatric Res 22: 201–207, 1987
28. Malhotra A, Siri FA, Aronson R: Cardiac contractile proteins in hypertrophied and failing guinea pig heart. Cardiovasc Res 26: 153–162, 1992
29. Dhalla NS, Dixon IMC, Beamish RE: Biochemical basis of heart function and contractile failure. J Appl Cardiol 6: 7–30, 1991
30. Hirzel HO, Tuchschmid CR, Schneider J, Krayenbuehl HP, Schaub MC: Relationship betwen myosin isoenzyme composition, hemodynamics and myocardial structure in various forms of human cardiac hypertrophy. Circ Res 57: 729–740, 1985
31. Hoffman U, Siegert E: Atrial and ventricular myosins from human hearts: I): Isoenzyme distribution during development and in the adults. Basic Res Cardiol 82: 348–358, 1987
32. Hoffman U, Axmann C, Palm N: Atrial and ventricular myosins from human hearts. II): Isoenzyme distribution after myocardial infarction. Basic Res Cardiol 82: 359–369, 1987
33. Eghbali M, Czaja MJ, Zeydel M, Weiner FR, Seifter S, Blumenfeld OO: Collagen chain mRNAs in isolated heart cells from young and adult rats. J Mol Cell Cardiol 20: 267–276, 1988
34. Kawahara E, Mukai A, Oda Y, Nakanishi I, Iwa T: Left ventriculotomy of the heart: tissue repair and localization of collagen types I, II, III, IV, V, VI and fibronectin. Virchows Archiv 417A: 229–236, 1990
35. Caspari PG, Gibbson K, Harris P: Changes in myocardial collagen in normal development and after beta blockade. In: Recent advances in studies on cardiac structure and metabolism. Academic Press, New York, vol. 7, 1976, pp 99–104
36. Kardami E, Fandrich RR: Basic fibroblast growth factor in atria and ventricles of the vertebrate heart. J Cell Biol 109: 1865–1874, 1989
37. Pelouch V, Ošt'ádal B, Procházka J, Urbanová D, Widimský J: Effect of high altitude hypoxia on the protein composition of the right ventricular myocardium. Prog Resp Res 20: 41–48, 1985
38. Pelouch V, Ošt'ádal B, Procházka J: Changes of contractile and collagenous protein induced by chronic hypoxia in myocardium during postnatal development of rat. Biomed Biochim Acta 46: S707–S711, 1987
39. Hollenber M, Honbo N, Samorodin AJ: The effect of hypoxia on cardiac growth in neonatal rat. Am J Physiol 231: 1445–1450, 1976
40. Holenber M, Honbo N, Samorodin AJ: Cardiac cellular responses to altered nutrition in neonatal rat. Am J Physiol 233: H356–H360, 1977
41. Laemmli UK: Cleavage of structural proteins during the assembly of the heaqd of bacteriophafe T4. Nature 227: 680–685, 1970
42. Factor SM: Pathological alteration of myocyte-connective tissue interaction in cardiovascular disease. In: T.F. Robinson, R.K.H. Kinne (eds). Cardiac myocyte-connective tissue interactions, in health and disease. Basel: 13, 1990, pp 130–146
43. Laurent GJ: Dynamic state of collagen: pathways of collagen degradation *in vivo* and their possible role in regulation of collagen mass. Am J Physiol 252: C1–C9, 1987
44. Schapper J, Froede R, Hein S, Buck A, Hashizume H, Speiser B, Friedl A, Bleese N: Impairment of the myocardial ultrastructure and changes of cytoskeleton in dilated cardiomyopathy. Circulation 83: 504–514, 1991
45. Weber KT, Janicki JS, Pick R, Capasso J, Anversa P: Myocardial fibrosis and pathological hypertrophy in the rat with renovascular hypertension. Am J Cardiol 65: G1–G7, 1990
46. Kivirikko KI: Collagens and their abnormalities in wide spectrum of diseases. Ann Med 2: 113–126, 1993
47. Price KM, Littler WA, Cummins P: Human atrial and ventricular myosin light chain subunits in the adult and during development. Biochem J 191: 571–580, 1980
48. Cummins P: Transitions in human atrial and ventricular myosin light chain isoenzymes in response to cardiac pressure overload-induced hypertrophy. Biochem J 205: 195–204, 1982
49. Schaub MC, Hirzel HO: Atrial and ventricular isomyosin composition in patients with different forms of cardiac hypertrophy. Basic Res Cardiol 82: (Suppl. 2) 357–367, 1987
50. Nakao K, Yasue H, Fujimoto K, Okumura K, Yamoto H, Hitoshi Y, Murohara T, Takatsu K, Miyamoto E: Increased expression of atrial myosin light chain 1 in the overloaded human left ventricle: possible expression of fetal type myocytes. Inter J Cardiol 36: 315–328, 1992
51. Trahair T, Yeoh T, Cartmill T, Keogh A, Spratt P, Chang V, dos Remedios CG, Gunning P: Myosin light chain gene expression associated with disease states of the human heart. J Mol Cell Cardiol 26: 577–585, 1993
52. Takeda N, Rupp H, Fenchel G, Hofmeister H-E, Jacob R: Relationship between the myofibrillar ATPase activity of human biopsy material and hemodynamic parameters. Jpn Heart J 26: 909–922, 1985
53. Bass A, Stejskalová M, Šamánek M, Ošt'ádal B, Hučin J, Procházka J: Comparison of energy-supply enzyme activities in the right atrium and ventricle of children with congenital heart disease. Physiol bohemoslov 37: 530–531, 1988
54. Morgan HF, Baker KM: Cardiac hypertrophy. Mechanical, neural and endocrine dependence. Circulation 83: 13–25, 1991

Molecular and Cellular Biochemistry **147**: 51–55, 1995.

Ventricular remodeling: insights from pharmacologic interventions with angiotensin-converting enzyme inhibitors

Sidney Goldstein, Victor G. Sharov, Jane M. Cook and Hani N. Sabbah

Department of Medicine, Division of Cardiovascular Medicine, Henry Ford Heart and Vascular Institute, Detroit, Michigan, USA

Abstract

Structural remodeling of the left ventricular (LV) myocardium develops in a time-dependent fashion following acute myocardial infarction and may be an integral component in the transition toward overt heart failure. Globally, the remodeling process is characterized by progressive LV enlargement and increased chamber sphericity. At the cellular level, the remodeling process is associated with myocyte slippage, hypertrophy, and accumulation of collagen in the interstitial compartment. In the present study, we examined the effects of early, long-term monotherapy with the angiotensin converting enzyme (ACE) inhibitor, enalapril, on the progression of LV remodeling in dogs with LV dysfunction (ejection fractions 30–40%) produced by multiple sequential intracoronary microembolizations. Dogs were randomized to 3 months oral therapy with enalapril (n = 7) or to no treatment (n = 7). In untreated dogs, LV end-systolic volume index (ESVI), end-diastolic volume index (EDVI) and chamber sphericity increased significantly during the 3 months follow-up period. In contrast, in dogs treated with enalapril ESVI, EDVI and chamber sphericity remained essentially unchanged. Treatment with enalapril attenuated myocyte hypertrophy and the accumulation of interstitial collagen in comparison to untreated dogs. These data indicate that early treatment with ACE inhibitors can prevent the progression of LV remodeling in dogs with LV dysfunction. Afterload reduction, inhibition of direct action of angiotensin-II and possibly the decrease in bradykinin degradation elicited by ACE inhibition may act in concert in preventing the progression LV chamber remodeling. (Mol Cell Biochem **147**: 51–55, 1995)

Key words: ventricular enlargement, myocyte hypertrophy, interstitial fibrosis, angiotensin converting enzyme inhibitors

Introduction

Structural and topographical remodeling of the left ventricle (LV) has long been recognized to develop following acute myocardial infarction. This remodeling process is progressive in nature in that it develops over a period of months or even years after the acute event. The factors that dictate the rate at which this process develops are not clear but are likely related to the extent of loss of viable myocardium. Thus, a larger infarction is likely to elicit a faster progression of LV remodeling in comparison to a smaller infarction. The term 'ventricular remodeling' includes several structural and topographical adaptations and/or maladaptations that occur in response to myocardial injury. Globally, these changes include LV chamber dilation and increased chamber sphericity [1]. At the cellular level, alterations in both the myocyte and non-myocyte compartment occur and include an increase in myocyte size and accumulation of collagen in the interstitium (reactive interstitial fibrosis) [1]. There is little doubt that this process of LV remodeling, if left unchecked, can lead to progressive LV dysfunction and ultimately to the syndrome of congestive heart failure. Despite the ominous association of LV remodeling with poor long-term prognosis, little is known about this process and the factors that dictate its development and progression. Our approach to probing the underlying factors that promote progressive LV remodeling has centered on the use of pharmacologic agents in a canine model of chronic heart failure which manifests progressive

Address for offprints: S. Goldstein, Henry Ford Hospital, 2799 West Grand Boulevard, Detroit, Michigan 48202, USA

LV remodeling [2, 3]. In the present study, we examined the effects of early, long-term, therapy with the angiotensin converting enzyme (ACE) inhibitor, enalapril, on the progression of LV remodeling in dogs with moderate heart failure. Specifically, we examined the effects of therapy with enalapril on LV chamber enlargement, LV chamber sphericity, cardiocyte hypertrophy and interstitial fibrosis. The cohort of animals used in this study represents a subset of a larger study which examined the hemodynamic effects of other pharmacologic agents including beta-blockers and digitalis preparations [4].

Methods

The animal model

Fourteen healthy mongrel dogs weighing between 18 and 31 kg were used in the study. Chronic LV dysfunction was produced by multiple sequential intracoronary embolizations with polystyrene latex microspheres (77–102 μm in diameter) as previously described [2]. Coronary microembolizations were performed during sequential cardiac catheterizations under general anesthesia and sterile conditions. Anesthesia consisted of intravenous injections of oxymorphone hydrochloride (0.22 mg/kg), diazepam (0.17 mg/kg) and sodium pentobarbital (150–250 mg to effect). In all dogs, coronary microembolizations were discontinued when LV ejection fraction, determined angiographically, was between 30–40%. To achieve this target ejection fraction, dogs underwent an average of five microembolization procedures performed 1–3 weeks apart. Three weeks after the last embolization procedure, all dogs underwent a pre-randomization left and right heart catheterization. One day after cardiac catheterization dogs were randomized to 3 months oral monotherapy with enalapril (10 mg twice daily, n = 7) or to no treatment at all (control, n = 7). Angiographic measurements were made at baseline, prior to any embolizations, and were repeated one day prior to randomization and initiation of therapy. Dogs were sacrificed after the final hemodynamic study namely, 3 months after initiating therapy and the hearts were removed and prepared for histologic evaluation. The study was approved by the Henry Ford Hospital Care of Experimental Animals Committee and conformed to the guiding principles of the American Physiological Society.

Ventriculographic measurements

Left ventriculograms were obtained during cardiac catheterization with the dog placed on its right side and were recorded on 35 mm cine at 30 frames/sec during the injection of 20 ml of contrast material (Hypaque meglumine 60%, Withrop Pharmaceuticals). Correction for image magnification was made with a radiopaque calibrated grid placed at the level of the LV. LV end-systolic and end-diastolic volumes were calculated from ventricular silhouettes using the area-length method [5] and were corrected for body surface area (end-diastolic volume index = EDVI and end-systolic volume index = ESVI). Global indexes of LV shape were used to quantitate changes in LV chamber sphericity. Left ventricular shape was quantified from angiographic silhouettes based upon the ratio of the major-to-minor axis at end-systole (ESR) and end-diastole (EDR) [6]. As these ratios decrease (approach unity), the shape of the LV deviates from that of a typical ellipsoid to one which approaches that of a sphere.

Immunohistochemical and morphometric assessments

From each dog, transmural tissue blocks were obtained from the LV free wall at the mid-ventricular level and rapidly frozen in isopentane cooled to –160°C in liquid nitrogen. Cryostat sections, 8–10 μm thick, were prepared and incubated at 4°C overnight in rabbit anti-human collagen type III polyclonal antibody (Chemicon International, Inc.). Sections were then stained with dichlorotriazinyl amino fluorescence (DTAF)-conjugated goat anti-rabbit IgG (Chemicon International, Inc.) to visualize interstitial collagen. Immunofluorescent staining was evaluated with an epifluorescent microscope optimized for DTAF. The same sections were used for quantitating myocyte size and volume fraction of interstitial collagen. From each section, five microscopic fields, each containing a minimum of 100 cardiocytes were selected at random for analysis. For each dog, the average myocyte cross sectional area (radial sections only) was calculated using computer-assisted planimetry (SigmaScan, Jandell Scientific). The volume fraction (%) of interstitial collagen, area occupied by collagen as a percent of total surface area, was quantified using computer-assisted video-densitometry (JAVA Video Analysis Software, Jandell Scientific). LV tissue specimen obtained from 5 normal dogs was processed in the same manner and used for comparison.

Data analysis

Comparisons of angiographic variable within each group were examined between measurements obtained just prior to the initiation of therapy and measurements made after completion of 3 months of therapy. For these comparisons, a Students paired t-test was used and a probability of 0.05 or less was considered significant. To ensure that angiographic parameters prior to randomization and initiation of therapy were similar between the untreated group and the enalapril treated group, comparisons were made using a t-statistic for two means. For this test, a probability of 0.05 or less was considered significant. Differences in morphometric measures between the control (untreated group) and the enalapril treated group were also examined using a t-statistic for two means. A probability of 0.05 or less was considered significant. All data are reported as the mean ± standard error of the mean.

Results

Angiographic findings

There were no significant differences in any of the pre-randomization angiographic parameters between dogs that were subsequently randomized to no treatment and dogs randomized to active treatment with enalapril. In dogs randomized to no treatment, LV ESVI was 44 ± 5 ml/m² and increased to 64 ± 7 ml/m² at the end of 3 months of follow-up compared to pre-randomization (71 ± 7 vs. 86 ± 9 ml/m², p = 0.007). During the 3 months follow-up period, in this untreated group of dogs, there was also associated significant reduction in ESR (1.56 ± 0.04 vs. 1.42 ± 0.04, p = 0.03) and EDR (1.43 ± 0.04 vs. 1.29 ± 0.05, p = 0.02) indicating increased LV chamber sphericity. In contrast, in dogs treated with enalapril, all angiographic parameters were not significantly different at the end of 3 months therapy compared to values obtained before randomization. In this cohort of treated dogs, ESVI was similar before and after therapy (44 ± 5 vs. 44 ± 5 ml/m², p = 0.94), as was EDVI (67 ± 8 vs. 72 ± 7 ml/m², p = 0.24), ESR (1.51 ± 0.08 vs. 1.46 ± 0.10, p = 0.67) and EDR (1.36 ± 0.07 vs. 1.27 ± 0.08, p = 0.036).

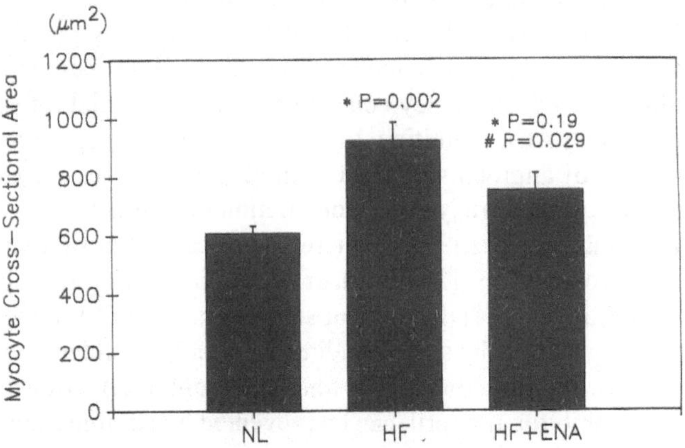

Fig. 1. Bar graph depicting values (mean ± SEM) of average myocyte cross-sectional area in normal dogs (NL), dogs with moderate heart failure that are untreated (HF) and dogs with moderate HF treated with enalapril (HF + ENA). * = p-value relative to NL; # = p-value relative to HF.

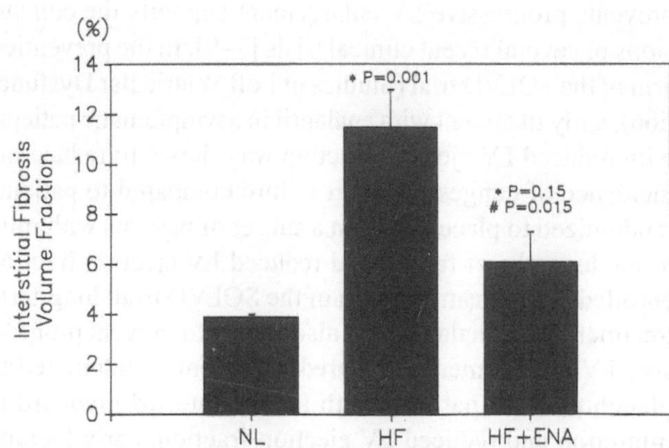

Fig. 2. Bar graph depicting values (mean ± SEM) of volume fraction of interstitial collagen in normal dogs (NL), dogs with moderate heart failure that are untreated (HF) and dogs with moderate HF treated with enalapril (HF + ENA). * = p-value relative to NL; # = p-value relative to HF.

Changes in myocyte size and volume fraction of interstitial collagen

In untreated dogs (control arm), the average LV myocyte cross-sectional area was substantially greater than in the LV of normal dogs (924 ± 63 vs. 608 ± 25 μm²) (p = 0.002). In dogs treated with enalapril, the average LV myocyte cross-sectional area (711 ± 58 μm²) was significantly smaller than untreated dogs (p = 0.029) and not significantly different than normal dogs (p = 0.19) (Fig. 1). In untreated dogs, the volume fraction of interstitial collagen was nearly 4-fold greater than that observed in normal dogs (11.5 ± 1.5 vs. 3.9 ± 0.1%) (p = 0.001). In dogs treated with enalapril, the volume fraction of interstitial collagen (6.1 ± 1.2%) was significantly lower than in untreated dogs (p = 0.015) and not significantly different than normal dogs (p = 0.15) (Fig. 2).

Discussion

Results of the present study indicate that in the absence of any drug interventions, dogs with moderate heart failure manifest considerable LV remodeling evidenced by progressive chamber enlargement, progressive chamber sphericity, increased cross-sectional area (hypertrophy) of residual cardiocytes and excessive accumulation of interstitial collagen. Long-term treatment with the ACE inhibitor, enalapril, on the other hand prevented or markedly attenuated all of these features of LV remodeling. Enalapril therapy prevented progressive LV dilation and attenuated the increase in chamber sphericity, myocyte size and interstitial fibrosis.

The observation that early long-term therapy with enalapril prevents progressive LV enlargement supports the conclusions of several recent clinical trials [7–9]. In the prevention arm of the SOLVD trial (Studies of Left Ventricular Dysfunction), early treatment with enalapril in asymptomatic patients with reduced LV ejection fraction was shown to reduce the incidence of congestive heart failure compared to patients randomized to placebo [7]. In a subset of patients with mild to moderate heart failure and reduced LV ejection fraction enrolled in the treatment arm of the SOLVD trial, long-term treatment with enalapril was also shown to prevent progressive LV enlargement compared to patients randomized to placebo [9]. In patients with a first anterior myocardial infarction and reduced LV ejection fraction, early therapy with the ACE inhibitor, captopril, was also shown to attenuate LV dilation [8]. Treatment with captopril in the first year after an anterior myocardial infarction was also shown to attenuate the increase in LV chamber sphericity [10]. The effects of other prototypical drugs used in the treatment of heart failure have also been examined in terms of their efficacy in preventing progressive LV chamber dilation. In dogs with moderate heart failure produced by multiple sequential intracoronary microembolizations, we showed that early, long-term therapy with the beta-blocker, metoprolol, can also prevent progressive LV dilation [4]. In the same dog model, however, early, long-term treatment with digoxin failed to prevent progressive LV dilation [4]. Consistent with these findings, studies in patients with dilated cardiomyopathy showed that long-term therapy with metoprolol was also effective in reducing LV chamber dimensions [11]. In patients with anterior myocardial infarction, treatment with digoxin initiated 7 to 10 days after the onset of symptoms failed to prevent the progressive increase in LV end-systolic and end-diastolic volume indexes after one year of therapy [12].

Although clear evidence exist to indicate that interference with the renin-angiotensin system in the form of ACE inhibition can modulate LV chamber enlargement in patients with chronic LV dysfunction and some evidence to suggest that this form of therapy can prevent progressive LV shape changes, there is no direct studies in patients which implicate ACE inhibition in the prevention or regression of myocyte hypertrophy or in the prevention of interstitial fibrosis. There are studies in animal models, however, that support the concept that ACE inhibitors can have a direct effect on myocyte hypertrophy and on the accumulation of interstitial collagen. In spontaneously hypertensive rats, for instance, treatment with captopril resulted in regression of LV hypertrophy [13]. In a rat model of renovascular hypertension, pretreatment with captopril was also shown to largely prevent the appearance of myocardial interstitial and perivascular fibrosis [14].

The mechanisms through which ACE inhibition and for that matter beta-adrenergic blockade elicit a beneficial effect on LV remodeling remain unclear. Certainly modulation of afterload must be taken into account when considering potential mechanisms of action of both ACE inhibitors and beta-blockers. Studies performed in this laboratory in dogs with moderate heart failure, from which the present animal cohort was selected, showed that therapy with either enalapril or metoprolol attentuated the progressive rise in systemic vascular resistance seen in untreated dogs; whereas monotherapy with digoxin did not [4]. These data provide compelling evidence that afterload reduction can prevent progressive LV enlargement in both patients and animals with chronic LV dysfunction. In spontaneously hypertensive rats, blood pressure reduction after therapy with captopril was also shown to be associated with a significant reduction in LV weight and with a significant reduction in total myocardial collagen content [15].

At present, it would be somewhat premature to suggest that afterload reduction is the sole factor responsible for the observed changes in the global and cellular feature of the LV remodeling process. Other factors should also be taken into account particularly with respect to therapy with ACE inhibitors. Angiotensin-II may have a direct effect on interstitial fibrosis through its action on fibroblasts which normally reside in the myocardium. Myocardial fibroblasts possess antiotensin-II receptor sites [16] and contain the mRNA which is responsible for gene expression of type-I and type-III collagens [17], the major fibrillar collagens of the myocardium [18]. Angiotensin-II formed locally, as a result of activation of a local renin-angiotensin system [19], may be a direct stimulus for cell growth [20] independent of its effect on afterload augmentation and may also be mitogenic to fibroblasts [16, 21]. Recent studies in rats with LV hypertrophy produced by aortic banding have suggested that specific blockade of angiotensin AT-1 receptors with LA sartan (Dup 753) was less effective than ACE inhibition in attenuating myocardial hypertrophy despite lowering of systemic blood pressure [22]. This observation would suggest that factors other than prevention of angiotensin-II formation and afterload reduction may contribute to the beneficial effects elicited by ACE inhibition. In addition to preventing the formation of angiotensin-II, ACE inhibitors also reduce the degradation of bradykinin. Potentiation of kinins following ACE inhibition can lead to increased release of nitric oxide and prostacyclin [23] both of which are thought to be antimitogenic [24]. In a recent study in rats with LV hypertrophy produced by aortic banding, Linz and Scholkens demonstrated that the beneficial antihypertrophic effect of the ACE inhibitor, ramipril, can be prevented by administration of a specific B-2 bradykinin receptor antagonist (HOE 140) in the absence of any reductions of blood pressure [25]. This observation, although unconfirmed, provides some evidence that ACE inhibition induced potentiation of bradykinin may contribute to the beneficial effects of ACE inhibitors on certain components of the LV remodeling process.

In conclusion, the results of the present study indicate that in dogs with moderate heart failure, early long-term monotherapy with enalapril prevents progressive LV remodeling as evidenced by prevention or attenuation of LV enlargement, LV chamber sphericity, myocyte hypertrophy and interstitial fibrosis. The exact mechanisms of this beneficial effect of ACE inhibitors remains uncertain. Additional studies are needed which probe the effects of angiotensin-II and bradykinin receptor blockade on LV remodeling in the setting of heart failure. In the absence of such studies, one can only conclude that the benefit of ACE inhibition on LV remodeling is likely derived from a combination of afterload reduction, direct inhibition of angiotensin-II formation and possibly from reduced degradation of bradykinin.

Acknowledgements

Supported, in part, by Grants from the American Heart Association of Michigan and National Heart, Lung and Blood Institute HL49090-01.

References

1. Sabbah HN, Goldstein S: Ventricular remodeling: consequences and therapy. Europ Heart J 14: 24–29, 1993
2. Sabbah HN, Stein PD, Kono T, Gheorghiade M, Levine TB, Jafri S, Hawkins ET, Goldstein S: A canine model of chronic heart failure produced by multiple sequential coronary microembolizations. Am J Physiol 260: H1379–H1384, 1991
3. Sabbah HN, Hansen-Smith F, Sharov VG, Kono T, Lesch M, Gengo PG, Steffen RP, Levine TB, Goldstein S: Decreased proportion of type I myofibers in skeletal muscle of dogs with chronic heart failure. Circulation 87: 1729–1737, 1993
4. Sabbah HN, Shimoyama H, Kono T, Gupta RC, Sharov VG, Scicli G, Levine TB, Goldstein S: Effects of long-term monotherapy with enalapril, metoprolol, and digoxin on the progression of left ventricular dysfunction and dilation in dogs with reduced ejection fraction. Circulation 84: 2852–2859, 1994
5. Dodge HT, Sandler H, Baxley WA, Hawley RR: Usefulness and limitations of radiographic methods for determining left ventricular volume. Am J Cardiol 18: 10–24, 1966
6. Sabbah HN, Kono T, Stein PD, Mancini GBJ, Goldstein S: Left ventricular shape changes during the course of evolving heart failure. Am J Physiol 263: H266–H270, 1992
7. The SOLVD Investigators: Effect of enalapril on mortality and the development of heart failure in asymptomatic patients with reduced left ventricular ejection fractions. N Engl J Med 327: 685–691, 1992
8. Pfeffer MA, Lamas GA, Vaughan DE, Parisi AF, Braunwald E: Effect of captopril on progressive ventricular dilatation after anterior myocardial infarction. N Engl J Med 319: 80–86, 1988
9. Konstam MA, Rousseau MF, Kronenberg, MW, Udelson JE, Melin J, Stewart D, Dolan N, Edens TR, Ahn S, Kinan D, Howe DM, Kilcoyne L, Metherall J, Benedict C, Yusuf S, Pouleur H: Effects of the angiotensin converting enzyme inhibitor enalapril on long-term progression of left ventricular dysfunction in patients with heart failure. Circulation 86: 431–438, 1992
10. Mitchell GF, Lamas GA, Vaughan DE, Pfeffer MA: Left ventricular remodeling in the year after first anterior myocardial infarction: a quantitative analysis of contractile segment lengths and ventricular shape. J Am Coll Cardiol 19: 1136–1144, 1992
11. Waagstein F, Caidahl K, Wallentin I, Bergh C-H, Hjalmarson A: Long-term beta-blockade in dilated cardiomyopathy. Effects of short and long-term metoprolol treatment followed by withdrawal and readministration of metoprolol. Circulation 80: 551–563, 1989
12. Bonaduce D, Petretta M, Arrichiello P, Conforti G, Montemurro MV, Attisano T, Bianchi V, Morgano G: Effects of captopril treatment on left ventricular remodeling after anterior myocardial infarction: comparison with digitalis. J Am Coll Cardiol 19: 858–863, 1992
13. Pfeffer JM, Pfeffer MA, Mirsky I, Braunweld E: Regression of left ventricular hypertrophy and prevention of left ventricular dysfunction by captopril in the spontaneously hypertensive rat. Proc Natl Acad Sci 79: 3310–3314, 1982
14. Jalil E, Janicki JS, Pick R, Weber KT: Coronary vascular remodeling and myocardial fibrosis in the rat with renovascular hypertension: response to captopril. Am J Hypertens 4: 51–55, 1991
15. Mukherjee D, Sen S: Collagen phenotypes during development and regression of myocardial hypertrophy in spontaneously hypertensive rats. Circ Res 67: 1474–1480, 1990
16. Schorb W, Booz GW, Dostal DE, Conrad KM, Chang KC, Baker KM: Angiotensin II is mitogenic in neonatal rat cardiac fibroblasts. Circ Res 72: 1245–1254, 1993
17. Eghbali M, Czaja MJ, Zeydel M, Weiner FR, Seifter S, Blumenfield OO: Collagen mRNAs in isolated adult heart cells. J Mol Cell Cardiol 20: 267–276, 1988
18. Medugorac I, Jacob R: Characterization of left ventricular collagen in the rat. Cardiovasc Res 17: 15–21, 1983
19. Dzau VJ: Circulating versus local renin-angiotensin system in cardiovascular homeostasis. Circulation 77: I-4–I-13, 1988
20. Khairallah PA, Robertson AL, Davila D: Effect of angiotensin-II on DNA, RNA and protein synthesis. In: J Genest, R. Koiw (eds) Hypertension. Springer-Verlag, New York, 1972, pp 212–220
21. Ganten D, Schelling P, Flugel RM, Ganten U: Effect of angiotensin and an angiotensin antagonist on iso-renin and cell growth in 3T3 mouse cells. Int Res Commun Med Sci 3: 327–332, 1975
22. Linz W, Henning R, Scholkens BA: Role of the angiotensin II receptor antagonism and converting enzyme inhibition in the progression and regression of cardiac hypertrophy in rats. J Hypertension 8:L S400–S401, 1991
23. Wiemer G, Scholkens BA, Becker RHA, Busse R: Ramilprilat enhances endothelial autacoid formation by inhibiting breakdown of endothelium-derived bradykinin. Hypertension 18: 558–563, 1991
24. Garg UC, Hassid A: Nitric oxide-generating vasodilators and 8-bromo-cyclic guanosine monophosphate inhibit mitogenesis and proliferation of cultured rat vascular smooth muscle cells. J Clin Invest 83: 1774–1777, 1989
25. Linz W, Scholkens BA: A specific B2-bradykinin receptor antagonist HOE 140 abolishes the antihypertrophic effect of ramipril. Br J Pharmacol 105: 771–772, 1992

Molecular and Cellular Biochemistry **147**: 57–68, 1995.

Factors involved in capillary growth in the heart

Olga Hudlická, Margaret D. Brown*, Helene Walter†, Jacqueline B. Weiss[1] and Anita Bate[1]

Department of Physiology, University of Birmingham Medical School and [1]Wolfson Angiogenesis Unit, Hope Hospital, University of Manchester, Salford M6, UK

Abstract

Growth of capillaries in the heart occurs under physiological circumstances during endurance exercise training, exposure to high altitude and/or cold, and changes in cardiac metabolism or heart rate elicited by modification of thyroid hormone levels. Capillary growth in all these conditions can be linked with increased coronary blood flow, decreased heart rate, or both. This paper brings evidence that, although increased blood flow due to long-term administration of coronary vasodilators results in capillary growth, a long-term decrease in heart rate induced by electrical bradycardial pacing in rabbits and pigs, or by chronic administration of a bradycardic drug, alinidine, in rats, stimulates capillary growth with little or no change in coronary blood flow. Decreased heart rate results in increased capillary wall tension, increased end-diastolic volume and increased force of contraction, and thus stretch of the capillary wall. This could lead to release of various growth factors possibly stored in the capillary basement membrane. Correlation was found between capillary density (CD) and the levels of low molecular endothelial cell stimulating angiogenic factor (ESAF) both in rabbit and pig hearts with CD increased by pacing. There was no relation between expression of mRNA for basic fibroblast growth factor and CD in sham-operated and paced rabbit hearts. In contrast, mRNA for TGFβ was increased in paced hearts, and the possible role of this factor in the regulation of capillary growth induced by bradycardia is discussed. (Mol Cell Biochem **147**: 57–68, 1995)

Key words: angiogenesis, capillaries, heart, exercise, bradycardia, vasodilation, mechanical factors, growth factors

Introduction

Growth of vessels in the heart is very intensive during early postnatal development. For example, capillary length and surface area increase 23× faster than myocardial mass during the first 11 days of life in the rat [1]. At the age of 45 days, capillary growth ceases in the rat heart [2], and growth of vessels is seldom observed in a normal adult heart. It can, however, occur under physiological circumstances in response to endurance training, exposure to high altitude hypoxia or cold [3], and, somewhat paradoxically, in response to increased [4] or decreased [5] levels of thyroid hormones. Exercise training by swimming is a more potent stimulus for capillary growth than training by treadmill running [see 6], which induces growth in young, but rarely in adult, animals [see 3].

It is not understood what factors are involved in growth of vessels in the normal adult heart. Although the participation of different growth factors in angiogenesis under pathological conditions [e.g. 7], and in ischaemic hearts [8], is clearly established, direct evidence for their involvement in vascular growth in physiological circumstances is, as yet, unavailable. On the other hand, while it is unlikely that mechanical factors play a significant role in the initiation of vessel growth in ischaemic hearts [9], there is circumstantial evidence for their involvement in normal hearts [10, 11].

The purpose of this study was to establish factors possibly involved in the initiation of capillary growth during endurance training. Training results in a decrease in heart rate, increase in maximal coronary blood flow and conductance, and increased force of cardiac muscle contraction. Increased blood flow increases the velocity of flow in capillaries [12]

**Present address:* School of Sport and Exercise Sciences, University of Birmingham, Birmingham B15 2TT; †Department of Clinical Chemistry, Wolfson Research Laboratory, Queen Elizabeth Hospital, Birmingham B15 2TT, UK
Address for offprints: O. Hudlická, Department of Physiology, University of Birmingham Medical School, Birmingham B15 2TT, UK

and thus calculated shear stress within vessels [13]. Decreased heart rate is linked with prolonged diastole and increased end-diastolic volume which may lead to increased capillary wall tension since capillary diameters are greater during diastole than during systole [14]. In addition, capillaries are stretched together with myocytes as a result of increased end-diastolic volume, and may be mechanically distorted during increased force of contraction.

Increased shear stress [15], increased stretch [16] and increased cyclic strain [17] all stimulate endothelial cell proliferation *in vitro*, and activate growth factors stored in the capillary basement membrane [18] or proteolytic enzymes such as plasminogen activators [19]. Degradation of the capillary basement membrane by proteases is a primary event in angiogenesis, enabling endothelial cells to migrate and subsequently form sprouts [20].

Each of the changes linked with exercise – increased coronary blood flow, decreased heart rate of increased stretch of myocytes and increased force of contraction – could result in either increased shear stress or capillary wall tension, and/or distortion of capillaries and subsequent activation of some growth factors and capillary growth. In order to differentiate which components are involved, we studied changes in capillary density in three different experimental models – long-term administration of coronary vasodilators, long-term bradycardia, and a long-term increase in inotropism induced by administration of a β agonist, dobutamine – and tried to link changes in capillary density with the presence of a small molecular angiogenic factor (ESAF), found in tissues with a high capillary supply [21], and with mRNA for basic fibroblast growth factor (bFGF) and transforming growth factor beta 1 (TGFβ1), the former a potent mitogenic factor for endothelial cells [see 7] present the heart [22], and the latter an angiogenic factor capable of modulating extracellular matrix also present in heart [see 23].

Materials and methods

Experiments were performed on male Sprague Dawley rats, initially 200 g body weight, New Zealand Red rabbits of either sex, 2.5–3.5 kg body weight, and female farm pigs, initially 23–29 kg. All procedures were carried out in accordance with Guidance under the 1986 Scientific Procedures for Animal Experimentation Act, issued by Her Majesty's Stationery Office. Initial surgery was performed under aseptic conditions and fluothane anaesthesia in rabbits, isofluorane/ N_2O/O_2 anaesthesia in pigs, and antibiotic cover and post-operative analgesia (buprenorphine) as appropriate.

Chronic instrumentation

a) Long-term coronary vasodilation was induced in rabbits by continuous administration of either adenosine (42 μmol.h^{-1}) or a methylxanthine derivative, propentofylline (HWA 285, Hoechst, Werk Albert, Wiesbaden, Germany; 57 μmol.h^{-1}), using portable i.v. infusion pumps [24]. Animals infused with vehicle (saline) served as controls. Drugs were infused for 24 h.day^{-1} for 3–5 weeks, the dosage having been estimated previously on the basis of preliminary experiments to increase coronary blood flow by approximately 40% without any marked change in mean arterial blood pressure.

b) Long-term bradycardia was produced in rats by i.p. injection of 3 mg.kg^{-1} alinidine (Boehringer Ingelheim) twice daily for 5 weeks. In rabbits and pigs, bradycardia was induced by transvenous atrial electrical pacing, as described previously by Wright and Hudlicka [25] and Brown *et al.* [26]. Briefly, rabbits had two teflon insulated stainless steel electrodes threaded through a polyethylene cannula introduced into the right atrium via the jugular vein. They were connected to an external portable pacemaker [27]. Heart rate was recorded using implanted subcutaneous ECG electrodes, and electrical stimuli applied within the mechanical refractory period eliminated every second heart beat, thus leading to a reduction in mechanical contraction frequency of approximately 50%.

In pigs, heart rate was monitored using ECG signals transmitted by a small telemetric recording device implanted under the skin of the chest wall at the same time as two atrial 'J' pacing electrodes were introduced into the right atrium via the jugular vein and connected to Medtronic® dual chamber telemetric pacemaker generators, sited subcutaneously in the neck region. One electrode was used to sense endogenous P-waves while the other paced the atrium with a stimulus 150–200 msec later. This double electrical event produced a single mechanical event in the ventricles. Sham-operated animals served as controls. Bradycardia was maintained by pacing for 4 weeks in both rabbits and pigs, with the exception of experiments where mRNA was estimated, when rabbits were paced for 2 weeks only.

c) Long-term administration of an inotrope, dobutamine (20 μg.kg^{-1}.min^{-1}), in the rabbit was via i.v. infusion using portable infusion pumps over a period of 2 weeks, 24 h.day^{-1} [28]. The dose was selected so as to give a significant increase in left ventricular dP/dt of 24 ± 6% ($p < 0.05$) with no change in coronary blood flow (2 ± 5%, n.s.), mean arterial blood pressure or heart rate.

Acute experiments to measure heart rate, coronary conductance and cardiac stroke work

Final experiments in rats receiving the bradycardic agent were performed 2–3 h after the last drug administration. In rabbits, experiments were performed 2–3 h after termination of the i.v. infusion in those receiving vasodilators or the inotrope, and 24 h after termination of pacing. In these groups, anaesthesia was by sodium pentobarbitone, initially 50 $mg.kg^{-1}$, i.p. in rats, i.v. in rabbits, supplemented as necessary via an intravenous cannula. Blood pressure was monitored by a femoral or brachial arterial catheter, heart rate being derived from this recording. To estimate coronary blood flow, the left ventricle was cannulated for injection of radio-labelled microspheres [29], with reference withdrawal from a brachial artery. Up to three differently-labelled (^{46}Sc, ^{57}Co, ^{113}Sn) microspheres were used to measure coronary flow under different conditions – resting, during maximal cardiac work in response to i.v. noradrenaline challenge (1–30 $\mu g.kg^{-1}$), or during maximal vasodilation by either adenosine in rats (1 $mg.kg^{-1}$) or propentofylline (3 $mg.kg^{-1}$) in rabbits. Tissue blood flow was calculated from radioactivity counts in samples of left ventricular free wall, kidneys and lung, normalized per gram of tissue, in relation to the reference blood flow activity. Adequacy of microsphere mixing was assessed by comparison of right and left kidney flows, and arteriovenous shunting was evaluated from lung flows, the criteria for acceptance being < 5% of total counts in the lung. Cardiac output was computed from the radioactivity levels in withdrawn blood relative to the total activity injected.

Estimation of capillary supply

At the end of experiments, hearts were removed, cleared of fat and weighed whole, or as portions – atria, septum, right and left ventricles – separately. Samples of left ventricular free wall were taken from subendo- and supepicardial regions, frozen in isopentane pre-cooled in liquid nitrogen, and 10 μm cryostat sections were stained for endothelial alkaline phosphatase using an indoxyl tetrazolium method [24] in rats and rabbits, and with biotinylated lectin, Griffonia simplicifolia I [30], in pigs. Capillaries were counted in each heart in a minimum of 10 fields per region, covering an area of 0.975 mm^2 at a magnification of 400 ×. Since counts in subendocardial and subepicardial regions did not differ in any of the species studied, they were pooled and expressed as mean left ventricular capillary density per mm^2, CD. In rabbits and rats, the number of myocytes, delineated by background eosin stain, was counted per sample field at the same time, and capillary supply was also calculated as capillary/fibre ratio, C/F.

Growth factor estimation

a) Endothelial cell stimulating angiogenic factor (ESAF)
Samples from the left and right ventricle of sham-operated and bradycardially-paced rabbits and pigs were weighed and promptly frozen in liquid nitrogen for later ESAF estimation. ESAF assay was performed according to the method of Cooper *et al.* [31]. Samples were homogenised in 2M $MgCl_2$, in order to release any ESAF bound to carrier, and centrifuged. The clear supernatant was passed through a filter with a 5,000 Mr exclusion limit, and the filtrates, now free of protein and higher Mr material, were desalted and lyophilised. ESAF was assayed by its ability to activate latent collagenase according to the method of Weiss *et al.* [32]. Results are expressed as units of ESAF where 1 unit is the percent activation of the enzyme per hour per mg of protein in the supernatant. Two determinations were done on each extract.

b) mRNA for bFGF and TGFβ1
Expression of mRNA for bFGF and TGFβ1 was studied in frozen samples of right and left ventricles from sham-operated and 2 week paced rabbits. Total cell RNA was extracted by either the one-step method [33] or a modification of that method using RNAzol (Biogenesis). Since Northern blot analysis with bFGF [34] and TGFβ1 [35] cDNA did not yield positive results, the radioprotection assay was used to detect specific hybridisation.

Statistics

All results are expressed as means ± s.e.m. Unpaired or paired, as appropriate, students *t*-test and single factor analysis of variance with significance at $p < 0.05$ were used.

Results

Effects of long-term increased blood flow on capillary supply

Acute administration of adenosine or the xanthine derivative propentofylline by i.v. infusion to rabbits increased coronary blood flow by 40% over a period of 3–4 h, but did not appreciably alter blood pressure [24]. When either drug was applied for 24 $h.day^{-1}$ for 3–5 weeks, mean left ventricular capillary density was significantly increased (Fig. 1), by 32 ± 5 and 30 ± 3% in the subepicardium and subendocardium respectively with adenosine, and by 18 ± 9 and 35 ± 9% in these regions with propentofylline. In contrast, the $alpha_1$ antagonist prazosin, which produces vasodilation in skeletal

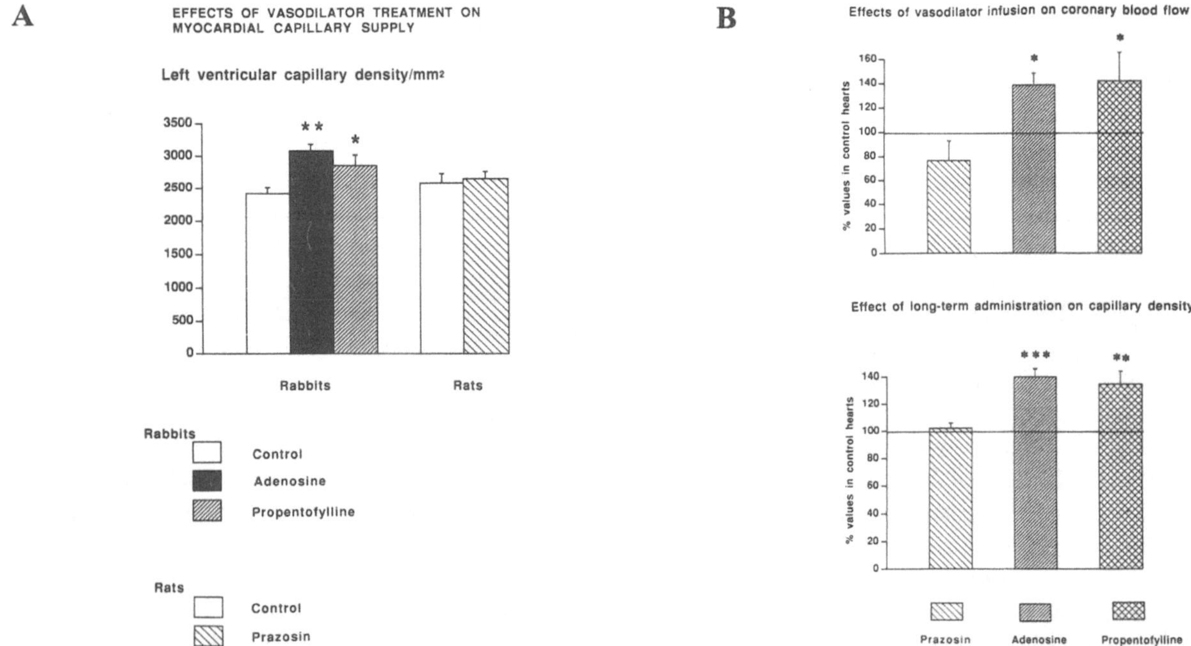

Fig. 1. Effect of long-term vasodilator administration on capillary density and coronary blood flow. A) Number of capillaries.mm^{-2} (ordinate) in the left ventricle in rats and rabbits; B) Top – % increase in coronary flow after 3–4 h i.v. infusion of 42 μmol.h^{-1} adenosine, 57 μmol.h^{-1} propentofylline or 30 μg.h^{-1} prazosin compared to preinfusion levels. Bottom – % increase in capillary density compared with hearts of similar weight from control animals after 3–4 weeks administration of the respective drugs. Adenosine and propentofylline were given i.v. to rabbits, prazosin in drinking water to rats – see methods. **p < 0.01, *p < 0.05 vs control.

muscles and induced capillary growth in this tissue [13], did not increase coronary blood flow on i.v. infusion, and did not alter capillary density after 5 weeks administration ([36]; Fig. 1).

Effects of long-term bradycardia on capillary supply

a) Heart rate

Bradycardia was induced by electrical pacing in rabbits and pigs, and by chronic administration of the bradycardic drug alinidine in rats. Heart rates in conscious rabbits (217 ± 2 beats.min^{-1}, n = 12) were decreased by around 50%, and maintained for up to 7 weeks, the minimum time necessary to result in an increase in capillary density above that found in sham-operated animals of similar heart weight being 2 weeks [25]. Heart rates in paced conscious pigs was 78 ± 4 beats.min^{-1} (n = 7), significantly lower (p < 0.001) than that in sham-operated animals (122 ± 6 beats.min^{-1}, n = 7). In final experiments under anaesthesia, heart rates were still significantly lower in paced pigs, and in rats treated with alinidine when compared to their respective controls (Table 1). Even when pacing was discontinued, resting heart rates in pigs remained low (91 ± 7 beats.min^{-1}, p < 0.01 vs. controls), and a similar observation was made in rabbits in which the underlying intrinsic heart rate decreased in twelve conscious-animals from 217 ± 2 beats.min^{-1} initially to 190 ± 3 beats.min^{-1} (p < 0.01) at the termination of electrical pacing.

Table 1. Heart rates recorded under anaesthesia in final experiments after chronically-induced bradycardia (4–5 weeks' duration). Numbers of animals in each group in parenthesis

	Sham-operated/controls	Bradycardic	
Pigs	117± 7 (12)	77± 5 (6)	P<0.0001
Rats	380±20 (7)	288±10 (9)	P<0.001

b) Heart–body weight ratios

Although decreased heart rate resulted in increased stroke volume, as cardiac output was not changed by the imposed bradycardia [25, 37], there were no signs of heart hypertrophy. Heart weights and heart–body weight ratios were similar in sham-operated/control rabbits, pigs and rats and in animals with bradycardia (Table 2). Consequently, capillary density, the number of capillaries.mm^{-2}, CD, could be used as an index of capillary growth [see 3]. Fibre diameters in rabbit hearts [38] and mean fibre areas in pig hearts [26] were also similar, but, based on the estimation of fibre density, were slightly larger in alinidine-treated rats (Table 3). This latter finding would seem to be due to some effect during tissue

Table 2. Heart and body weights and their ratios in sham-operated/control and chronic bradycardic groups. Numbers of animals in each group in parenthesis

	Sham-operated/controls	Bradycardic	
Rabbits			
Body wt. Kg	2.700 ± 0.080 (9)	2.890 ± 0.180 (8)	N.S.
Heart wt. g	6.410 ± 0.300 (9)	6.800 ± 0.360 (8)	N.S.
Heart/body wt. × 100	0.237 ± 0.007 (9)	0.238 ± 0.010 (8)	N.S.
Pigs			
Body wt. Kg	37.4 ± 2.9 (12)	39.4 ± 3.3 (8)	N.S.
Heart wt. g	154.2 ± 10.8 (12)	175.5 ± 13.1 (8)	N.S.
Heart/body wt. × 100	0.420 ± 0.022 (12)	0.449 ± 0.012 (8)	N.S.
LV free wall wt. g	70.1 ± 6.2 (12)	68.9 ± 5.4 (8)	N.S.
Rat			
Body wt. g	406 ± 18 (7)	423 ± 15 (7)	N.S.
Heart wt. g	1.31 ± 0.03 (7)	1.35 ± 0.06 (7)	N.S.
Heart/body wt. × 100	0.325 ± 0.023 (7)	0.320 ± 0.010 (7)	N.S.
LV free wall wt. g	0.730 ± 0.028 (5)	0.732 ± 0.041 (4)	N.S.

Table 3. Capillary supply in the left ventricle of different species after chronic bradycardia induced by electrical pacing (rabbits, pigs) or by pharmacological means (rats). Numbers of animals in each group in parenthesis

	Sham-operated/ controls	Bradycardic	
Capillary density/mm²			
Rabbits	1963 ± 168 (9)	2456 ± 194 (8)	$p < 0.05$
Pigs	1445 ± 67 (7)	1709 ± 104 (6)	$p < 0.05$
Rats	1971 ± 97 (7)	1903 ± 21 (7)	N.S.
Fibre density/mm²	1518 ± 39 (7)	1240 ± 21 (7)	$p < 0.05$
Capillary/fibre ratio			
Rabbit	1.26 ± 0.07 (12)	1.46 ± 0.07 (5)	$p < 0.05$
Rat	1.30 ± 0.04 (7)	1.54 ± 0.02 (7)	$p < 0.05$

sampling rather than actual fibre hypertrophy since there was no relationship between heart weight and fibre density in rat hearts (data not shown).

c) Capillary supply

Capillary supply was assessed on the basis of staining of capillary endothelium using alkaline phosphatase in rabbits and rats, and lectin Griffonia simplicifolia I in pigs, since all capillaries do not stain for alkaline phosphatase in the latter species. Due to different staining protocols, capillary supply could be assessed as CD in all three species, but as capillary/fibre ratio, C/F, only in rabbits and rats in which the outlines of myocytes were clearly delineated by background staining. The data on capillary supply are presented in Table 3. Left ventricular capillary density was similar in sham-operated rabbits and control rats and was lower in pigs. Long-term bradycardia resulted in a significant increase in CD in pigs and rabbits, while the values in alinidine-treated rats were similar to those on control animals. However, fibre density, which was independent of heart or left ventricular weight, was lower in bradycardic rats than in controls and thus C/F ratio was significantly higher. C/F ratio was also significantly increased in paced rabbit hearts. Since all anatomically-present capillaries were counted, a significant increase in either CD or C/F ratio can only be interpreted as capillary growth.

d) Coronary conductance, calculated shear stress and wall tension

As indicated in the introduction, bradycardia may be linked with increasing coronary blood flow which could lead to increased shear stress and/or vessel wall tension. Coronary blood flow per beat was indeed significantly higher during acute bradycardial pacing [39]. However, coronary conductance measured during maximal vasodilation with i.v. infusion of propentofylline in rabbits or adenosine in rats, was not significantly different in either species compared to their respective controls (Fig. 2), so it is unlikely that increased blood flow *per se* would initiate capillary growth induced by long-term bradycardia.

However, since a higher proportion of coronary flow to the heart occurs during diastole than during systole, and the duration of the latter is prolonged less than that of the former during bradycardia, it is possible that capillary wall shear stress could be increased. Calculations of shear stress during systole and diastole using the data on capillary diameters and red blood cell velocities from Tillmans *et al.* [14] show (Fig. 3) that shear stress during systole is much higher than during diastole, and it is unlikely that the prolongation of diastole during bradycardia could compensate for this difference. Thus, shear stress does not seem to be an important factor initiating capillary growth in the case of long-term

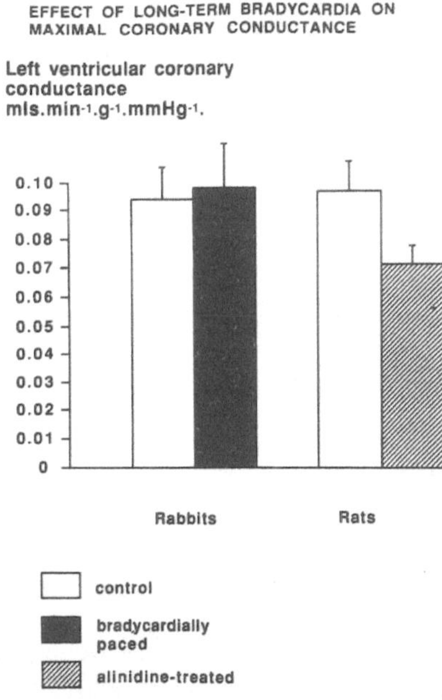

Fig. 2. Effect of long-term bradycardia (induced by electrical pacing in rabbits and long-term administration of alinidine in rats) on maximal coronary conductance (administration of 10^{-4} mol adenosine in rats and 3 $mg.kg^{-1}$ propentofylline in rabbits).

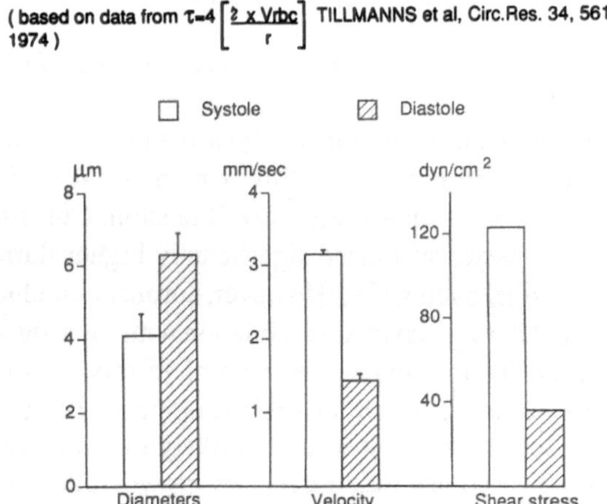

Fig. 3. Diameters of capillaries, velocity of red blood cells and calculated shear stress during systole and diastole based on data from dog hearts.

bradycardia, although it could play a role in capillary growth elicited by chronic administration of vasodilating drugs. On the other hand, capillary wall tension, again calculated from vessel pressure [40] and diameter data [14], is significantly higher during diastole than during systole (Fig. 4), and therefore, stretch of the capillary basement membrane due to increased wall tension may be an important triggering factor in angiogenesis during long-lasting bradycardia.

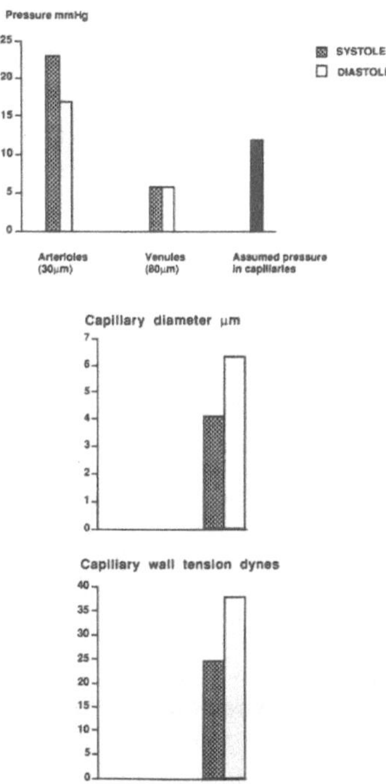

Fig. 4. Calculated capillary wall tension based on data on capillary diameters (Tillmans *et al.* [14]) and pressures (Nellis and Liedtke [51]) during systole and diastole.

Stretch of myocytes

Long-term bradycardia results in increased end-diastolic volume and thus increased stroke work [25], which is actually apparent before any signs of capillary growth [41]. Increased end-diastolic volume stretches myocytes and their aligned capillaries which, under normal circumstances, are slightly tortuous (Fig. 5). It also results in an increased force of

Fig. 5. Schematic arrangement of capillaries in the heart. The arrowheads show capillary interconnections. Modified from Hossler *et al.*, Scan Electron Microscop Pt 4, 1469–1479, 1986

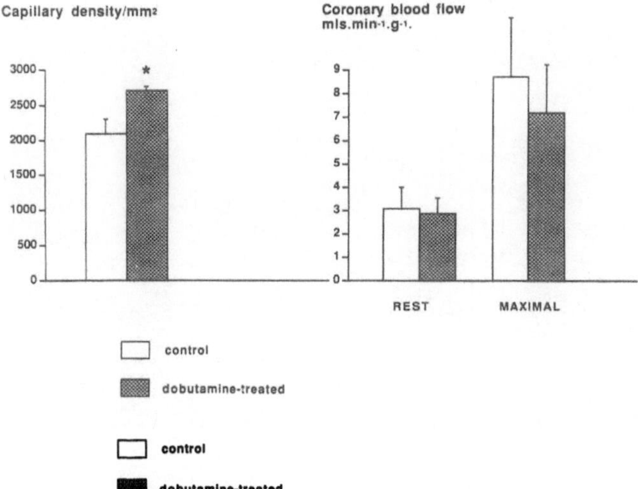

Fig. 6. Capillary density (number of capillaries.mm^{-2}) and coronary blood flow in hearts of control and dobutamine-treated rabbits. Maximal coronary blood flow measured at the peak cardiac output achieved by i.v. infusion of noradrenaline. $^{*}p < 0.05$ vs control.

contraction. If the latter factor is important in the initiation of capillary growth, it should be possible to elicit it by drugs which have an inotropic action. A β agonist, dobutamine, which had positive inotropic effects in rabbits with minimal negative chronotropic effects during long-term administration [28], indeed stimulated capillary growth without altering maximal coronary blood flow (Fig. 6). It also increases stroke work to a slightly greater extent than bradycardial pacing (Fig. 7). This increased force of contraction would distort the

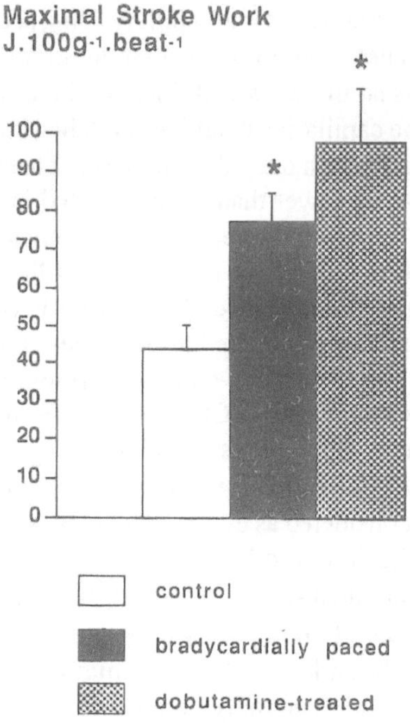

Fig. 7. Maximal stroke work (joules.100 g^{-1} heart weight.beat^{-1} (during i.v. infusion of noradrenaline) in control, bradycardially-paced and dobutamine-treated rabbit hearts. $^{*}p < 0.05$ vs control.

capillary basement membrane and could act in a similar way as described for cyclic strain, which, when imposed upon endothelial cells in tissue culture, induced an increase in DNA [17].

Are growth factors involved in capillary growth during long-term bradycardia?

Increased capillary wall tension as well as increased stretch due to increased force of myocyte contraction could result in a mechanical disturbance of the capillary basement membrane, releasing some of the growth factors, such as bFGF, which are stored in it [7]. It is possible that this mitogen could also be released during increased capillary pressure, since Acevedo *et al.* [42] described its release from cultured endothelial cells in response to elevated hydrostatic pressure.

Another growth factor possibly involved in angiogenesis and stored in the extracellular matrix is TGFβ1 [43]. In assays for mRNA for bFGF in rabbit hearts bradycardially-paced for 2 weeks, and in hearts from appropriate sham-operated controls, there was either none or a very weak expression in the left ventricle, where CD was moderately increased, while positive mRNA expression in the right ventricle was not related to any changes in CD (Fig. 8, Table 4). There is therefore no indication that this factor is involved in capillary growth due to chronic bradycardia. mRNA for TGFβ1, however, was expressed in paced but not sham-operated left ventricle, and not at all in the right ventricle (Table 4).

Table 4. Capillary density (CD) and mRNA expression for bFGF and TGFβ1 in rabbit hearts after two weeks' bradycardial pacing

	Sham-operated		Bradycardic	
	Left ventricle	Right ventricle	Left ventricle	Right ventricle
CD	1423 ± 88	1655 ± 6	1780 ± 148*	1601 ± 62
mRNA expression				
bFGF	±/0	+/±	±/0	+/±
TGFβ1	±/0	–/–	+/+	–/±

0 = no expression
± = weak expression
+ = positive expression
$^{*}p < 0.05$ bradycardiac left ventricle vs. sham-operated

There was, however, a good relationship between capillary density and the levels of activity of a small molecular endothelial cell stimulating angiogenic factor (ESAF) in both rabbits and pigs, with a tendency towards higher levels in paced than sham-operated hearts (Fig. 9).

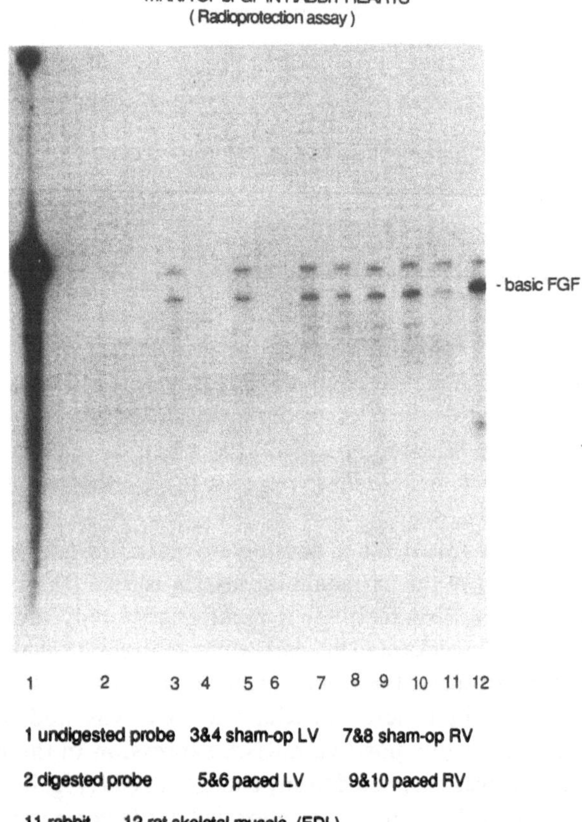

Fig. 8. Example of original gel for mRNA bFGF in rabbit hearts and skeletal muscles.

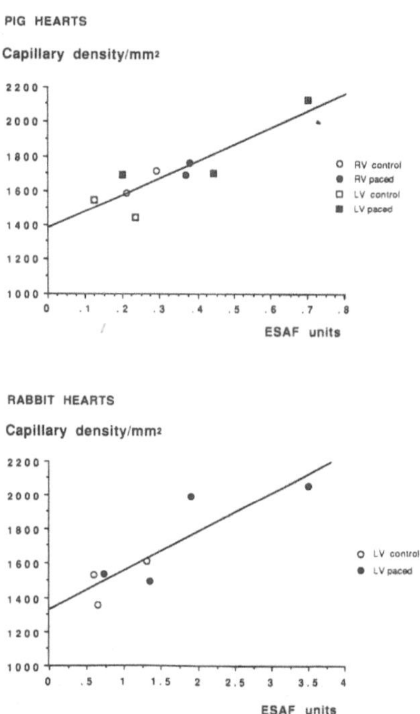

Fig. 9. Relationship between endothelial cell stimulating angiogenic factor, ESAF, (abscissa) and number of capillaries.mm^{-2} (ordinate) in pig (top) and rabbit (bottom) hearts.

Discussion

Assessment of capillary growth

Capillary growth is best assessed by the presence of capillary sprouts [44]. These are, however, almost impossible to identify in the heart since the capillary network is very dense and inter-connecting (see Fig. 5), and the tissue is not transparent. Incorporation of ^{3}H-thymidine into endothelial cells is another reliable method for estimation of the number of endothelial cells entering mitosis. Tornling [45] found a good correlation between the labelling index and increased number of capillaries in hearts of dipyridamole-treated animals. Provided that capillaries are counted in tissue sections where myocytes are in good cross-section, and that there is no hypertrophy (whereby larger myocytes would separate capillaries further, thus decreasing CD), CD or C/F based on either counts from low power electron micrographs or from light microscopy after staining specifically for capillary endothelium are a good estimate of capillary growth. The only error could result from a high capillary tortuousity which could entail one and the same capillary being sectioned twice, but this does not seem to be the case in normal hearts [46], and even less so in hearts with increased end-diastolic volume when the capillaries would be stretched and tortuousity diminished. Although our values for CD in control rabbit, pig and rat hearts are lower than those reported in the literature [see 3], it is important to note that all samples from control hearts and those after long-term bradycardia in all species were taken under identical conditions. This is very important since samples from contracted hearts in our experiments, i.e. maximally contracted in ice-cold saline prior to freezing and sectioning, will show lower CD than those from hearts arrested in diastole in which fibres would be relaxed and thus thinner. Increased CD and C/F ratio in the present study could therefore be considered as evidence for capillary growth. This growth can be initiated by factors acting either from the luminal or abluminal side of the vessel, and must be preceded by derangement of the capillary basement membrane to create space for the endothelial cells to migrate outside, divide, form sprouts and eventually connect to other pre-existing capillaries.

Role of mechanical factors in capillary growth acting from the luminal side

One mechanical factor which could initiate capillary growth from the luminal side is increased shear stress, as a result of either increased velocity of flow or decreased diameter of vessels. In the present experiments, increased capillary density was induced by long-term administration of adenosine or a xanthine derivative, propentofylline. Other authors have induced capillary growth by chronic treatment with dipyridamole [45] or ethanol [47]. Although adenosine may stimulate proliferation of endothelial cells directly [48], dipyridamole has no such action [49]. All the above substances are coronary vessel dilators and increase blood flow, and hence the velocity of flow in myocardial capillaries. Data on the effects of dipyridamole on the coronary microcirculation indicate that it increases velocity of red blood cells but does not alter capillary diameters [12], so that it is very likely that the increased flow elevates shear stress within these vessels.

There is some evidence to suggest that increased shear stress can perturb endothelial cells in such a way that might encourage them to undergo division. In endothelial cells in tissue culture, it caused activation of mRNA for tissue plasminogen activator [19], which might assist in degradation of the basement membrane. It disrupted the luminal glycocalyx layer in capillaries in skeletal muscle just prior to active capillary growth in response to long-term electrical stimulation [50]. Shear stress is known to increase DNA content in endothelial cell cultures [15], and the possible mechanism of his action could be interpreted on the basis of Berthiaume *et al.* [51] finding that shear stress causes perturbation of phospholipids in the upper leaflet of the cell membrane, thus possibly initiating the cascade of transduction signals leading to cell mitotis [see 41]. However, shear stress does not seem to be involved in the hearts where capillary growth was elicited by imposition of bradycardia, since neither coronary flow nor conductance were increased in this situation.

Another factor acting from the luminal side is capillary pressure and/or wall tension. Although we do not have data on capillary pressure in the heart during systole and diastole, and there are no data on how these parameters might vary with heart rate changes, Nellis and Liedtke [40] showed that during systole and diastole, venular pressures were the same and arteriolar pressures were only slightly different, so that, by derivation, capillary pressures would vary little during the phases of the cardiac cycle (Fig. 4). However, capillary diameters have been observed to be larger during diastole than systole [14], and thus capillary wall tension would be considerably higher for a proportionately longer period of time during long duration bradycardia. Increased tension was shown to stimulate uptake of ^{3}H-thymidine and proline into the wall of arterial segments *in vitro* [52]. It would also flatten endothelial cells and thus make them more susceptible to growth factors, as has been demonstrated for endothelial cells in tissue culture by Folkman and Greenspan [53].

Mechanical factors involved in capillary growth acting from the abluminal side

Increased capillary wall tension, although initiated by factors acting from the luminal side, stretches the basement membrane in addition to the endothelial cells. Basement membrane injury [54] or stretch can lead to the release of growth factors which are stored in it [54]. Furthermore, in tissue culture, stretched endothelial cells which had a larger area had a higher incorporation of ^{3}H-thymidine [55]. Stretch also activates Ca^{2+} channels, leading to increased intracellular Ca^{2+} concentration [56, 57], an essential step in cell proliferation [see 41].

In addition to stretch of the endothelium due to increased vessel wall tension, capillaries can be exposed to increased myocyte stretch in connection with increased force of contraction. This could be as a result of either increased end-diastolic volume via Starling's law during long-term bradycardia, or from increased inotropism in experiments where dobutamine was administered long-term. In either case, the extracellular matrix would be subjected to enhanced mechanical distortion. From tissue culture studies, it is known that the extracellular matrix, consisting largely of a fibronectin-collagen network, decreases the mobility of endothelial cells [58], and its disturbance is the first step in angiogenesis [20]. Cyclic stretch inhibits the production of collagen by endothelial cells [59], which would weaken the basement membrane integrity. Physical/mechanical forces also activate matrix metalloproteinases [60], which are essential for the lysis of the basement membrane.

Role of growth factors in capillary growth during long-term bradycardia

A positive relationship was shown in both pig and rabbit hearts after long-term bradycardia between levels of ESAF activity and capillary density. ESAF is a unique physiological activator of pro-matrix-metalloproteinase 2, which is highly specific for the degradation of Type IV collagen, a major component of the basement membrane. ESAF levels are higher in tissues with active capillary growth [21], and were increased in skeletal muscles subjected to increased activity by chronic electrical stimulation [61], which would be linked, as in paced hearts, to distortion of the capillary basement membrane.

In contrast, expression of mRNA for bFGF was very weak and did not correlate with capillary density. The reports on the presence of bFGF in the heart and supplying vessels are at variance, with some authors describing its presence in myocytes [43], other in blood vessels [62, 22], either during development or during ischaemia. There is no data on the involvement of bFGF in capillary growth in normal hearts under physiological circumstances. Parker and Schneider [43] did not find altered expression of bFGF or TGFβ1 in hearts exposed to increased haemodynamic load, but we found an increased expression of mRNA for TGFβ1 in the left ventricle of hearts subjected to bradycardia by pacing, at the time when capillary growth would be expected to be active.

TGFβ1 accentuates extracellular matrix proliferation [63] and is assumed to inhibit angiogenesis, possibly by mediating an inhibitory effect of pericytes on endothelial cell proliferation [64]. However, recent reports on the role ofTGFβ1 in the regulation of angiogenesis indicate that it is involved in tube formation [65], and that it induces formation of sprouts *in vitro* [66]. It also induced growth of vessels in the rabbit cornea [67]. Low concentrations (less than 1 ng.ml^{-1}) potentiated the effect of bFGF, while concentrations at 5–10 ng.ml^{-1} were inhibitory [68]. In addition, TGFβ1 promotes the differentiation of endothelial cells into smooth muscle cells [69], and could thus contribute to transformation of capillaries to arterioles.

In conclusion, capillary growth during long-term bradycardia appears to be due to mechanical factors such as increased wall tension and stretch of surrounding myocytes, which may activate stretch-dependent Ca^{2+} channels in the endothelium or stretch the basement membrane, releasing either TGFβ1 or ESAF. Bradycardia, together with increased inotropic action, is therefore likely to be an important factor initiating capillary growth during endurance training, possibly to a greater extent than the increase in coronary blood flow which is associated with training.

Acknowledgement

This work was supported by the Wellcome Trust, the British Heart Foundation and the Jean Shanks Foundation.

References

1. Olivetti G, Anversa P, Melissari M, Loud AV: Morphometric study of early postnatal development of the thoracic aorta in the rat. Circ Res 47: 417–424, 1980
2. Rakusan K: Cardiac growth, maturation and aging. In: R. Zak (ed). Growth of the Heart in Health and Disease. Raven Press, New York, 1984, pp 131–164
3. Hudlicka O, Brown MD, Egginton S: Angiogenesis in skeletal and cardiac muscle. Physiol Rev 72: 369–417, 1992
4. Chilian WH, Wangler RD, Peters KG, Tomanek RJ, Marcus ML: Thyroxine-induced left ventricular hypertrophy in the rat: anatomical and physiological evidence for angiogenesis. Circ Res 57: 591–598, 1985
5. Tomanek RJ, Barlow PA, Connell PM, Chen Y, Torry RJ: Effects of hypothyroidism and hypertension on myocardial perfusion and vascularity in rabbits. Am J Physiol 265: H1638–1644, 1993
6. Laughlin MH, McAllister RM: Exercise training-induced coronary vascular adaptation. J Appl Physiol 73: 2209–2225, 1992
7. Folkman J, Klagsbrun M: A family of angiogenic peptides. Nature 329: 671, 1987
8. Schaper W, Sharma HS, Quinkler W, Markert T, Wünsch M, Schaper J: Molecular biologic concepts of coronary anastomoses. J Am Coll Cardiol 15: 513–518, 1990
9. Schaper W, Görge G, Winkler B, Schaper J: The colateral circulation of the heart. Prog Cardiovasc Dis 31: 57–77, 1988
10. Hudlicka O: Capillary growth: role of mechanical factors. NIPS 3: 117–120, 1988
11. Tomanek RJ: Response of the coronary vasculature to myocardial hypertrophy. J Am Coll Cardiol 15: 528–533, 1990
12. Tillmans H, Steinhausen M, Leinberger H, Thederan H, Kübler W: The effect of coronary vasodilators on the microcirculation of the ventricular myocardium. In: H. Tillmans, W. Kübler, H. Zebe (eds). Microcirculation of the Heart. Springer Verlag, Berlin, 1982, pp 305–312
13. Dawson JM, Hudlicka O: Can changes in microcirculation explain capillary growth in skeletal muscle? Int J Exp Path 74: 65–71, 1993
14. Tillmans TH, Ikeda S, Hansen H, Sarma JS, Fauvel JH, Bing RJ: Microcirculation in the ventricle of the dog and turtle. Circ Res 34: 561–569, 1974
15. Ando J, Nomura H, Kamiya A: The effect of fluid shear stress on the migration and proliferation of cultured endothelial cells. Microvasc Res 33: 62–70, 1987
16. Ingber DE, Folkman J: Regulation of endothelial growth factor action – solid state control by extracellular matrix. Progress in Clin Biol Res 249: 273–284, 1987
17. Sumpio BF, Banes AJ, Levin LG, Johnson G: Mechanical stress stimulates aortic endothelial cells to proliferate. J Vasc Surg 6: 252–256, 1987
18. Malek AM, Gibbons GH, Dzau VJ, Izumo S: Fluid shear stress differentially modulates expression of genes encoding basic fibroblast growth factor and platelet-derived growth factor-B chain in vascular endothelium. J Clin Invest 92: 2013–2021, 1993
19. Diamond SL, Sharefkin B, Dieffenbach C, Frasier-Scott K, McIntyre LV, Eskin SG: Tissue plasminogen activator messenger RNA levels increase in cultured human endothelial cells exposed to laminar shear stress. J Cell Physiol 173: 364–371, 1990
20. Ausprunck DH: Tumor angiogenesis. In: J.C. Houck (ed). Handbook of Inflammation. Vol 1, Elsevier/North Holland, Amsterdam, 1979, pp 317–351
21. Odedra R, Weiss JB: Low molecular weight angiogenesis factors. Pharmac Ther 49: 111–124, 1991
22. Casscells W, Speier E, Sasse J, Klagsbrun M, Allen P, Lee M, Calvo B, Cjiba M, Haggroth L, Folkman J, Epstein SE: Isolation, characterization and localization of heparin-binding growth factors in the heart. J Clin Invest 85: 433–441, 1990
23. Sharma HS, Zimmerman R: Growth factors and development of coronary collaterals. In: P. Cummins (ed). Growth Factors and the Cardio-vascular System. Kluwer Academic Publishers, Boston, 1993, pp 119–148
24. Ziada AMAR, Hudlicka O, Tyler KR, Wright AJA: The effect of long-term vasodilation on capillary growth and performance in rabbit heart and skeletal muscle. Cardiovasc Res 18: 724–732, 1984
25. Wright AJA, Hudlicka O: Capillary growth and changes in heart per-

formance induced by chronic bradycardial pacing in the rabbit. Circ Res 49: 469–478, 1981

26. Brown MD, Davies MK, Hudlicka O, Townsend P: Long-term bradycardia in the conscious pig produced by electrical pacing: effects on myocardial capillary supply. J Physiol 475: 62P, 1994
27. Tyler KR, Wright AJA: Lightweight portable stimulators for stimulation of skeletal muscles at different frequencies and for cardiac pacing. J Physiol 307: 8–9P, 1980
28. Brown MD, Hudlicka O: Cardiac performance *in vivo* and anatomical capillary supply in the rabbit after prolonged dobutamine infusion. Cardiovasc Res 25: 909–915, 1991
29. Schrock GD, Krahmer RL, Ferguson JL: Coronary flow by left atrial and left ventricular microsphere injection in the rat. Am J Physiol 259: H635–638, 1990
30. Alroy J, Goyal V, Skutelsky E: Lectin histochemistry of mammalian endothelium. Histochemistry 86: 603–607, 1987
31. Cooper RG, Taylor CM, Choo JJ, Weiss JB: Elevated endothelial-cell-stimulating-angiogenic factor activity in rodent glycolytic skeletal muscles. Clin Sci 81: 267–270, 1991
32. Weiss JB, Hill CR, Davis RJ, McLaughlin B: Activation of mammalian procollagenase and basement membrane degrading enzymes by low-molecular weight angiogenesis factors. Agents and Actions 15: 107–108, 1984
33. Chomcyznski P, Sacchi N: Single step method of RNA isolation by Guanidinium thiocyanate-Phenol-Chloroform extraction. Anal Biochem 162: 156–159, 1987
34. Shimasaki S, Emoto N, Koba A, Mercado M, Shibata F, Cooksey K, Baird A, Ling N: Complementary DNA cloning and sequencing of rat ovarian basic fibroblast growth factor and tissue distribution study of its mRNA. Biochem Biophys Res, Commun 157: 256–263, 1988
35. Quian SW, Kondaiah P, Roberts AB, Sporn MB: cDNA cloning by PCR of rat transforming growth factor beta 1. Nucl Acids Res 18: 3059–3063, 1990
36. Ziada AMAR, Hudlicka O, Tyler KR: The effect of long-term administration of α_1-blocker prazosin on capillary density in cardiac and skeletal muscle. Pflügers Arch 415: 355–360, 1989
37. Brown MD, Cleasby MJ, Hudlicka O: Capillary supply of hypertrophied rat hearts after chronic treatment with the bradycardic agent alinidine. J Physiol 427: 40P, 1990
38. Hudlicka O, Wright AJA, Hoppeler H, Uhlmann E: The effect of chronic bradycardial pacing on the oxidative capacity in rabbit hearts. Resp Physiol 72: 1–12, 1988
39. Hudlicka O, West D, Kumar S, El Khelly F, Wright AJA: Can growth of capillaries in the heart and skeletal muscle be explained by the presence of an angiogenic factor? Br J Exp Path 70: 237–246, 1989
40. Nellis SH, Liedtke AJ: Pressures and dimensions in the terminal vascular bed of the myocardium determined by a new free motion technique. In: H. Tillmans, W. Kübler, H. Zebe (eds). Microcirculation of the Heart. Springer Verlag, Berlin, 1982, pp 61–74
41. Hudlicka O, Brown MD: Physical forces and angiogenesis. In: G.M. Rubanyi (ed). Mechanoreception by the Vascular Wall. Futura Publishing Co, Mount Kisco, NY, 1993, pp 197–241
42. Acevedo AD, Bowser SS, Gerritsen ME, Bizios R: Morphological and proliferative responses of endothelial cells to hydrostatic pressure – role of fibroblast growth factor. J Cell Physiol 157: 603–614, 1993
43. Parker TG, Schneider MD: Growth factors, proto-oncogenes and plasticity of the cardiac phenotype. Ann Rev Physiol 53: 179–200, 1991
44. Schoefl GI, Majno G: Regeration of blood vessels in wound healing. In: Advances in Biology of Skin, Vol V, Pergamon, Oxford, UK, 1964, pp 73–193
45. Tornling G: Capillary neoformation in the heart of dipyridamole treated rats. Acta Pathol Microbiol Scand, sec A 90: 269–271, 1982
46. Poole DC, Batra S, Mathieu-Costello O, Rakusan K: Capillary geometrical changes with fiber shortening in rat myocardium. Circ Res 70: 697–706, 1992
47. Mall G, Mattfeldt T, Reiger P, Volk B, Frolov VA: Morphometric analysis of the rabbit myocardium after chronic ethanol feeding – early capillary changes. Basic Res Cardiol 77: 57–67, 1982
48. Meininger CJ, Schelling ME, Granger HJ: Adenosine and hypoxia stimulate proliferation and migration of endothelial cells. Am J Physiol 255: H554–H562, 1988
49. Jakob W, Zipper J, Savolvy SB, Siems W-E, Jentzsch KD: Is dipyridamole an aniogenic agent? Exp Pathol 22: 217–224, 1982
50. Brown MD, Egginton S, Hudlicka O: Changes in capillary endothelial cell glycocalyx in rat skeletal muscles during chronic electrical stimulation. Int J Microcirc Clin Exp 11: 447, 1992
51. Berthiaume F, Frangos JA: Fluid flow causes membrane perturbation in cultured human umbilical vein endothelial cells (HUVECS). FASEB J 4: A835, 1990
52. Hume WR: Proline and thymidine uptake in rabbit ear artery segments *in vitro* increased by chronic tangential load. Hypertension 2: 738–743, 1980
53. Folkman J, Greenspan HP: Influence of geometry on control of cell growth. Biochem Biophys Acta 417: 211–231, 1975
54. Folkman J, Klagsbrun M, Sasse J, Wadzinski M, Ingber DE, Vlodavsky I: A heparin-binding angiogenic protein – based fibroblast growth factor – is stored within basement membrane. Am J Pathol 130: 393–400, 1988
55. Ingber DE, Folkman J: Regulation of endothelial growth factor action – solid state control by extracellular matrix. Progr Clin Biol Res 249: 273–284, 1987
56. Lansman JB, Hallam TJ, Rink TJ: Single stretch-activated ion channels in vascular endothelial cells as mechanotransducers? Nature 235: 811–812, 1987
57. Naruse K, Sokabe M: Involvement of stretch activated ion channels in Ca^{2+} mobilization to mechanical stretch in endothelial cells. Am J Physiol 264: C1037–1044, 1993
58. Vlodavsky LK, Johnson D, Gospodarowicz R: Appearance in confluent vascular endothelial cell monolayers of a specific cell surface protein (CSP-60) not detected in actively growing in multiple layers. Proc Natl Acad Sci USA, 76: 2306–2310, 1979
59. Sumpio BF, Banes AJ, Link GW, Iba T: Modulation of endothelial cell phenotype by cyclic stretch: inhibition of collagen production. J Surg Res 48: 415–429, 1990
60. Woessner FJ Jr: Matrix metalloproteinases and their inhibitors in connective tissue remodelling. FASEB J 5: 2145–2154, 1991
61. Brown MD, Hudlicka O, Fakhoury R, Weiss JB: Low molecular cell angiogenesis stimulating factor (ESAF) and capillary growth in skeletal muscles. Int J Microcirc Clin Exp 10: 401, 1991
62. Kardami E, Fandrich RR: Basic fibroblast growth factor in atria and ventricles of the vertebrate heart. J Cell Biol 109: 1865–1871, 1989
63. Davidson JM, Zoia O, Liu JM: Modulation of transforming growth-factor-beta-1 stimulated elastin and collagen production and proliferation in porcine vascular smooth muscle cells and skin fibroblasts by basic fibroblast growth factor alpha and insulin-like growth factor-1. J Cell Physiol 155: 149–156, 1993
64. Antonelli-Orlidge A, Saunders KB, Smith SR, D'Amore PA: An activated form of TGFbeta is produced by co-cultures of endothelial cells and pericytes. Proc Natl Acad Sci USA 86: 4544–4548, 1989
65. Iruela-Arispe ML, Sage EH: Endothelial cells exhibiting angiogenesis in vitro proliferate in response to TGF beta 1. J Cell Biochem 52: 414–430, 1993
66. Gajdusek CM, Luo Z, Mayberg MR: Basic fibroblast growth factor and transforming growth factor beta-1: synergistic mediators of angiogenesis in vitro. J Cell Physiol 157: 133–144, 1993

67. Phillips GD, Whitehead RA, Stone AM, Ruebel MW, Goodkin ML, Knighton DR: Transforming growth factor beta (TGF-β) stimulation of angiogenesis: an electron microscopic study. J Submicrosc Cytol Pathol 25: 149–155, 1993
68. Pepper MS, Vassalli LD, Orci J, Montesano R: Biphasic effect of transforming factor beta-1 on in vitro angiogenesis. Exp Cell Res 204: 356–363, 1992
69. Arciniegas E, Sutton AB, Allen TD, Schor AN: Transforming growth factor beta 1 promotes the differentiation of endothelial cells into smooth muscle-like cells in vitro. J Cell Sci 103: 521–529, 1992

Molecular and Cellular Biochemistry **147**: 69–73, 1995.

Biomechanical signals in the coronary artery triggering the metabolic processes during cardiac overload

Mária Gerová, Olga Pecháňová, Venceslav Stoev, Margita Kittová[1], Iveta Bernátová, Milan Juráni[2] and Svatoplute Doležel

Institute of Normal and Pathological Physiology, [1]Department of Physiology, Medical School, Comenius University; [2]Institute of Animal Biochemistry and Genetics, Slovak Academy of Sciences, Bratislava, Slovakia

Abstract

Peculiarities in structure and deformability of epicardial conduit coronary arteries are described. The thin wall of animal coronary artery contrasts the human coronary artery in which the remarkable wall thickness is due namely by the intima thickness. Deformation in length and diameter of conduit coronary arteries, due to the left and right ventricle volume increase, has been defined in non-beating canine heart. Ramus interventricularis anterior being firmly tethered to the myocardium undergoes about 3 times larger deformation than ramus circumflexus. In anaesthetized dogs a 30% increase in blood pressure, elicited by aortic constriction, induces an increase in diameter of coronary artery, in segment length, in blood flow and consequently in shear stress which represents a load for circumferentially running smooth muscle bundles, longitudinally running smooth muscle bundles, as well as for the endothelium. The above load lasting 4 h is already reflected by an increase in total RNA content and [^{14}C] leucin incorporation in the left ventricle myocardium in the wall of ramus interventricularis anterior, not in ramus circumflexus. The findings fit completely with the different range of deformation of both the above coronary branches and indicates an increase in proteosynthesis not only in myocardium, but in ramus interventricularis anterior as well. An increase in ornithindecarboxylase activity in coronary wall leading to an increase in biogenic polyamines, is present in the case only, when blood pressure increase is induced by infusion of noradrenaline. (Mol Cell Biochem **147**: 69–73, 1995)

Key words: diameter, length, wall thickness, shear stress, protein synthesis, ornithine decarboxylase

Conduit coronary arteries are unique in the structure: in animals they are extremely thin wall vessels. Figure 1 represents the cross section of ramus interventricularis anterior of an adult dog. Wall thickness/diameter ratio represents 1:20. For comparison in Fig. 1 is the cross section of a similar sized vessel from the hind extremity – dorsal pedal artery, which is characterized by a wall thickness/diameter ratio 1:5 [1]. Sims [2] confirmed the thin wall of the coronary artery in 10 other animal species used in laboratories: from mouse to baboon and beef.

However, a completely different situation was described in human coronary arteries, where wall/lumen ratio was 1:7.4 in adults [3, 4]. The thickness of the human coronary wall is caused mainly by the thickness of the intima. The contribution of the intima to the wall thickness increases with age (Fig. 2). Looking at the post birth development of the human coronary artery it follows: while media increases from 50 μ in newborn to 200 μ in adults, e.g. about 4 times, intima grows in the same period from 6–300 μ in adults e.g. 500 times, and continues in growth further up to 450 μ [3, 4]. The factors which underlie the growth of the coronary wall, and the intima in particular, are not well understood.

Coronary arteries are unique among the vessels also in that point, that they are more complex deformed. To the radial and longitudinal deformation due to each stroke volume as it is obvious in other vessels, deformation in length and diameter

Address for offprints: M. Gerová, Institute of Normal and Pathological Physiology, Sienkiewiczova 1, 813 71 Bratislava, Slovakia

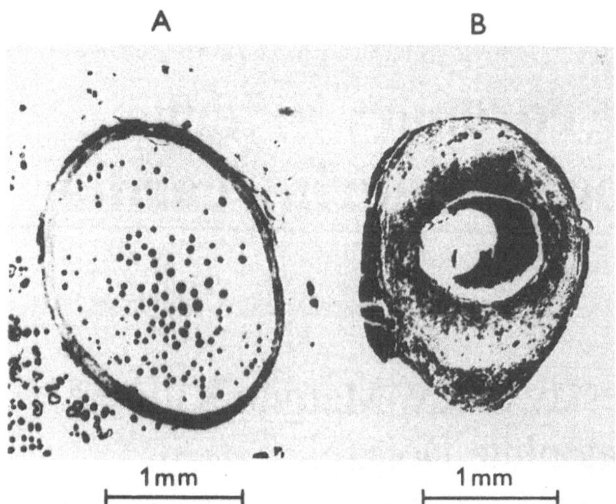

Fig. 1. Cross section of ramus interventricularis anterior (A) and dorsal pedal artery (B), both fixed under normal systemic blood pressure. From Gerová *et al.* [1].

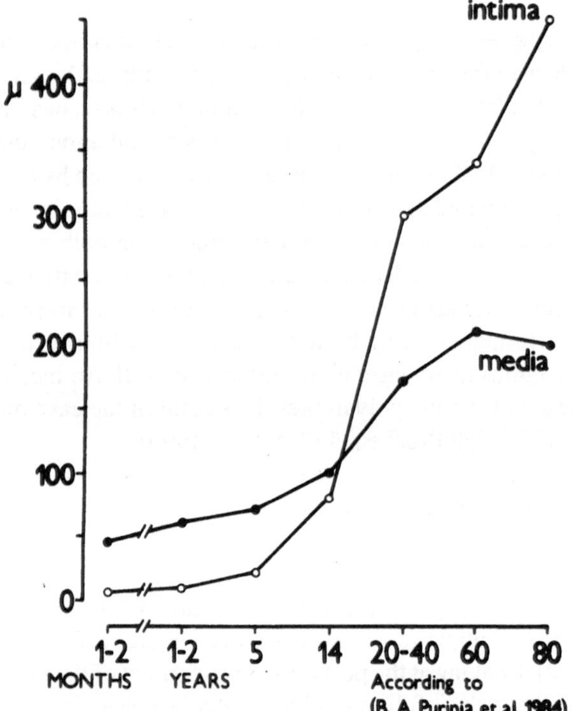

Fig. 2. Growth of the intima and media of the human coronary artery after birth.

does access due to the underlying continuous changes in actual geometry of the ventricles [5].

The deformation of contractile cells, and cardiomycytes inclusively, was proven by Mann *et al.* [7], Kent *et al.* [8] and Watson [9] as a stimulus for the increase in proteosynthesis. In coronary wall, however, the elastin-collagen skeleton

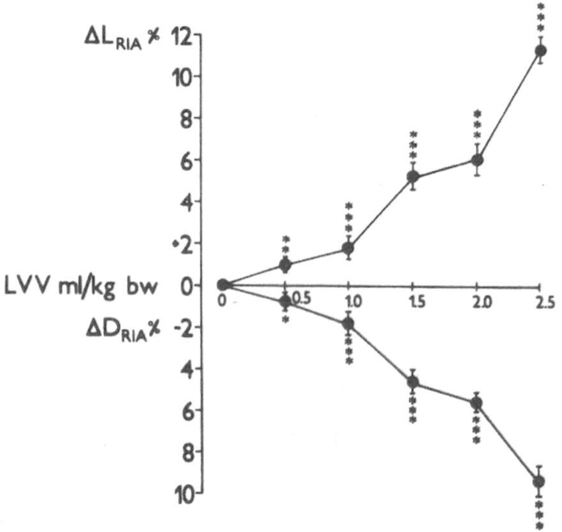

Fig. 3. Mean values ± SE of segment length increase in percent of resting value (L_{RIA}, top ordinate) and diameter decrease in percent of resting value (D_{RIA}, bottom of ordinate) of ramus interventricularis anterior during stepwise increase of left ventricle volume.

damp in certain limits the deformation of smooth muscle. A question arises: can a current hemodynamical load which induces an increase in proteosynthesis in myocardium and induces a deformation of coronary artery affect the proteosynthesis in coronary wall?

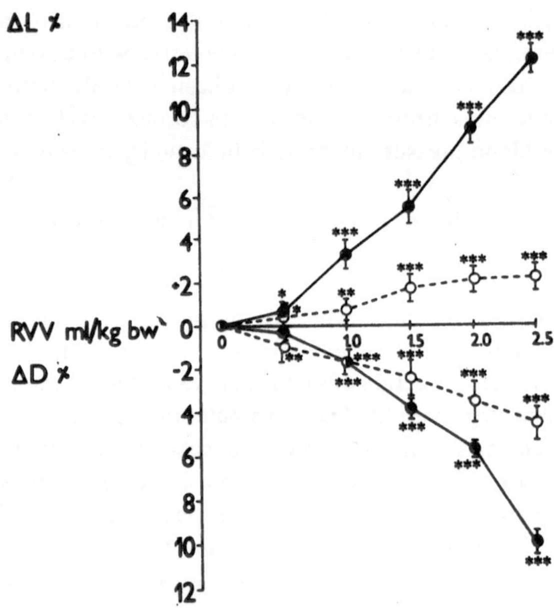

Fig. 4. Mean values ± SE of segment length increase in percent of resting value (L, top ordinate) and diameter decrease in percent of resting value (D, bottom of ordinate) of ramus interventricularis anterior (full line) and ramus circumflexus (interrupted line) during stepwise increase of right ventricle volume.

The first part of the question addressed the range of *deformation in length and diameter* of the *conduit coronary artery* brought about by the changes in volume of the ventricles. In the *non-beating canine heart* placed in a bath, and using the ultrasound technique for measuring segment length and diameter, an increase in left ventricle volume by 100–150% induces a segment length increase by about 12%, and diameter decrease by about 10% as demonstrated in Fig. 3 [5].

Qualitatively similar, quantitatively, however, by far smaller changes were found measuring the ramus circumflexus (Fig. 4). Radial and longitudinal deformation of this vessel represented only about one third of changes observed in ramus interventricularis anterior [6]. This finding is important and will be related to the proteosynthesis during overloading of the heart. The change in *geometry of coronary artery in a beating heart* during a shortlasting blood pressure increase was studied in anaesthetized dogs. An increase of blood pressure was induced by constriction of abdominal aorta above renal arteries. Segment length and diameter of ramus interventricularis anterior were monitored by ultrasound technique. Transducing crystals were sewn into the adventitia of the coronary artery on a beating heart. Moreover, diameter and blood flow in the ramus circumflexus was measured by electromagnetic flowmeter and shear stress was calculated according to Hagen–Poisseuille equation

$$\tau = \frac{4\eta \quad Q}{\pi \, r_i^3} \ \text{dyn / cm}^3$$

Shear stress was used as an index of endothelium deformation [10]

Figure 5 represents mean values of parameters monitored:

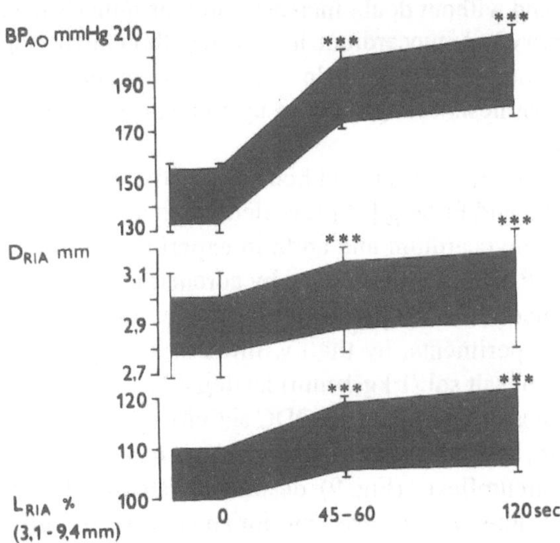

Fig. 5. Geometry of coronary artery during increase of blood pressure in anaesthetized dog. Blood pressure in the root of aorta (BP_{AO}), diameter (D_{RIA}) and segment length (L_{RIA}) of ramus interventricularis anterior.

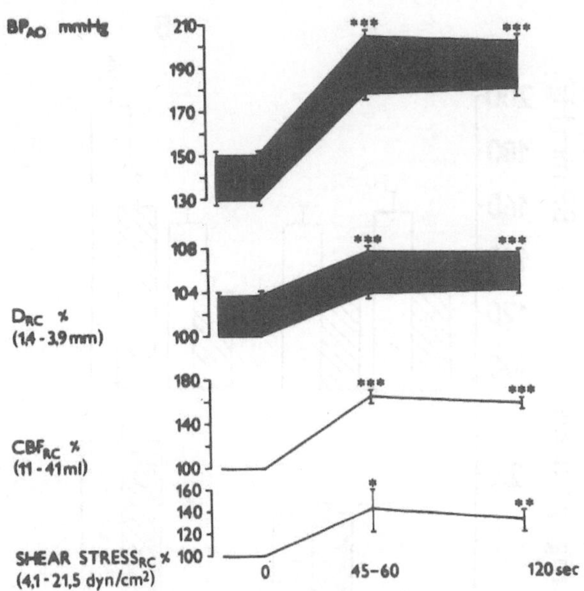

Fig. 6. Blood flow (BF_{RC}), diameter (D_{RC}) and shear stress in ramus circumflexus during increase of blood pressure (BP_{AO}) in anaesthetized dog.

increasing the blood pressure by about 30%, the coronary diameter increased by 4.6%, representing a load for circumferentially and/or spirally running smooth muscle bundles and segment length increased too, by 8%, representing the load for longitudinally running smooth muscle bundles.

In addition: a 30% increase in BP induced an increase of blood flow in ramus circumflexus by 50% and calculated shear stress increased by about 35% (Fig. 6).

Increasing the systemic blood pressure, the load on, and/or deformation of, both circumferentially and longitudinally running smooth muscle cells increased [11] and also load and deformation, of endothelial cells effected by shear stress, increased.

The main goal of the study was, as stated in the introduction, the *metabolic process in coronary wall* as a consequence of the pressure overloading lasting a short period of 4 h.

RNA content estimated according to Canev and Markov [12], and labelled ^{14}C-leucin incorporation into the coronary wall estimated according to Nagai [13], were used as indicators of proteosynthesis. As controls, shame operated animals were used, lying under anaesthesia for 4 h.

A 30% increase in blood pressure lasting 4 h induced a significant increase in RNA content in ramus interventricularis anterior and myocardium (Fig. 7). No change was found in ramus circumflexus. The no response of ramus circumflexus is noteworthy, in particular, if compared to the different range of deformation of ramus interventricularis anterior and ramus circumflexus, as mentioned above. The later was proven to deform during 100–150% increase in ventricle volume only about 3%, e.g. one third of the deformation of RIA.

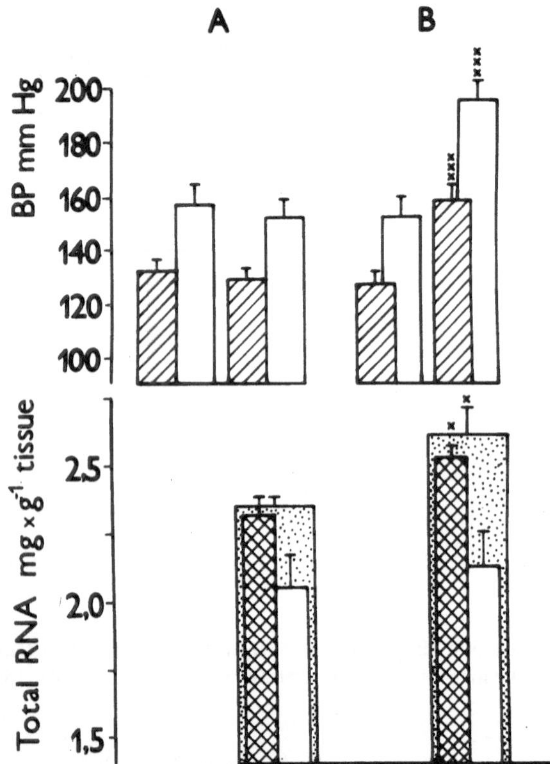

Fig. 7. Total RNA content in ramus interventricularis anterior (crossed hatched columns), ramus circumflexus (white columns) and myocardium (dotted columns) in the control group of animals (shame operated and 4 h lasting anaesthesia) (A), and in the group of animals with 4 h lasting increase of blood pressure (B). Upper row of columns represent blood pressure, lower row the RNA content.

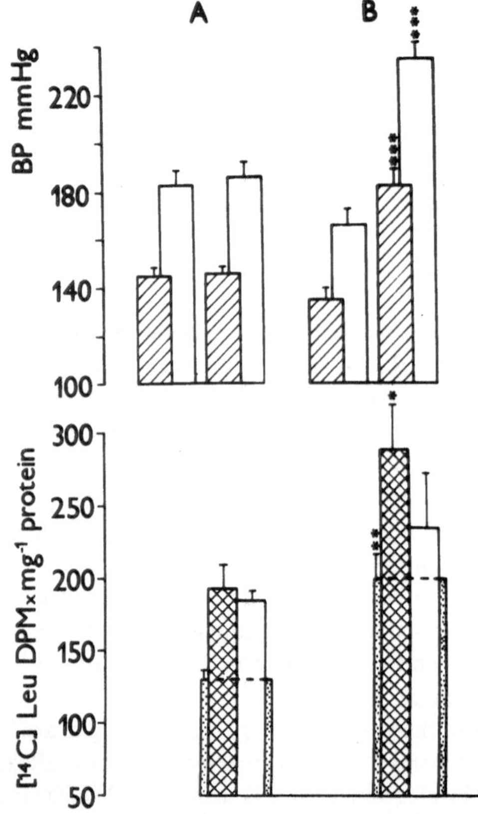

Fig. 8. ¹⁴C-leucin incorporationin the wall of ramus interventricularis (crossed hatched columns), ramus circumflexus (white columns) and myocardium (dotted columns) in the control group of animals (shame operated and 4 h lasting anaesthesia) (A), and in the group of animals with 4 h lasting increase of blood pressure (B). Upper row of columns represent blood pressure, lower row the RNA content.

The same conclusion could be drawn from the other series of experiments where labelled ^{14}C-leucin was estimated after the cardiovascular system was exposed to a 30% increase in BP (Fig. 8).

The process was reversible and 2 h after setting free the aortic constriction, with decrease of BP the RNA content decreased too. Thus a clear-cut correlation between the BP and proteosynthesis was found in ramus interventricularis anterior in myocardium, not in ramus circumflexus.

Studying further the individual links in proteosynthesis, biogenic polyamines (spermin, spermidin, putrescin) were taken into consideration. Biogenic polyamines were shown to precede the proteosynthesis in myocardium and aorta by Johnson and Schanberg [14], Majesky [15], Thomson and Friberg [16]. The source for polyamines being amino acid orinithine and enzyme starting the process is ornithin decarboxylase (ODC). In the above model, after the 4 h lasting 30% increase of blood pressure, the samples of coronary arteries and myocardium were analysed for the ODC activity using the method of Bartolomé *et al.* [17].

Surprisingly, no change of ODC activity was found either in myocardium or in ramus intraventricularis anterior or in ramus circumflexus. Thus a 4 h lasting pressure-overload evidently and without doubt increases proteosynthesis in coronary artery and myocardium, however, without involving the polyamines. Polyamines do not play any control role in proteosynthesis brought about by this short hypertensive episode.

The data by Johnson and Schanberg [14], Majesky [15], Thomson and Friberg [16] revealed the increased ODC activity in myocardium and aorta in experimental models in which BP increase was elicited by adrenergic stimuli. Thus, we induced a blood pressure increase comparable to the above experiments, by the i.v. infusion of noradrenaline (10 g/1 ml salt sol./1 kg/1 min) lasting 4 h.

To a positive surprise, the ODC activity increased in myocardium, in ramus interventricularis anterior and also in ramus circumflexus (Fig. 9), despite the rate of deformation of ramus interventricularis anterior and ramus circumflexus being quantitatively different, and despite this difference being reflected by a different rate of proteosynthesis.

As far as myocardium is concerned, Zimmer already in 1986 [18] stated that in proteosynthesis and consequent cardiac

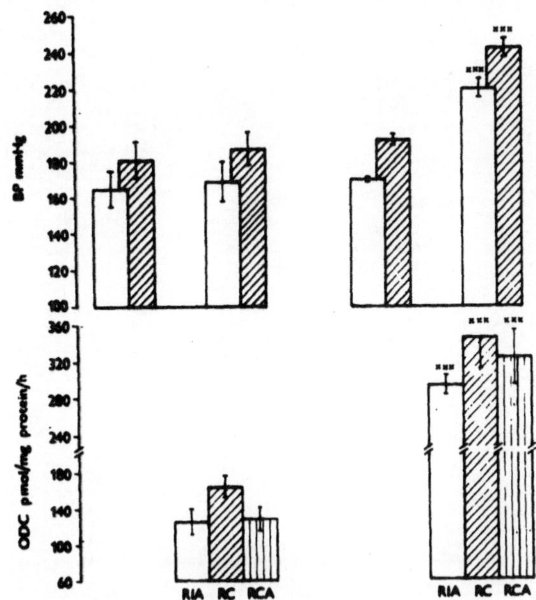

Fig. 9. Ornithine decarboxylase activity in ramus interventricularis anterior, ramus circumflexus, in the right coronary artery and in the myocardium during 4 h lasting blood pressure increase induced by i.v. administration of noradrenaline.

hypertrophy in certain experimental models ODC is involved and in others it is not involved. Present experiments include the conduit coronary artery into the paradigma concerning the metabolic response to loading the cardiovascular system.

The present experiments concerned the early changes in proteosynthesis in coronary wall, and the results indicated heterogenous pathways dependent on heterogenous stimuli. Long-term cardiac overloading by pressure or volume resulted in a complete remodelling of coronary wall characterized by an increase in wall thickness and increase in extracellular matrix, as it was described by Wiener and Giacomelli [19] in pressure overloading and Gerová *et al.* [20] in volume overloading of the heart. The contribution of endothelium and smooth muscle cells to the increase in proteosynthesis and finally to the wall thickness remains for further investigation.

References

1. Gerová M, Gero J, Barta E, Dole§el S, Smieško V, Levický V: Neurogenic and myogenic control of conduit coronary a.: a possible interference. Basic Res Cardiol 76: 503–507, 1981
2. Sims FH: A comparison of structural features of the walls of coronary arteries from 10 different species. Pathology 21: 115–124, 1989
3. Rabe D: Kalibermessung an den Herzkranzarterien und dem Sinus Coronarius. Basic Res Cardiol 68: 356–379, 1973
4. Purinja B, Kasjanov BA: Biomechanics of large conduit arteries in man (in Russian). Zinatne, Riga, 1984
5. Gerová M, Bárta E, Stolárik M, Gero J: Geometry of the conduit coronary artery in diastole is determined by the volume of the left and right ventricles. Basic Res Cardiol 84: 583–590, 1989
6. Gerová M, Bárta E, Stolárik M, Gero J: Heterogeneity in geometrical alterations of two main branches of left coronary artery induced by increase in left and right ventricle volume. Am J Physiol 262: H1049–H1053, 1992
7. Mann DL, Kent AL, Cooper G: Load regulation of the properties of adult feline cardiocytes: growth induction by cellular deformation. Circ Res 64: 1079–1090, 1989
8. Kent RL, Hoober JK, Cooper G: Load responsiveness of protein synthesis in adult mammalian myocardium: role of cardiac deformation linked to sodium influx. Circ Res 64: 74–85, 1989
9. Watson PA: Function follows form: generation of intracellular signals by cell deformation. Faseb J 5: 2013–2019, 1991
10. Gerová M, Stoev V, Kittová M, Koska J: Deformation of conduit coronary artery during afterload increase. Physiol Res 42: 9, 1993
11. Meyer WW, Walsh SZ, Lind J: Functional morphology of human arteries during fetal and postnatal development. In: C.J. Schwartz, N.T. Werthessen, S. Wolf (eds). Structure and Function of the Circulation. Plenum Press, New York, 1981, pp 95–380
12. Canev RG, Markov GG: Quantitative estimation of nucleic acids by spectrophotometric method (in Russian). Biochimija 25: 151–159, 1960
13. Nagai R, Low RB, Stirewalt WS, Alpert NR, Litten RZ: Efficiency and capacity of protein synthesis are increased in pressure overload cardiac hypertrophy. Am J Physiol 255: H325–H328, 1988
14. Johnson MD, Grognolo A, Kuhn CM, Schanberg SM: Hypertension and cardiovascular hypertrophy during chronic catecholamine infusion in rats. Life Sciences 33: 169–180, 1983
15. Majesky MW, Yang HYL, Juchau MR: Interaction of alpha and beta adrenergic stimulation on aortic ornithine decarboxylase activity. Life Sciences 36: 153–159, 1985
16. Thomson KE, Friberg P, Adams MA: Vasodilators inhibit acute alpha-adrenergic receptor-induced trophic responses in the vasculature. Hypertension 20: 809–815, 1992
17. Bartolome J, Huguenard J, Slotkin TA: Role of ornithine decarboxylase in cardiac growth and hypertrophy. Sciences 210: 793–794, 1980
18. Zimmer HG, Peffer H: Metabolic aspects of the development of experimental cardiac hypertrophy. Basic Res Cardiol 81 (S1): 127–137, 1986
19. Wiener J, Giacomelli F: Structural characterization of coronary arteries and myocardium in renal hypertensive hypertrophy. In: R.C. Tarazzi, J.B. Dunbar (eds). Perspectives in Cardiovascular Research. Raven Press, New York, 1983, 8, pp 60–72
20. Gerová M, Holécyová A, Kristek F, Fízel' A, Fízel'ová: Remodelling and functional alterations of the rabbit coronary artery in volume overloaded heart. Cardiovasc Res 27: 2005–2010, 1993

1. [illegible]

[illegible]

11. [illegible] Wolf [illegible]

17. [illegible] Role of [illegible] growth and [illegible]
18. [illegible]

20. [illegible] remodeling [illegible]

PART II

MYOCYTIC ADAPTATION AND MYOCARDIAL INJURY

Molecular and Cellular Biochemistry **147**: 77–81, 1995.

Oxidative stress and heart failure

Neelam Singh, Arvinder K. Dhalla*, Charita Seneviratne and Pawan K. Singal

Division of Cardiovascular Sciences, St. Boniface General Hospital Research Centre and Department of Physiology, Faculty of Medicine, University of Manitoba, Winnipeg, Canada

Abstract

Various abnormalities have been implicated in the transition of hypertrophy to heart failure but the exact mechanism is still unknown. Thus heart failure subsequent to hypertrophy remains a major clinical problem. Recently, oxidative stress has been suggested to play a critical role in the pathogenesis of heart failure. Here we describe antioxidant changes as well as their significance during hypertrophy and heart failure stages. Heart hypertrophy in rats and guinea pigs, in response to pressure overload, is associated with an increase in 'antioxidant reserve' and a decrease in oxidative stress. Hypertrophied rat hearts show increased tolerance for different oxidative stress conditions such as those imposed by free radicals, hypoxia-reoxygenation and ischemia-reperfusion. On the other hand, heart failure under acute as well as chronic conditions is associated with reduced antioxidant reserve and increased oxidative stress. The latter may have a causal role as suggested by the protection seen with antioxidant treatment in acute as well as in chronic heart failure. It is becoming increasingly apparent that, anytime the available antioxidant reserve in the cell becomes inadequate, myocardial dysfunction is imminent. (Mol Cell Biochem **147:** 77–81, 1995)

Key words: antioxidants, redox state, lipid peroxidation

Introduction

Free radicals are significantly important in the maintenance of normal physiological function. However, an increased production of these toxic chemicals through an uncontrolled chain reaction is potentially lethal. Therefore, in normal conditions, there is a balance between free radicals and their quenchers, antioxidants. All enzymatic as well as non-enzymatic antioxidants available in the cell constitute what is known as the 'antioxidant reserve' [1]. During disease states or pathological conditions, the balance shifts in favour of a relative increase in free radicals resulting in increased oxidative stress [2]. Generally, the redox state, the ratio of reduced glutathione to oxidized glutathione (GSH/GSSG), is used as a sensitive index of oxidative stress [3, 4]. An increase in the redox state indicates reduced oxidative stress and vice versa.

It has been shown that certain oxidative stress conditions are also associated with changes in the antioxidant defense mechanisms. Myocardial antioxidant enzyme activities have been reported to change under acute as well as chronic stress conditions [4–8]. Therefore, antioxidant status of the heart is suggested to be a dynamic function adjusting to various physiological as well as pathological conditions [1, 9–11]. Hypertrophy, an adaptive response of the heart to increased workload, is associated with increased antioxidant activities [7, 9]. Heart failure, apparently a common clinical end point of many cardiac conditions including hypertrophy, is associated with depressed antioxidant reserve [1, 7]. Although transition from hypertrophy to heart failure is determined by many factors, increased oxidative stress also appears to play a role [2]. In this concise review, we focus on changes in the antioxidant status during hypertrophy and heart failure as well as their probable significance in these two cardiac conditions.

Address for offprints: P.K. Singal, St. Boniface General Hospital Research Centre, 351 Tache Avenue, Rm.# R3022, Winnipeg, MB, R2H 2A6, Canada
**Current address:* Department of Internal Medicine, Division of Cardiology, University of Missouri, Columbia, Missouri, USA

Antioxidant changes and their significance in heart hypertrophy

An increase in myocardial antioxidant capacity was observed in rat hearts subsequent to chronic pressure overload [9]. There was a significant increase in myocardial superoxide dismutase (SOD) and glutathione peroxidase (GSHPx) activities up to 12 weeks of hypertrophy due to banding of the abdominal aorta in rats. At 48 weeks of hypertrophy, SOD activity in hypertrophied heart was the same as in control hearts, whereas, GSHPx activity was still higher when compared to control. Lipid peroxidation, another index of oxidative stress and cell injury, was found to be significantly lower in the hypertrophied hearts. Increase in antioxidants and a reduction in lipid peroxidation were also reported at 10 weeks of hypertrophy in rats [12–14]. During this period, the hemodynamic function, as compared with controls, was better and stable as indicated by increased aortic pressure, left ventricular systolic pressure (LVSP) and its first derivative (dP/dt) with no change in left ventricular end diastolic pressure (LVEDP) [9]. Heart hypertrophy due to exercise-training in rats was also accompanied by improved antioxidants [11]. Myocytes isolated from hypertrophied rat hearts also showed increased antioxidants and reduced lipid peroxidation [14].

Guinea pigs subjected to chronic pressure overload also showed significant increase in myocardial SOD as well as GSHPx activities and a decrease in lipid peroxidation at 10 weeks following constriction of the ascending aorta [7]. The GSH/GSSG redox state was higher at 10 weeks suggesting reduced oxidative stress [7]. Heart hypertrophy in these animals was indicated by 50% increase in ventricle wt/body wt ratio as well as a 38% increase in left ventricular wall thickness [15]. Hemodynamic assessment showed increased LVSP as well as +dP/dt and no change in LVEDP.

In order to explore the significance of these antioxidant changes, rat hearts with increased antioxidant capacity during the hypertrophy stage were subjected to various forms of exogenous oxidative stress conditions including exposure to free radicals, ischemia-perfusion and hypoxia-reoxygenation [9, 12–14]. Perfusion of rat hearts with xanthine-xanthine oxidase, an exogenous source of free radicals, resulted in contractile failure and caused a significant rise in resting tension in controls. However, a better maintenance of contractile function and significantly less rise in resting tension was seen in the hypertrophied hearts [9]. It is likely that increased antioxidant reserve in the hypertrophied hearts is responsible for a better maintenance of the cardiac function upon exposure to free radicals. Under ischemic conditions, contractile failure and resting tension in both control and hypertrophied hearts were comparable; however, upon reperfusion, hypertrophied hearts showed better recovery of the developed force and resting tension as well as reduced incidence of arrhythmias [13]. In addition, supplementation of the perfusion medium with antioxidants, SOD and catalase, significantly attenuated the ischemia-reperfusion injury in control group. The injury seen in control hearts supplemented with antioxidants was comparable to that seen in the hypertrophied hearts having a higher antioxidant reserve. These findings suggest the significance of increased myocardial endogenous antioxidants in offering protection against ischemia-reperfusion injury [13].

Experiments with hypertrophied hearts subjected to hypoxia-reoxygenation provided further evidence that antioxidants might be protective [12]. Hypoxia for 15 min resulted in complete failure of the developed tension and about 200% increase in resting tension in both hypertrophied and control hearts. However, upon reoxygenation, hypertrophied hearts recovered developed tension to 60% and resting tension was higher by only 80%. Control hearts, on the other hand, showed a poor recovery of developed as well as resting tensions. Both SOD and GSHPx activities were significantly higher in the hypertrophy group than the corresponding reoxygenated control hearts. Even though reoxygenation resulted in an increase in MDA content in both groups, increase in hypertrophied group was significantly less than that seen in the control group. These findings suggested that a higher endogenous antioxidant reserve in hypertrophied hearts might be useful in offering resistance against reoxygenation injury [12]. Exercise-induced hypertrophied hearts with increased myocardial endogenous antioxidants were also found to be more resistant to adriamycin induced cardiotoxic effects [11]. Adriamycin, an antitumor drug, is suggested to cause its cardiotoxic effect through free radical mechanisms [16, 17]. Hypertrophied myocytes, upon reoxygenation, showed better preservation of endogenous antioxidant enzyme activity and lower levels of lipid peroxidation as compared to control myocytes [14]. These studies on hypertrophied rat myocytes subjected to hypoxia-reoxygenation established that protection of the hypertrophied hearts was occurring at the cardiomyocyte level.

Antioxidant changes and their significance in heart failure

Significant changes in endogenous antioxidants have also been reported in acute and chronic conditions of myocardial dysfunction as well as failure. In this regard, acute failure in rat hearts due to 10 min hypoxia at 37°C was accompanied by a decrease in SOD and GSHPx activities and the function recovery upon reoxygenation was poor [6]. However, when hypoxia was induced at 22°C for 10 min or at 37°C for only 5 min, there were no changes in these antioxidants and the recovery upon reoxygenation was complete. Hypoxic injury at 37°C for 10 min was also diminished by the addition of catalase in the medium, suggesting that reduced antioxidant reserve during hypoxia did contribute to the injury upon reoxygenation and that better maintenance of endogenous antioxidant levels during hypoxia is important for recovery [6]. This point was further supported by studies on the cardiac myocyte isolated from normal adult hearts and exposed to hypoxia and reoxygenation. Hypoxia for 15 min resulted in a reduction in MnSOD and GSHPx activities with no change in CAT activity and lipid peroxidation levels [18]. Upon reoxygenation for 15 min, recovery of MnSOD was seen but there was no change in GSHPx activity. The increase in lipid peroxidation was accompanied by changes in cell morphology and function. These latter changes due to hypoxia-reoxygenation were modulated by the addition of catalase in the medium [18].

Another study in isolated and perfused rabbit hearts showed that ischemia-induced decline of developed pressure and increase in diastolic pressure were restored upon reperfusion after 30 min [19]. These changes were also accompanied by a decline in tissue redox state (GSH/GSSG) and recovery upon reperfusion. However, reperfusion after 60 and 90 min did not re-establish normal function and the changes correlated with decreased redox state and increased oxidative stress [19]. The clinical relevance of this phenomenon was also established by monitoring the arterial coronary sinus differences in GSH and GSSG in patients with coronary artery disease during coronary by-pass surgery [19]. During cardioplegia in surgery, when the heart was globally ischemic for 30 min, the arteriovenous difference for GSH and GSSG remained constant. However, upon reperfusion, there was a release of GSH and GSSG. Reperfusion after longer duration of ischemia caused more loss of tissue glutathione resulting in increased oxidative stress [19]. Perfusion of the rabbit heart with N-acetyl-cysteine, a sulphydryl group donor, resulted in recovery of developed pressure and increased tissue GSH content with no change in GSSG content, suggesting a therapeutic role of glutathione in improving the redox state and the efficacy of myocardial reperfusion [19].

There is now evidence available suggesting that in chronic cardiac dysfunction, oxidative stress may increase and antioxidant reserve may be depressed [7, 17, 20]. In this regard, banding of the ascending aorta in guinea pigs, resulted in heart failure at 20 weeks post-surgery duration. Hemodynamic data revealed decreased systolic aortic pressure, depressed LVSP, increased LVEDP along with clinical signs of heart failure such as dyspnea. These failing hearts showed decrease in myocardial SOD and GSHPx activity compared to control groups. The redox state was found to be depressed [7] and the change was mainly due to increased GSSG accumulation. This suggested increased oxidative stress condition in the 20 week failing group may be due to depressed antioxidant activities or increased free radical production [7]. In this regard, increased free radical production has been shown in mitochondria isolated from failing hearts [21, 22]. Myocardial lipid peroxidation was also reported to increase in cardiomyopathic hamsters [22]. Depressed antioxidant reserve has also been reported in congestive heart failure due to adriamycin [17, 20]. Although these reports demonstrated increased oxidative stress in heart failure, role of such a change in the pathogenesis of heart failure was suggested by studies employing treatment with antioxidants.

Fig. 1. Ultrastructure of a failing guinea pig heart. Ascending aorta in the animal was banded for 20 weeks and the animal showed all signs and symptoms of congestive heart failure as described elsewhere (Randhawa and Singal 1992). Intracellular edema (*) and mitochondrial abnormalities (arrows) are apparent. Magnification line is one micron.

Catalase and SOD have been shown to offer protection against loss of cardiac function due to ischemia-reperfusion injury in a variety of experimental conditions [23]. Therapeutic potential of vitamin E against myocardial injury has also been documented [24–27]. The lipophilic nature of vitamin E is responsible for its excellent antioxidant properties [25, 28]. In cardiac membranes as well as in isolated cardiomyocytes, vitamin E offers protection against oxidative damage [29, 30]. In guinea pigs, heart failure due to banding of the ascending aorta was accompanied by significant deposition of collagen [15] and myocyte damage (Fig. 1). However, in animals treated with vitamin E, by implanting slow release tablets, at the time of surgery to induce pressure overload, myocardial ultrastructure was significantly maintained (Fig. 2) and the animals also showed a better hemodynamic function.

An association between a high intake of vitamin E and a reduced risk of coronary heart disease has been shown both in men and women [31, 32]. Trolox, a water soluble analogue

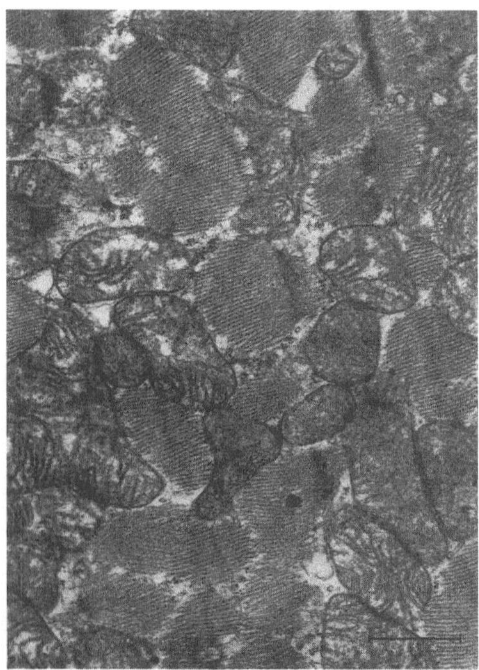

Fig. 2. Ultrastructure of a guinea pig heart. Animal, same as in Fig. 1, was treated with vitamin E for 20 weeks. Myocyte structure is better maintained. Magnification line is one micron.

of α-tocopherol and ascorbic acid, were found to be effective in protecting isolated myocytes from free radical-induced injury [33]. Trolox and ascorbic acid reduced the area of infarction produced by occlusion of the anterior descending coronary artery in dogs [33]. Probucol, another antioxidant, offered protection against adriamycin-induced congestive heart failure [17, 20].

In addition to heart failure, the above observations may also suggest a potential therapeutic value of a chronic antioxidant treatment in other disease conditions associated with increased oxidative stress. In this regard, SOD has been administered safely to human neonates for the prevention of bronchopulmonary displasia in infant respiratory distress syndrome with no evidence of allergy or toxicity to hepatic or renal systems [34]. In hypertensive patients, a reduction in SOD and GSHPx in the blood was reported [35] and antioxidant therapy with vitamin C, thiopronine and glutathione was found to improve these antioxidants which was also accompanied by vasodilatory influence [36].

Conclusion

Studies on the antioxidant changes and their significance during hypertrophy and heart failure have provided a new insight about the pathogenesis of heart disease. During the development of heart hypertrophy in response to a chronic increase in cardiac workload there appears to be a phase of increased antioxidant reserve. However, during prolonged durations of stress, some deficit in these compensatory antioxidant changes can occur. Thus, a reduction in antioxidants in failing hearts and a better maintenance of cardiac function with antioxidant therapy has provided strong evidence that antioxidant deficit under chronic conditions may contribute to the pathogenesis of heart failure.

Acknowledgements

This work was supported by the Medical Research Council Group Grant in Experimental Cardiology (PKS). Ms N Singh was supported by a fellowship from the University of Manitoba. Dr AK Dhalla was supported by a studentship from the Manitoba Health Research Council and is now a post-doctoral fellow of the Medical Research Council.

References

1. Singal PK, Kirshenbaum LA: A relative deficit in antioxidant reserve may contribute in cardiac failure. Can J Cardiol 6: 47–49, 1990
2. Kaul N, Siveski-Iliskovic N, Hill M, Slezák J, Singal PK: Free radicals and the heart. J Pharmacol Toxicol Meth 30: 55–67, 1993
3. Curello S, Ceconi C, Bigoli C, Ferrari R, Albertini A, Guarnieri C: Change in the cardiac glutathione after ischemia and reperfusion. Experimentia 41: 42–43, 1985
4. Ferrari R, Ceconi C, Curello S, Guarnieri C, Caldarera CM, Albertini A, Visioli O: Oxygen-mediated myocardial damage during ischemia and reperfusion: role of the cellular defenses against oxygen toxicity. J Mol Cell Cardiol 17: 937–945, 1985
5. Guarnieri C, Flamigni F, Caldarera CM: Role of oxygen in cellular damage induced by reoxygenation of hypoxic heart. J Mol Cell Cardiol 12: 797–808, 1980
6. Dhaliwal H, Kirschenbaum LA, Randhawa AK, Singal PK: Correlation between antioxidant changes during hypoxia and recovery upon reoxygenation. Am J Physiol 261: H632–H638, 1991
7. Dhalla AK, Singal PK: Antioxidant changes in hypertrophied and failing guinea pig hearts. Am J Physiol 266 (Heart Circ Physiol): H1280–H1285, 1994
8. Singal PK, Dhalla AK, Hill M, Thomas TP: Endogenous antioxidant changes in the myocardium in response to acute and chronic stress conditions. Mol Cell Biochem 129: 179–186, 1993
9. Gupta M, Singal PK: Higher antioxidative capacity during a chronic stable heart hypertrophy. Circ Res 64: 398–406, 1989
10. Higuchi M, Cartier LJ, Chen M, Holloszy JO: Superoxide dismutase and catalase in skeletal muscle: adaptive response to exercise. J Gerontology 40: 281–286, 1985
11. Kanter MM, Hamlin RL, Unverferth DV, Davis HW, Merola AJ: Effect of exercise training on antioxidant enzymes and cardiotoxicity of doxorubicin. J Appl Physiol 59: 1298–1303, 1985
12. Kirshenbaum LA, Singal PK: Antioxidant changes in heart hypertrophy: significance during hypoxia-reoxygenation injury. Can

J Physiol Pharmacol 70: 1330–1335, 1992
13. Kirshenbaum LA, Singal PK: Increase in endogenous antioxidant enzymes protects the heart against reperfusion injury. Am J Physiol 265: H484–H493, 1993
14. Kirshenbaum LA, Hill M, Singal PK: Endogenous antioxidants in isolated hypertrophied cardiac myocytes and hypoxia-reoxygenation injury. J Mol Cell Cardiol 27: 263–272, 1995
15. Randhawa AK, Singal PK: Pressure overload-induced cardiac hypertrophy with and without dilation. J Am Coll Cardiol 20: 1569–1575, 1992
16. Singal PK, Deally CMR, Weinberg LE: Subcellular effects of adriamycin in the heart: a concise review. J Mol Cell Cardiol 19: 817–828, 1987
17. Siveski-Iliskovic N, Kaul N, Singal PK: Probucol promotes endogenous antioxidants and provides protection against adriamycin-induced cardiomyopathy in rats. Circulation 89: 2829–2835, 1994
18. Kirshenbaum LA, Singal PK: Changes in antioxidant enzymes in isolated cardiac myocytes subjected to hypoxia-reoxygenation. Lab Invest 67 (6): 796–803, 1992
19. Ferrari R, Curello S, Ceconi C, Cargnoni A, Condorelli E, Albertini A: Alterations of glutathione status during myocardial ischemia and reperfusion. In: P.K. Singal (ed). Oxygen Radicals in the Pathophysiology of Heart Disease. Kluwer Academic Press, Massachussets, 1988 pp 145–160
20. Siveski-Iliskovic N, Hill M, Chow DA, Singal PK: Probucol protects against adriamycin cardiomyopathy without interfering with its antitumor effect. Circulation 1995 (in press)
21. Guarnieri C, Muscari C, Caldarera CM, Slefanelli C, Pretolani E: The effect of treatment with coenzyme Q_{10} on the mitochondrial function and superoxide radical formation in cardiac muscle hypertrophied by mild stenosis. J Mol Cell Cardiol 19: 63–71, 1985
22. Kobayashi A, Yamashita T, Kaneko M, Nishiyama T, Hayashi H, Yamaszaki N: Effect of verapamil on experimental cardiomyopathy in the bio 14.6 Syrian hamster. J Am Coll Cardiol 10: 1128–1134, 1987
23. Werns SW, Lucchesi BR: The role of the polymorphonuclear leukocyte in mediating myocardial reperfusion injury. In: P.K. Singal (ed). Oxygen Radicals in the Pathophysiology of Heart Disease. Kluwer Academic Press, Massachussets, 1988, pp 123–144
24. Janero DR: Therapeutic potential of vitamin E against myocardial ischemia-reperfusion injury. Free Rad Biol Med 10: 315–324, 1991
25. van Acker SABE, Koymans LMH, Bast A: Molecular pharmacology of vitamin E: structural aspects of antioxidant activity. Free Rad Biol Med 15: 311–328, 1993
26. Dhalla AK, Singal PK: Vitamin E delays pathogenesis of heart failure due to chronic pressure overload. Can J Cardiology 10(A): 72A, 1994
27. Dhalla AK, Singh N, Singal PK: Antioxidant therapy associated with the reversal of oxidative stress, delays the pathogenesis of heart failure. Circulation 1995 (in press)
28. Packer L: Vitamin E is nature's master antioxidant. Scientific American Science and Medicine 1: 54–63, 1994
29. Massey KD, Burton KP: Alpha-tocopherol attenuates myocardial membrane-related alterations resulting from ischemia and reperfusion. Am J Physiol 256: H1192–H1199, 1989
30. Ferrari R, Visiolo O, Guarnieri C, Caldarera M: Vitamin E and the heart: possible role as antioxidant. Acta Vitaminol Enzymol 5: 11–22, 1983
31. Rimm EB, Stampfer MJ, Ascherio A, Giovannucci E, Colditz GA, Willett WC: Vitamin E consumption and the risk of coronary disease in men. N Eng J Med 328: 1450–1456, 1993
32. Stampfer MJ, Hennekens CN, Manson JE, Colditz GA, Rosner B, Willett WC: Vitamin E consumption and the risk of coronary disease in women. N Eng J Med 328: 1444–1449, 1993
33. Mickle DAG, Li RK, Weisel RD, Birnbaum PL, Wu TW, Jackawski G, Madonik MM, Burton GW, Ingold KU: Myocardial salvage with Trolox and ascorbic acid for an acute evolving infarction. Ann Thorac Surg 47: 553–557, 1989
34. Rosenfeld W, Evans H, Jhaveri R, Moainie H, Vohra K, Georgatos E, Salazar JD: Safety and plasma concentrations of bovine superoxide dismutase administered to human premature infants. Dev Pharmacol Ther 5: 151–161, 1982
35. Iarema NI, Konovalova GG, Lankin VZ: Changes in the activity of antioxidant enzymes in patients with hypertension. Kardiologia 32(3): 46–48, 1992
36. Ceriello A, Giugliano D, Quatraro A, Lefebvre PJ: Antioxidants show an antihypertensive effect in diabetic and hypertensive subjects. Clin Sci Colch 81(6): 739–742, 1991

Molecular and Cellular Biochemistry **147**: 83–88, 1995.

Structural and biochemical remodelling in catecholamine-induced cardiomyopathy: comparative and ontogenetic aspects

Bohuslav Ošťádal, Václav Pelouch, Ivana Ošťádalová and Olga Nováková[1]

Institute of Physiology, Academy of Sciences of the Czech Republic and [1]Department of Physiology and Developmental Biology, Faculty of Science, Charles University, Prague, Czech Republic

Abstract

Excessive release or administration of beta-mimetic catecholamines may induce cardiomegaly, necrotic lesions and accumulation of connective tissue in the heart of adult homoiotherms. It was examined here whether similar changes can also be observed at different stages of evolution of the cardiovascular system, i.e. in poikilotherms and in homoiotherms during embryonic life.

Sensitivity of the poikilothermic hearts (carp, frog, turtle) to isoproterenol (IPRO) was significantly lower than in the homoiotherms. Necrotic lesions, if present, were localized in the inner spongious musculature which has no vascular supply but which exhibits higher activities of enzymes connected with aerobic oxidation. Moreover, the IPRO-induced decrease of the phospholipid content was also significantly more expressed in the spongious layer. IPRO treatment did not influence the total weight of the fish heart but the proportion of the outer compact layer was significantly higher. These changes were accompanied by an increase of collagen, higher water content and an increase of isomyosin with a lower ATPase activity. The response of the poikilothermic heart to IPRO-induced overload thus differs significantly from that in the homoiotherms.

The administration of IPRO during embryonic life of homoiotherms (chick) induces serious cardiovascular disturbances, including cardiomegaly and cellular oedema. Necroses of myofibrils, characteristic of IPRO-induced lesions of adults, were, however, rather exceptional. IPRO did not elevate the concentration of ^{85}Sr (as a calcium homologue) in the immature myocardium; it seems, therefore, that IPRO-induced changes of the embryonic heart are not necessarily due to an intracellular calcium overload.

It may be concluded that the character of catecholamine-induced cardiomyopathy is not uniform and depends strictly on the stage of cardiac development. (Mol Cell Biochem 147: 83–88, 1995)

Key words: catecholamine-induced cardiomyopathy, isoproterenol, cardiotoxicity, poikilothermic heart, embryonic heart, necrotic lesions, cardiac overload

Introduction

It is now well established that the administration of high doses of beta-mimetic catecholamines may induce cardiomegaly, necrotic lesions and accumulation of connective tissue in the heart of adult homoiotherms [for review, see 1, 2]. Since the discovery of Rona *et al.* [3], isoproterenol (IPRO)-induced cardiac lesions have been demonstrated not only in a number of adult mammals, such as dogs [4], hamsters [5], rabbits [6], cats [7], monkeys [8], and mice [9], but also in adult birds [10]. The pathogenesis of catecholamine-induced myocardial lesions is multifactorial; the major hypotheses include a relative hypoxia, coronary microcirculatory effects, altered membrane permeability, myofilament overstimulation, high energy phosphate deficiency, catecholamine-induced formation of oxidation products and calcium overload [2].

Address for offprints: B. Ošťádal, Institute of Physiology, Academy of Sciences of the Czech Republic, Vídeňská 1083, 142 20 Prague 4, CZ Czech Republic

Whereas abundant data are available concerning the cardiotoxicity of catecholamines in adult homoiotherms, much less is known about the possible toxic effect of these substances on the immature heart. It is the aim of this short review to summarize some of our experimental data on the effect of high doses of IPRO at different stages of evolution of cardiovascular system, i.e. in poikilothermic vertebrates and in homoiotherms during embryonic life.

Effect of beta-mimetic catecholamines on the heart of adult poikilotherms

a) General characteristics of the poikilothermic heart

While the heart of adult homoiotherms consists entirely of a compact musculature with coronary blood supply, the ventricular myocardium of most species of cold-blooded vertebrates is formed by two different muscular layers: the inner avascular spongious musculature supplied by diffusion from the ventricular lumen is covered by an outer compact layer with a coronary blood supply [11–13].

We have observed [14] that in the carp and turtle, the compact musculature comprises 37 and 55%, respectively, of the total cardiac weight. In both species the weight of the compact layer increases with increasing heart weight. The water content, as well as the protein composition of both layers is not different. The myosin ATPase activity (Ca activated) of the compact layer of the carp heart is significantly higher than that of the spongious musculature. No such difference was found in the turtle. On the other hand, the activities of enzymes connected with aerobic oxidation (citrate synthase, malate dehydrogenase) and glucose phosphorylation (hexokinase) are higher in the spongious musculature as compared with the compact layer. Similarly, the content of phospholipids is higher in the spongious musculature, the greatest difference being in the content of diphosphatidylglycerol DPG – [15]. Maresca *et al.* [16] and Greco *et al.* [17] have demonstrated that differences in enzyme activities are accompanied by different mitochondrial populations in both layers.

The heterogenous heart of cold-blood animals thus offers a unique opportunity to compare the sensitivity of two defined types of musculature to catecholamines under identical experimental conditions.

b) Isoproterenol-induced necrotic lesions

The administration of large doses of IPRO (2 × 80 mg.kg^{-1}) induced similar myocardial lesions in adult fish [10] and reptiles [18] to those as described in adult homoiotherms. The changes were typical, with myolysis of the muscle fibers, accompanied by marked inflammatory cell infiltration, mainly of a mononuclear character. Unlike reptiles (turtle) and homoiotherms the fish reacts to the action of IPRO with some delay as demonstrated by the fact that we did not detect any necrotic lesions unless the observation period was extended to 7 days. In this connection it is interesting to note that in poikilothermic animals the IPRO-induced necrotic lesions were localized exclusively in the inner spongy-like musculature which has no vascular supply but which has higher activities of enzymes connected with aerobic oxidation [14, 19]. This suggests that the formation of toxic changes is likewise independent of the presence of vascularization and that the substance acts directly on the myocardium.

The sensitivity of the poikilothermic heart to the cardiotoxic effects of IPRO was, however, significantly lower as compared with homoiotherms (incidence of necrotic lesions was 100% in rats, pigeons and hens, 3% in tench and 30% in turtle); the frog's heart was resistant to the necrotizing effect of IPRO [10, 20].

c) Adaptation to isoproterenol-induced overload

The administration of a single or two consecutive high doses of IPRO did not affect either the total heart weight or weight proportions of the two layers of the poikilothermic (carp) myocardium. Repeated administration of lower doses (15 × 5 mg.kg^{-1}) did not influence the total weight of the fish (carp) heart but the proportion of the outer compact layer was significantly higher [21, 22] (Figs 1 and 2). In both layers the changes were accompanied by a higher water content, an increase of isomysin with a lower ATPase activity and an increase of collagenous proteins (type I and III) (Fig. 3).

The response of the poikilothermic heart to catecholamine-induced overload thus differs significantly from the homoiotherms, where IPRO induces significant increase of the absolute right and left ventricular weight [23–25]. We have

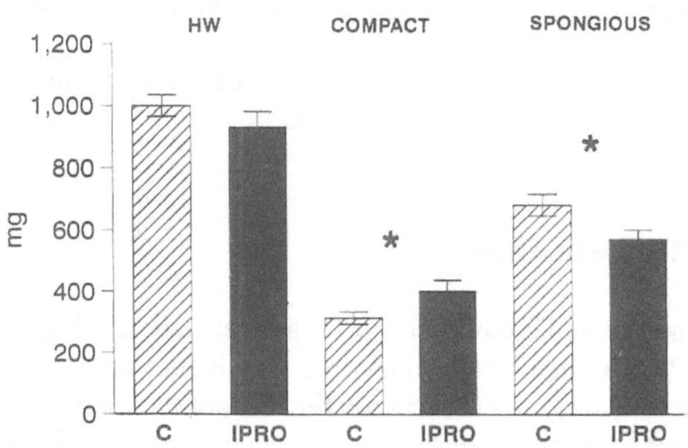

Fig. 1. Total heart weight (HW) and absolute weights of the compact and spongious musculature in control (C) and isoproterenol (IPRO)-treated (15 × 5 mg.kg^{-1}) carps. * – $p < 0.01$. Data from [21, 22].

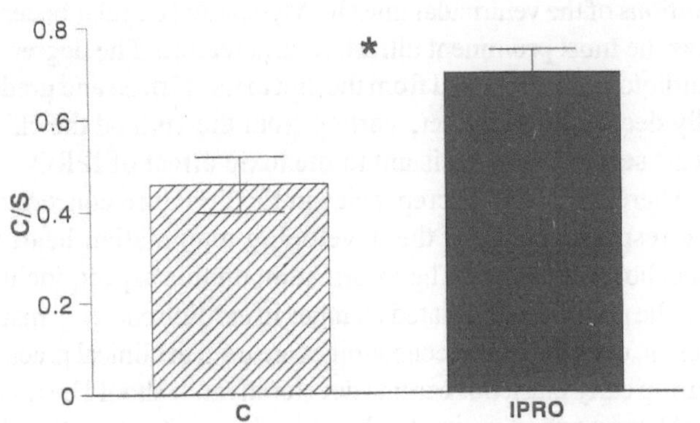

Fig. 2. Compact weight/spongious weight in control (C) and isoproterenol (IPRO)-treated (15 × 5 mg.kg^{-1}) carps. * – $p < 0.01$. Data from [21, 22].

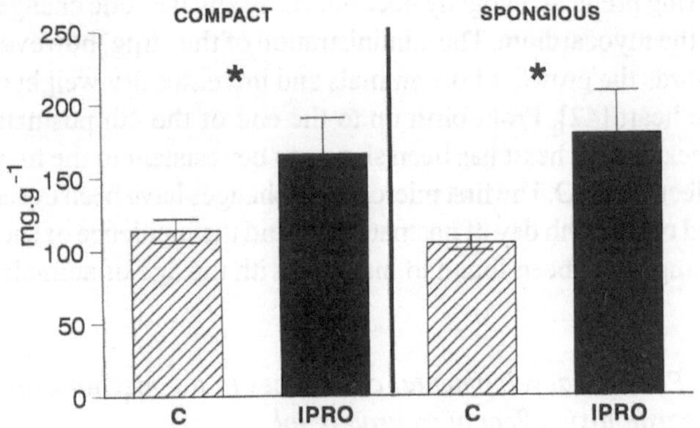

Fig. 3. Collagenous proteins (mg.g^{-1}) in compact and spongious musculature in control (C) and isoproterenol (IPRO)-treated (15.5 mg.kg^{-1}) carps. * – $p < 0.01$. Data from [22].

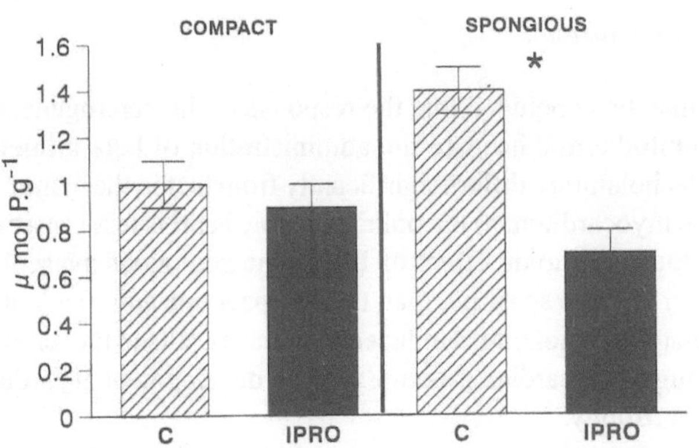

Fig. 4. Phosphatidylinositol (umol P.g^{-1}) in compact and spongious musculature in control (C) and isoproterenol (IPRO)-treated (15.5 mg.kg^{-1}) carps. * – $p < 0.01$. Data from [21].

observed, however [14, 26], that the size of the compact layer increased even with increasing weight of the intact poikilothermic heart, both during ontogeny and in different species varying in body weight. All these results support the hypothesis that the development of the compact musculature is necessary for the maintenance of balanced blood pressure conditions in the larger hearts (law of Laplace). Whether the described increase of the compact/spongious ratio is the first step or the only mechanism of the adaptation of the poikilothermic heart to the overload remains a matter of speculation.

d) Effect of isoproterenol on the phospholipid content of the compact and spongious musculature

It has been shown that administration of IPRO leads in adult homoiotherms to a change in the phospholipid metabolism, suggesting that changes in heart membranes in catecholamine-induced cardiomyopathy are of crucial importance in determining the functional and structural status of the myocardium [27–29]. Okumura *et al.* [30] found a decrease of total phospholipid content in the rat heart 24 h after a single dose (40 mg.kg^{-1}) of IPRO.

In contrast to the rat heart the carp ventricle does not respond to a high (40 mg.kg^{-1}) dose of IPRO by diminishing the total phospholipid content 24 h after treatment [22]. Whereas no differences were found in the content of individual phospholipids in the compact musculature, significant (40%) decrease of phosphatidylinositol (PI) occurred in the spongious layer (Fig. 4). Changes in PI metabolism are generally ascribed to activation of alfa$_1$ receptors in the heart [31]. However, Kiss and Farkas [32] report a slowed down conversion of phosphatidic acid to PI in the IPRO-treated rat heart which is in accordance with our findings. This observation confirms the previous results that spongious musculature is more sensitive to IPRO than the compact one.

Much more impressive differences occur after a repeated administration of lower doses (5 mg.kg^{-1}) of IPRO. The decrease of the mitochondrial phospholipid DPG was the most significant change; this change was more pronounced in the spongious layer and did not reach the original level even after 15 doses, while in the compact musculature a distinct overshoot occurred after 10 doses. The tissue content of ethanolamine and choline phosphoglycerides in both layers increased which probably compensates for the decrease in DPG and thus contributes to the constancy of the total phospholipid content. Similarly as after a single high dose, the PI content in the spongious layer decreased, but this change was only transient and PI content returned to the control values after 15 doses.

A decrease in DPG signals undoubtedly damages and possibly breaks down mitochondria. During a repeated administration of IPRO the tissue is, however, capable of a considerable repair of this injury as judged from the recovery of DPG content to control values. To our knowledge nothing is known about how the cardiac adaptation to IPRO treatment is reflected in the phospholipid metabolism. The changes observed apparently result from such an adaptation.

e) Conclusion

It may be concluded that the response of the heterogeneous poikilothermic heart to the administration of beta-mimetic catecholamines differs significantly from that in the mammalian myocardium: a) the poikilothermic heart is less sensitive to the cardiotoxic effect of IPRO; the spongious avascular layer is more sensitive than the compact one and b) adaptation to catecholamine-induced overload results in the remodelling of the cardiac structure without development of cardiac hypertrophy.

Cardiotoxicity of beta-mimetic catecholamines during prenatal development of homoiotherms

a) Isoproterenol-induced cardiac lesions in chick and rat immature heart

While abundant data are available concerning the cardiotoxicity of catecholamines in adult homoiotherms, very little is known about the toxic effect of these substances in the immature heart. The interest in this field has recently been stimulated by an increasing clinical use of catecholamines during early phases of ontogenetic development, particularly during pregnancy as well as in neonates [33–35]. Developmental changes in cardiac sensitivity to catecholamines and possible risks of their clinical use in obstetrics and pediatric cardiology were recently summarized [36, 37]. In this short review we would like to focus on the different response of the immature and adult homoiothermic heart to the necrogenic doses of beta-mimetic catecholamine, IPRO.

As mentioned above, the administration of high doses of IPRO produces necrotic lesions in the myocardium of adult homoiotherms. However, the administration of the same drug to chick embryos induces different nonnecrotic cardiovascular disturbances [38–40]. The type of changes was shown to depend on the time at which the beta agonist was administered during embryogenesis. Defects of the interventricular septum occurred on the 2nd embryonic day (ed), malformations of the large vessels were observed on the 3rd to 6th ed. Starting from the 5th ed, cardiomegaly occurred, partly as a result of the increased water content in the myocardium. The slight increase in cardiac dry weight was obviously caused, as in adult hearts, by the combination of the direct proteosynthetic effect of catecholamines and the increased workload induced by their stimulation of beta receptors. From the 7th to the 15th ed a block in the development of coronary vascularization, always associated with the persistence of an evolutionary older type of blood supply (diffusion from the ventricular cavity) occurs in the nonvascularized, spongious portions of the ventricular muscle. Myocardial cellular oedema was the most prominent ultrastructural feature. The degree of cardiotoxicity increased from the 2nd to the 12th ed and gradually decreased thereafter; starting from the 16th ed the chick heart seemed to be resistant to the toxic effect of IPRO.

There are many discrepancies in the literature concerning the responsiveness of the developing mammalian heart to catecholamines [36]. The information on this aspect, including the possible age-related changes in cardiotoxicity is insufficient, even though catecholamines are used in clinical practice during early phases of cardiac development. It should be noted that beta-sympathomimetic drugs are known to cross the placental barrier and reach the fetus, and their concentration in the fetal blood is nearly equal to that in the mother [41].

The acute administration of high doses of IPRO in rats during prenatal ontogeny does not cause any necrotic changes in the myocardium. The administration of this drug, however, retards the growth of the animals and increases dry weight of the heart [42]. From birth up to the end of the 4th postnatal week, the rat heart has been shown to be resistant to the toxic effect of IPRO. The first microscopic changes have been observed on the 30th day of postnatal life, and the incidence of such changes has been found to increase with the age of animals.

b) Possible developmental differences in the mechanisms of cardiotoxic effect of isoproterenol

According to Fleckenstein [43], the main pathogenetic mechanism of the cardiac damage in adults is an excess of intracellular calcium, followed by depletion of high-energy phosphates and injury to the mitochondria. The development of IPRO-induced cardiac lesions can be quantified by the measurement of ^{45}Ca uptake into the myocardial cells [43, 44]. However, the pathogenesis of IPRO-induced disturbances of the immature heart is poorly understood. Some of the effects may be attributed to the hemodynamic changes induced by catecholamine stimulation of beta receptors [45, 46].

One of the possible mechanisms could be an excess of intracellular calcium, similarly as in adult myocardium. We have tested this hypothesis by using strontium – the homologue element of calcium. The reason is the methodological advantage: ^{85}Sr is measurable as a gamma emitter; therefore, samples need no modification and work with them is quicker and simpler [47–49].

In 15-day-old rats, which were resistant to the necrogenic action of IPRO, the accumulation of strontium was not different from controls. IPRO stimulates ^{85}Sr uptake beginning from the 30th day of postnatal life and that is connected with the development of cardiac necroses (Fig. 5). On the other hand, IPRO did not influence ^{85}Sr uptake in the chick embryonic heart in any of the investigated developmental periods (Fig. 6). These findings are in agreement with the results that

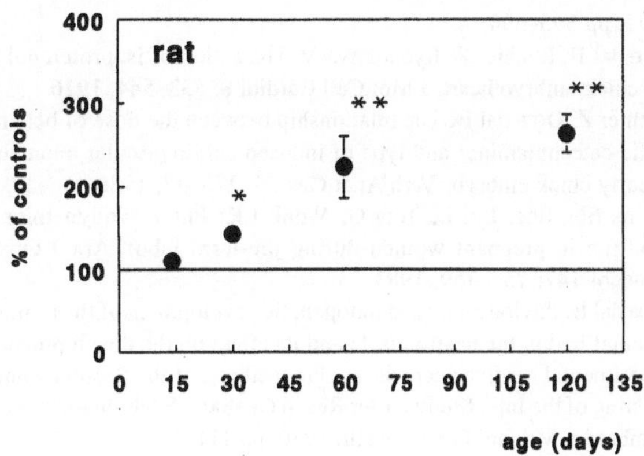

Fig. 5. Isoproterenol (IPRO)-stimulated ^{85}Sr accumulation in the rat heart during postnatal development (% of control values). ** – $p < 0.01$. Data from [49].

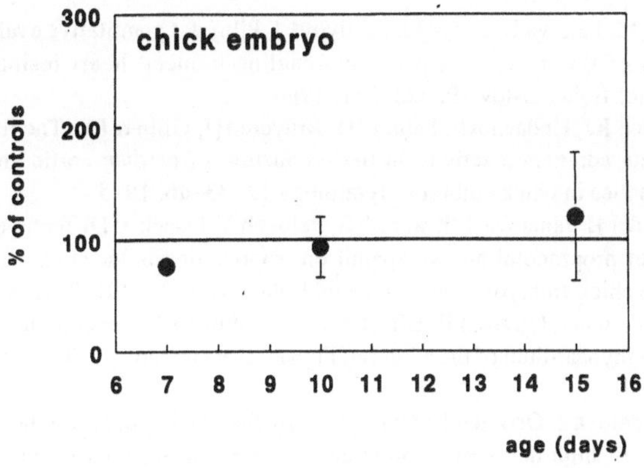

Fig. 6. Isoproterenol (IPRO)-stimulated ^{85}Sr accumulation in chick embryonic heart (% of control values). ** – $p < 0.01$. Data from [48].

verapamil, a calcium antagonist, was ineffective in preventing the toxic effects of high doses of catecholamines in chick embryonic hearts [46]. The exact mechanisms for the prenatal (chick) as well as early postnatal (rat) subsensitivity to necrogenic doses of catecholamines should be the subject of further analyses.

References

1. Rona G: Catecholamine cardiotoxicity. J Mol Cell Cardiol 17: 291–306, 1985
2. Dhalla NS, Yates JC, Naimark B, Dhalla KS, Beamish RE, Ošťádal B: Cardiotoxicity of catecholamines and related agents. In: D. Acosta (ed). Cardiovascular Toxicology. New York, Raven Press, 1992, pp 239–283
3. Rona G, Chappel CI, Balasz T, Gaudry R: An infarct-like myocardial lesions and other toxic manifestations produced by isoproterenol in the rat. Arch Pathol 67: 443–455, 1959a
4. Rona G, Zsoter T, Chappel C, Gaudry R: Myocardial lesions, circulatory and electrographic changes produced by isoproterenol in the dog. Rev Can Biol 18: 83–94, 1959b
5. Handforth CP: Myocardial infarction and necrotizing arteritis in hamsters, produced by isoproterenol (Isuprel). Med Serv J Can 18: 506–512, 1962
6. Amelin AZ, Anshelevich JV, Malzobar MJ: Eksperimentalnyje izmenenija miokarda pri vozdějstvii izadrinom (isopropylnoradrenalinom). Arkh Pathol 25: 25–30, 1963
7. Rosenblum I, Wohl A, Stein AA: Studies in cardiac necrosis. I Production of cardiac lesion with symphathomimetic amines. Toxicol Appl Pharmacol 7: 1–8, 1965
8. Maruffo CA: Fine structural study of myocardial changes induced by isoproterenol in Rhesus monkeys (Macaca mulatta). Am J Pathol 50: 27–37, 1967
9. Zbinden G, Moe RA: Pharmacologial studies on heart muscle lesions induced by isoproterenol. Ann NY Acad Sci 156: 294–308, 1969
10. Ošťádal B, Rychterová V: Effect of necrogenic doses of isoproterenol on the tench (Tinca tinca – Osteoichthyes), the frog (Rana temporaria – Anura) and the pigeon (Columbia livia – Aves). Physiol Bohemoslov 20: 541–547, 1971
11. Benninghoff A Herz: In: L. Bolk, E. Göppert, E. Killius, W. Lubosch (eds). Handbuch vergleich. Anat Wirbeltiere, VI Band. Wien, Berlin, 1933
12. Poupa O, Ošťádal B: Experimental cardiomegalies and 'cardiome – galies' in free living animals. Ann NY Acad Sci 156: 445–468, 1969
13. Ošťádal B, Rychter Z, Poupa O: Comparative aspects of the development of the terminal vascular bed in the myocardium. Physiol Bohemoslov 19: 1–7, 1970
14. Bass A, Ošťádal B, Pelouch V, Vítek V: Differences in weight parameters, myosin-ATPase activity and the enzyme pattern of energy supplying metabolism between the compact and spongious cardiac musculature of carp (Cyprinus carpio) and turtle (Testudo Horsfieldi). Pflügers Arch 343: 65–77, 1973
15. Drnková J, Nováková O, Pelouch V, Ošťádal B, Kubišta V: Phospholipid content in the compact and spongious musculature of the carp heart (Cyprinus carpio). Physiol Bohemoslov 34: 381–384, 1985
16. Maresca B, Modigh M, Servillo L, Tota B: Different temperature dependences of oxidative phosporylation in the inner and outer layers of tuna heart ventricle. J Comp Physiol 105: 167–172, 1976
17. Greco G, Martino G, Tota B: Further characterization of two mitochondrial populations in tuna heart ventricle. Comp Biochem Physiol 71B: 71–75, 1982
18. Ošťádal B, Rychterová V, Poupa O: Isoproterenol-induced acute experimental cardiac necrosis in the turtle (Testudo Horsfieldi). Am Heart J 76: 645–649, 1968
19. Gemelli L, Martino G, Tota B: Oxidation of lactate in the compact and spongy myocardium of tuna fish (Thunnus thynnus thynnus L.). Comp Biochem Physiol 65B: 231–236, 1980
20. Karlsten A, Akholm LE, Poupa O: Ultrastructural and functional aspects of frog heart lesions after isoproterenol. Acta Path Microbiol Scand 85: 251–267, 1977
21. Pelouch V, Ošťádal B, Červený M, Procházka J: Isoprenaline-induced cardiac enlargement in fish. J Mol Cell Cardiol 15: 221, 1983
22. Drnková J, Nováková O, Pelouch V, Ošťádal B, Kubišta V: The effect of isoprenaline on the phospholipid content of the compact and spongious musculature of the carp ventricular myocardium. Comp Biochem Physiol 90C: 257–261, 1988
23. Rakušan K, Tietzová H, Turek Z, Poupa O: Cardiomegaly after repeated application of isoprenaline in the rat. Physiol Bohemoslov 14: 456–459, 1965
24. Stanton HC, Brenner G, Mayfield ED: Studies on isoproterenol-induced cardiomegaly in rats. Am Heart J 77: 72–80, 1969
25. Čihák R, Kolář F, Pelouch V, Procházka J, Ošťádal B, Widimský J: Functional changes in the right and left ventricle during the development of cardiac hypertrophy and after its regression. Cardiovasc Res 26: 845–850, 1992

26. Oçœ dal B, Schiebler TH: Uber die terminale Strombahn in Fischherzen. Z Anat Entwickl-Gesch 133: 288–304, 1971
27. Vorbeck ML, Malewski FF, Erhart LS, Martin AP: Membrane phospholipid metabolism in the isoproterenol-induced cardiomyopathy in the rat. In: A. Fleckenstein, G. Rona (eds). Recent Advances in the Study of Cardiac Structure and Metabolism. University Press, Baltimore, 1975, pp 175–181
28. Dhalla NS, D§urba A, Pierce GN, Tregaskis MG, Panagia V, Beamish RE: Membrane changes in myocardium during catecholamine-induced pathological hypertrophy. In: N.R. Alpert (ed). Perspectives in Cardiovascular Research. Raven Press, New York, 1983, pp 527–534
29. Dhalla NS, Ganguly PK, Panagia V, Beamish RE: Catecholamine-induced cardiomyopathy: alterations in Ca transport systems. In: C. Kawai, W.H. Abelman (eds). Pathogenesis of Myocarditis and Cardiomyopathy. University of Tokyo Press, Tokyo, 1987, pp 135–147
30. Okumura K, Ogawa K, Satake T: Pretreatment with chlorpromazine prevents phospholipid degradation and creatine kinase depletion in isoproterenol-induced myocardial damage in rats. J Cardiovasc Pharmac 5: 983–988, 1983
31. Brown JH, Buxton IL, Bruton LL: Alfal-adrenergic and muscarinic cholinergic stimulation of phosphoinositide hydrolysis in adult rat cardiomyocytes. Circ Res 57: 532–537, 1985
32. Kiss Z, Farkas T: The effect of isoproterenol on the metabolism of phosphatidylinositol by rat heart in vitro. Biochem Pharmac 24: 999–1002, 1975
33. Nuwayhid B, Rajabi M: Beta-sympathomimetic agents: Use in perinatal obstetrics. Clin Perinatol 14: 757–782, 1987
34. Driscoll DJ: Use of inotropic and chronotropic agents in neonates. Clin Perinatol 14: 931–948, 1987
35. Papp J Gy: Autonomic responses and neurohumoral control in the early antenatal heart. Basic Res Cardiol 83: 2–9, 1988
36. Oçœ dal B, Beamish RE, Barwinsky J, Dhalla NS: Ontogenic development of cardiac sensitivity to catecholamines. J Appl Cardiol 4: 467–486, 1989
37. Oçœ dal B, Rychter Z, Oçœ dalová I: Ontogenic development of cardiotoxicity for catecholamines. In: M. Nagano, T. Takeda, N.S. Dhalla (eds). The Cardiomyopathic Heart. Raven Press, New York, 1994, pp 395–400
38. Oçœ dal B, Rychter Z, Rychterová V: The action of isoproterenol on the chick embryo heart. J Mol Cell Cardiol 8: 533–544, 1976
40. Rychter Z, Oçœ dal B: The relationship between the dose of beta mimetic catecholamines and type of induced cardiovascular anomalies in early chick embryo. Verh Anat Ges 80: 275–27, 1986
41. Caritis SN, Shei Lin L, Toig G, Wonk LK: Pharmacodynamics of ritodrine in pregnant women during pre-term labor. Am J Obstet Gynecol 147: 752–759, 1983
42. Oçœ dal B: Phylogenetic and ontogenetic development of the terminal vascular bed in the heart muscle and its effect on the development of experimental cardiac necrosis. In: Proceedings of the Second Annual Meeting of the Inter Study Gr for Res in Cardiac Metabolism, Instituto Lombardo, Milano. Fond Baselli, 1970, pp 111–132
43. Fleckenstein A: Specific inhibitors and promoters of calcium action in the excitation-contraction coupling of heart muscle and their role in the prevention or production of myocardial lesions. In: P. Harris, L.H. Opie (eds). Calcium and the Heart. Academic Press, London, 1971, pp 135–188
44. Mráz M, Faltová E, æedivý J, Protivová J, Pilný J: Quantitative evaluation of the development of isoprenaline-induced heart lesions. Physiol Bohemoslov 29: 323–331, 1980
45. Hodach RJ, Hodach AE, Fallon JD, Bruyere HJ, Gilbert EF: The role of beta adrenergic activity in the production of cardiac aortic arch anomalies in chick embryos. Teratology 12: 33–46, 1975
46. Oçœ dal B, Janatová T, Krause EG, Pelouch V, Dušek J: Different effect of propranolol and verapamil on isoprenaline-induced changes in the chick embryonic heart. Physiol Bohemoslov 36: 301–311, 1987
47. Oçœ dalová I, Oçœ dal B: Effect of isoprenaline on ^{85}Sr accumulation in the myocardium of the adult rat. Physiol Bohemoslov 37: 351–353, 1988
48. Oçœ dalová I, Oçœ dal B: ^{85}Sr uptake by the chick embryonic heart: effect of high doses of isoproterenol. Can J Physiol Pharmacol 70: 959–962, 1992
49. Oçœ dalová I, Oçœ dal B: Ontogenic differences in isoproterenol-induced ^{85}Sr uptake in the myocardium. In: M. Nagano, N. Takeda, N.S. Dhalla (eds). The Cardiomyopathic Heart. Raven Press, New York, 1993, pp 395–400

Molecular and Cellular Biochemistry **147**: 89–97, 1995.

Angiotensin converting enzyme inhibitors, left ventricular hypertrophy and fibrosis

Wolfgang Linz, Gabriele Wiemer, Jutta Schaper[1], René Zimmermann[1], Kazushige Nagasawa[1], Peter Gohlke[2], Thomas Unger* and Bernward A. Schölkens

PGU Cardiovascular Agents, Hoechst AG, D-65926 Frankfurt/Main; [1]Max-Planck-Institute, Experimental Cardiology, D-61231 Bad Nauheim, Germany; [2]Christian Albrechts University of Kiel, Department of Pharmacology, D-24105 Kiel, Germany

Abstract

From pharmacological investigations and clinical studies, it is known that angiotensin converting enzyme (ACE) inhibitors exhibit additional local actions, which are not related to hemodynamic changes and which cannot be explained only by interference with the renin angiotensin system (RAS) by means of an inhibition of angiotensin II (ANG II) formation. Since ACE is identical to kininase II, which inactivates the nonapeptide bradykinin (BK) and related kinins, potentiation of kinins might be responsible for these additional effects of ACE inhibitors.

a) In rats made hypertensive by aortic banding, the effect of ramipril in left ventricular hypertrophy (LVH) was investigated. Ramipril in the antihypertensive dose of 1 mg/kg/day for 6 weeks prevented the increase in blood pressure and the development of LVH. The low dose of ramipril (10 μg/kg/day for 6 weeks) had no effect on the increase in blood pressure or on plasma ACE activity but also prevented LVH after aortic banding. The antihypertrophic effect of the higher and lower doses of ramipril, as well as the antihypertensive action of the higher dose of ramipril, was abolished by coadministration of the kinin receptor antagonist icatibant. In the regression study the antihypertrophic actions of ramipril were not blocked by the kinin receptor antagonist. Chronic administration of BK had similar beneficial effects in a prevention study which were abolished by icatibant and N^G-nitro-L-arginine (L-NNA).

In a one year study the high and low dose of ramipril prevented LVH and fibrosis. Ramipril had an early direct effect in hypertensive rats on the mRNA expression for myocardial collagen I and III, unrelated to its blood pressure lowering effect.

b) In spontaneously hypertensive rats (SHR) the preventive effects of chronic treatment with ramipril on myocardial LVH was investigated. SHR were treated in utero and, subsequently, up to 20 weeks of age with a high dose (1 mg/kg/day) or with a low dose (10 μg/kg/day) of ramipril. Animals on a high dose remained normotensive, whereas those on a low dose developed hypertension in parallel to vehicle-treated controls. Left ventricular mass was reduced only in high-dose-treated, but not in low-dose treated animals but both groups revealed an increase in myocardial capillary length density. In SHR stroke prone animals cardiac function and metabolism was improved by ramipril and abolished by coadministration of icatibant.

In contrast to the prevention studies, in a regression study ramipril reduced cardiac hypertrophy also by low dose treatment.

c) In rats chronic nitric oxide (NO) inhibition by N^G-nitro-L-arginine-methyl ester (L-NAME) treatment induced hypertension and LVH. Ramipril protected against blood pressure increase and partially against myocardial hypertrophy.

These experimental findings in different models of LVH characterise ACE inhibitors as remarkable antihypertrophic and antifibrotic substances. (Mol Cell Biochem 147: 89–97, 1995)

Key words: left ventricular hypertrophy, fibrosis, ramipril, autocrine-paracrine actions, ACE inhibitors, bradykinin, prostacyclin

Address for offprints: W. Linz, Hoechst AG, PGU Cardiovascular Agents (H821), D-65926 Frankfurt/Main, Germany

Introduction

Left ventricular hypertrophy (LVH) is regarded – beside high blood pressure as an independent risk factor for cardiovascular diseases especially with respect to the sequels of ischemia, arrhythmias and left ventricular dysfunction [1].

The renin angiotensin system (RAS) has been implicated in the development and maintenance of hypertension and cardiac hypertrophy [2]. Angiotensin converting enzyme (ACE) inhibitors may partly suppress the cardiac hypertrophic response by reducing the formation of angiotensin II (ANG II), which stimulates hypertrophy, matrix protein and collagen synthesis [3, 4].

From new insights into the molecular biology of the RAS we know that ANG II is not only synthesised in the blood stream but also locally in tissues [5–11]. Thus, the traditional endocrine concept has evolved into a concept of autocrine-paracrine functions of the RAS [12]. Consequently ACE inhibitors may exert part of their pharmacological effects via these autocrine-paracrine mechanisms including not only the RAS but also the kallikrein-kinin-system [13, 14].

ACE inhibitors attenuate the formation of ANG II and accumulate kinins by inhibition of their degradation, namely bradykinin (BK). Thereby, they prevent the systemic and local actions of ANG II and potentiate the local and/or cardiovascular and metabolic effects of BK [15, 16]. Especially the effects of kinins had been underestimated for long time.

ACE inhibitors seem to have a more pronounced antihypertrophic effect pointing to the importance of interference with the RAS and the kallikrein kinin system to prevent or regress this target organ damage [17].

Antihypertrophic effect of ACE inhibitors in rats with aortic constriction and pressor overload hypertrophy

In our studies renal hypertensive rats with LVH following aortic constriction between the origin of renal arteries were used. After aortic banding during development of pressure overload hypertrophy the circulating RAS is markedly activated. However, once LVH is established at 6 weeks following aortic constriction, plasma values of renin activity and aldosterone are in the range of sham operated animals. Immediately after aortic banding ACE activity and ACE mRNA levels within the myocardium as well as intracardiac ANG I to ANG II conversion rates were increased [18, 19].

In rats subjected to abdominal aortic constriction and LVH an increase in vulnerability to arrhythmias was found [20]. This is in line with observations in patients with LVH [21].

To investigate the possible involvement of locally formed ANG II by the cardiac RAS and its possible trophic properties [22, 23], long-term administration of an ACE inhibitor was compared with other antihypertensive agents in the prevention and regression of LVH [24]. The effects of equipotent oral antihypertensive doses of the ACE inhibitor ramipril (1 mg/kg/day), the calcium antagonist nifedipine (30 mg/kg/day), and the arterial vasodilator dihydralazine (30 mg/kg/day) on cardiac mass in rats subjected to constriction of the abdominal aorta were compared. Daily oral treatment over 6 weeks was started immediately following acute aortic constriction (prevention experiments) or 6 weeks after aortic banding, when hypertension and cardiac hypertrophy were established (regression experiments). Groups of sham operated animals and untreated animals with aortic banding served as controls. In the regression experiments an additional group received ramipril in a low dose of 10 μg/kg/day.

All three drugs lowered the blood pressure to a similar level with the exception of the low dose of ramipril, which was without effect on high blood pressure. Only the ACE inhibitor induced a significant and complete prevention or regression of cardiac hypertrophy compared to control normotensive rats which were similar to the sham-operated normotensive rats. Surprisingly the low dose of ramipril showed the same complete regression of cardiac hypertrophy as seen with the antihypertensive dose of the ACE inhibitor [24].

A comparable antihypertrophic effect was observed in a recent one year study in rats [25]. The aim of this study was to separate local cardiac effects using a low dose from those effects on systemic blood pressure when using an antihypertensive dose of ramipril. After one year, treatment with both doses the antihypertensive and the low dose which had no effect on blood pressure had prevented LVH. Plasma ACE activity was inhibited in the high but not in the low dose group, whereas the conversion of ANG I–ANG II in isolated aortic segments was suppressed in both treated groups. Plasma catecholamines were increased in the vehicle control group but treatment with either dose of the ACE inhibitor normalised the values. The myocardial phosphocreatine/ATP ratio as an indicator for the energy state of the heart was reduced in the vehicle control group, whereas the hearts from treated animals showed a normal ratio comparable to hearts from sham operated animals.

After one year from each group 7 animals were separated, treatment stopped and housed for additional 6 month (withdrawal experiments). Withdrawal of the treatment did not change left ventricular weight to body ratio in the different groups and in the earlier group with high ACE inhibitor treatment blood pressure did not reach the value of the stenosis vehicle group.

These experiments showed that long-term treatment with an ACE inhibitor effectively prevented cardiac hypertrophy even in the presence of high blood pressure. This protective effect was still present after 6 month treatment, withdrawal. Local ACE inhibition involving decreased ANG II formation, an increased kinin accumulation and an attenuation of sympathetic activities should be considered as factors evoking these long term beneficial cardiac effects of ACE inhibitors.

The dissociation between the effects of ramipril on blood pressure in a high dose and on cardiac mass already in a low dose stresses the role of factors other than blood pressure and afterload in the development of hypertensive cardiac hypertrophy.

From other series of experiments using the same model it was known however, that losartan was more active to regress an already established LVH than to prevent the development of cardiac hypertrophy [26, 27]. Therefore during the development of LVH other factors than ANG II seem to play a role. Since inhibition of ACE besides reducing ANG II formation also increases kinin levels, kinins might contribute via generation of nitric oxide (NO) and prostacyclin (PGI_2) to the prevention of the hypertrophic response.

Patients (n:115) with essential hypertension and LVH [28], after a selection period of 4–6 weeks under antihypertensive therapy with 20 mg furosemide daily, were randomized in a double blind manner to receive either placebo, the subhypotensive dose of 1.25 mg or the antihypertensive dose of 5 mg ramipril daily for 6 months. Treatment with furosemide was continued during this period. Ramipril at both treatment regimens for 6 months induced LVH regression, independent of changes in ambulatory blood pressure in patients under antihypertensive therapy.

Contribution of kinins to the antihypertrophic effect of ramipril

To evaluate the role of BK and related kinins in the antihypertrophic effect of ACE inhibitors the influence of the kinin receptor antagonist icatibant on the effects of ramipril on LVH in rats with aortic banding was investigated [29]. Ramipril in the antihypertensive dose of 1 mg/kg per day p.o. for 6 weeks prevented the increase in blood pressure and the development of LVH. Plasma ACE activity was significantly inhibited. The low dose of ramipril (10 μg/kg/day p.o. for 6 weeks) had no effect on the increase in blood pressure and on plasma ACE activity but also prevented LVH after aortic banding. The antihypertrophic effect of the high and the low dose ramipril as well as the antihypertensive action of the high dose of ramipril were abolished by icatibant. However, when treatment (high and low) was started 6 weeks after aortic constriction (regression experiments) the kinin receptor antagonist was not able to reverse the antihypertrophic effects of the ACE inhibitor.

Furthermore, chronic administration of BK in a dose without effect on blood pressure via osmotic minipumps prevented development of LVH, however did not induce regression of LVH. The preventive effect of BK was abolished by coadministration of icatibant or of the NO synthase inhibitor L-NNA [30].

These data suggest that kinins are involved in the antihypertrophic effects of ACE inhibitors in the developmental phase of LVH in rats with aortic constriction. NO releasing vasodilators and cyclic GMP are known to be antimitogenic and antiproliferative, *in vitro* [31, 32]. Similar effects were found for PGI_2 and cyclic AMP [33]. Both NO and PGI_2 when increased by kinin accumulation following ACE inhibition may contribute to these antihypertrophic effects of ACE inhibitors [34].

From these experimental studies in rats with pressure overload LVH one can assume that kinin accumulation induced by ACE inhibitors may contribute to the antihypertrophic action during the prevention phase, whereas attenuation of ANG II formation by the ACE inhibitor may be more important during the regression period.

Role of ACE inhibition on myocardial fibrosis

ANG II has been demonstrated to act as a growth factor in a variety of tissues including cardiac fibroblasts. Cardiac fibroblasts can mediate ANG II induced cardiac myocyte hypertrophy through a paracrine mechanism stimulating the production of a transferable growth factor or factors in cardiac fibroblasts [35]. ACE inhibition might at least in part via reduced ANG II formation positively interfere with these mitogenic signaling pathways in cardiac fibroblasts.

In line with the values for LVH obtained in the one year study are the observations on the occurrence of myocardial fibrosis which was evaluated by staining the left ventricular tissue for fibronectin. Myocardial fibrosis was not seen in hearts from animals treated with the high as well as the low dose of ramipril, whereas in hearts from vehicle treated rats with aortic banding, myocardial fibrosis occurred.

Myocardial fibrosis did not recur after 6 months withdrawal of ACE inhibitor treatment.

In the same model cardiac mRNA levels of the collagens α_1(I) [col I] and α_1(III) [col III] was isolated from control and aortic banded-hypertensive rats as well as from rats treated with either the high or the low dose ramipril after 2 and 6 weeks aortic constriction. Banded hypertensive rats with vehicle treatment showed increased cardiac col mRNA levels. Low dose ramipril treatment led to normal col mRNA levels but blood pressure was still elevated, whereas high dose ramipril treatment reduced col and blood pressure below control values. At 6 weeks blood pressure and col were at control levels in animals with high dose ramipril treatment. These results indicate that ramipril has a direct and early effect in hypertensive rats on the mRNA expression of col I and col III, unrelated to its blood pressure lowering effect. Influences of transcriptional control of collagen gene expression by ramipril are thus able to reduce an increased left ventricular weight to body weight ratio and, possibly hypertrophy [36].

Determination of the fibronectin content in ACE inhibitor treated animals showed in the prevention study lower values than in the regression study. Fibrosis was seen until 12 weeks after aortic constriction (regression study).

Beyond 4 weeks of renovascular hypertension, an accumulation of fibrillar collagen is seen within the adventitia of intramyocardial coronary arteries. From their perivascular location fibrillar collagen begins to radiate outward into neighboring intramuscular spaces [37–39]. Eight weeks later, a progressive perivascular and interstitial fibrosis has developed [38, 40]. After 12 weeks, foci replacement fibrosis secondary to myocyte necrosis, appear predominantly within the endomyocardium of the rat [41, 42]. At 20 weeks of renovascular hypertension, the complete pattern of diffuse interstitial and perivascular fibrosis has developed. The reparative fibrosis becomes more pronounced at 32 weeks or more, where overall collagen volume fraction accounts of the myocardial structural space [41] as shown in the one year study.

In our regression study the fibronectin values in both ramipril treated groups (low and high) were comparable to values found in sham operated animals (~ 23%) whereas in banded vehicle treated rats the values ranged about 43%. Cotreatment with icatibant did not abolish this antifibrotic effect of ramipril. These observations would imply that endothelium derived kinins increased by ACE-inhibition in this phase might act more on myocytes than on fibroblasts in this model of renovascular hypertension. Thus, when one considers the presence or absence of the remodeling of the interstitial space, hypertrophy is a heterogenous process and myocyte and nonmyocyte compartments appear to have independent regulatory controls. Local concentrations of stimulators (e.g. ANG II, endothelin, aldosteron, norepinephrine) and inhibitors (e.g. kinins, PGI_2, NO, atrial natriuretic peptide) may regulate fibroblast collagen turnover and the healing response [43].

Effects of ACE inhibitors on cardiac and vascular hypertrophy in Spontaneously Hypertensive Rats (SHR)

Earlier chronic studies in SHR had shown that oral administration of ramipril in doses of 0.1, 1, and 10 mg/kg/day resulted in a dose-dependent antihypertensive effect, with a threshold antihypertensive dose of 0.1 mg/kg/day [44, 45]. Measurements of ACE activity in homogenates of hearts from normotensive rats pretreated with single doses of 1, 10 and 100 µg/kg of ramipril demonstrated a long-lasting inhibition of ACE activity at all doses [46].

SHR and stroke prone SHR (SHRSP), animals with genetic hypertension associated with normal to low plasma renin levels, were treated with different ACE inhibitors at antihypertensive high doses (1 mg/kg per day) and low doses (0.01 mg/kg per day). Prevention studies were begun before hypertension developed (prenatally) and were continued for 20 weeks. The effects of chronic ACE inhibitor treatment on myocardial LV weight and on capillary length density as well as on structural alterations in mesenteric arteries were investigated [47–49].

Early-onset treatment with high doses of the ACE inhibitors ramipril and zabicipril prevented or attenuated the development of hypertension and prevented the development of cardiac LVH. These effects were not altered by chronic kinin receptor blockade with icatibant demonstrating that kinins do not contribute to the antihypertensive and antihypertrophic actions in genetically hypertensive rats.

The development of LVH is associated with a diminished capillary density leading to relative ischemia. Therefore, the effect of chronic ACE inhibitor treatment on cardiac capillary length density was determined. The results revealed an increase in the length of capillaries per volume of the left ventricle in animals treated with an antihypertensive dose (high dose) of the ACE inhibitors indicating an improved oxygen supply of the heart [47, 50].

In addition, high dose treatment with ramipril and zabicipril affected the development of vascular structural alterations. This effect was demonstrated by the decrease in the number of smooth muscle cell layers in the vascular media and the media to lumen and wall to lumen ratios of mesenteric arteries [50, 51].

In contrast to the antihypertensive dose, cardiac hypertrophy as well as vascular structural alterations were not affected by chronic early-onset treatment with low doses of the ACE inhibitors ramipril, zabicipril and perindopril. Therefore, in genetically hypertensive animals the effects of the ACE inhibitors on the development of cardiac and vascular hypertrophy appear to be related to their antihypertensive actions [47–51]. On the other hand, low-dose ACE inhibitor treatment like high-dose treatment improved myocardial capillary length density [48]. This suggests that capillary proliferation is independent of blood pressure and of structural alterations in the myocardium. The underlying mechanism for the ACE inhibitor induced myocardial capillary growth is not known. One possible explanation resides in the kinin potentiating effect of the ACE inhibitor. BK has been shown to improve myocardial blood flow, even at very low concentrations [52]. An enhanced myocardial blood flow on the other hand appears to be the common denominator of all experimental conditions associated with myocardial capillary proliferation [53, 54]. In addition, long term ACE inhibitor treatment of SHRSP improved cardiac function, increased coronary flow and myocardial tissue concentrations of glycogen and the energy-rich phosphates ATP and creatine phosphate as will be outlined below [49]. These effects could be prevented by chronic kinin receptor blockade with icatibant. In addition, these effects are comparable with the known cardiac metabolic effects of BK to enhance myocardial glucose uptake in normoxic isolated rat hearts [55]. Interestingly, in the aging mouse ACE inhibition was found to decrease renal and myocardial sclerosis and to increase the number of mitochondria in heart and liver cells, which was associated with a significant increase in survival [56]. Further studies comparing specific ANG II and BK receptor antagonists will particularly have to address the possible effect of an ACE inhibitor-induced kinin potentiation on myocardial capillary growth and mitochondrial density in more detail.

The observations, that low-dose ACE inhibitor treatment did not affect the development of LVH in SHR and SHRSP is at variance with the results reported in the coarctation model of renal hypertension mentioned above (Fig. 1). The discrepancy between these studies could be explained by the fact that the coarctation model represents a highly renin-dependent model of experimental hypertension, which may respond to ACE inhibition, more marked than the SHR and SHRSP, models with normal to low plasma renin.

In a regression study adult 16 week old SHR with established hypertension and cardiac and vascular hypertrophy were treated for 16 weeks with the ACE inhibitors ramipril and zabicipril at doses of 1 mg/kg per day and 0.01 mg/kg per day. Treatment with the high dose of both drugs normalized blood pressure and reduced cardiac hypertrophy, but had no effect on morphometric parameters in the mesenteric arteries [47, 50]. Thus, mesenteric vascular hypertrophy could only be prevented by early-onset high-dose treatment with ACE inhibitors but not once hypertrophy has been established. In contrast, cardiac hypertrophy was significantly reduced by low dose treatment with ramipril (Fig. 1), but not with zabicipril. It should be noted that the hypertension-induced increase in vascular mass of SHR mensenteric arteries appear to be mainly due to hyperplasia, that is an increase in the number of cells. On the other hand, the increase in cardiac mass is mainly a result of an increase in cell size (hypertrophy). Most likely, a regression of an increased number of cells is more difficult to achieve any antihypertensive treatment than a regression of an increased cell size.

These results demonstrate, that in SHR early-onset treatment with ramipril can induce myocardial capillary growth, even at doses too low to antagonise the development of hypertension or LVH. This ability of ramipril to induce capillary growth might be of great importance for induction of coronary collateral vessels in humans with coronary artery disease and heart failure [57].

Meanwhile other investigators found similar beneficial

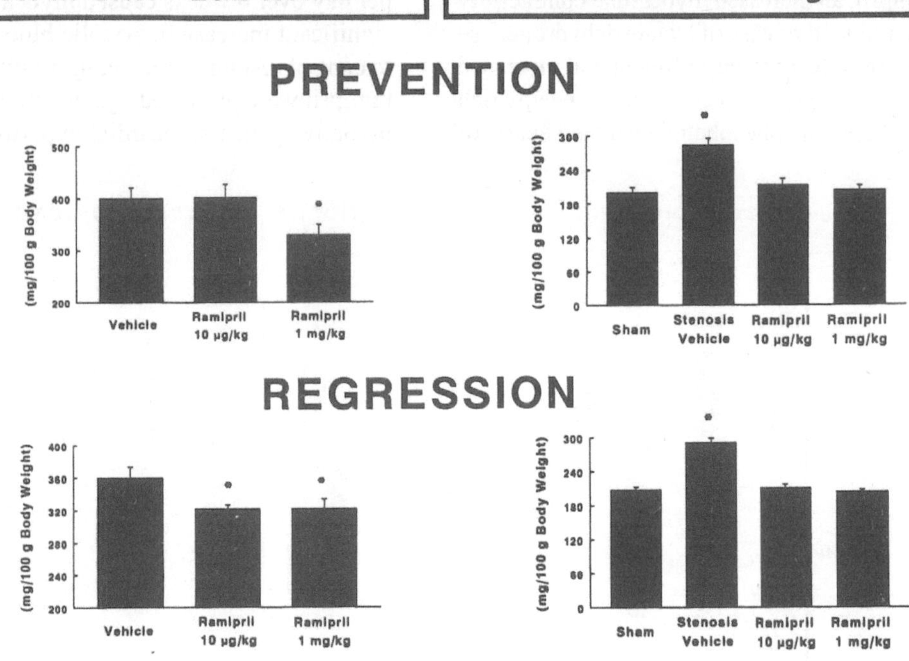

Fig. 1. Effect of long term oral treatment with ramipril, high (1 mg/kg/day) and low (10 µg/kg/day) dose, on left ventricular hypertrophy (LVH) in rats. Left hand side: Spontaneously hypertensive rats (SHR) with low plasma renin levels. In a prevention study rats were treated *in utero* and continued for 20 weeks. In a regression study adult 16 week old SHR with established hypertension and cardiac hypertrophy were treated for 16 weeks with the ACE inhibitor. Right hand side: In a prevention study rats with aortic constriction and high plasma renin levels were treated immediately after operation for 6 weeks. In a regression study treatment started 6 weeks – after aortic constriction after LVH hypertrophy has been established – for 6 weeks. *$p < 0.05$ vs. vehicle and sham respectively.

effects on LVH and/or fibrosis by ACE inhibition without blood pressure reduction [58–61], whereas others could not observe beneficial effects on LVH by low dose ACE inhibitor treatment [62].

Cardiac and vascular function in SHR

Low- and high-dose treatment with ramipril inhibited vascular ACE activity ex vivo demonstrated by the inhibition of aortic vasoconstrictor responses to ANG I but not to ANG II. Early onset treatment with high-dose ramipril increased aortic vasodilatory responses to acetylcholine and decreased vasoconstrictor responses to noradrenaline. Treatment of adult SHR for 16 weeks with high-dose ramipril (regression study) had similar effects on vascular function, but did not affect vascular hypertrophy (see above). Low-dose ramipril, although having no effect on blood pressure, significantly decreased the aortic vasoconstrictor responses to noradrenaline in both the prevention and the regression study. This regimen further increased the vasodilatory responses to acetycholine in the regression study and to a more limited extent in the prevention study. Low and high dose ACE inhibitor treatment resulted in a significant increase in aortic cyclic GMP by 98 and 160% respectively [63].

In a more recent study in stroke prone SHR by long term treatment with ramipril, an increased myocardial contractility and coronary flow, reduced release of lactate dehydrogenase and creatine kinase into the coronary effluent and increased myocardial tissue levels of glycogen as well as the energy rich phosphates ATP and creatine phosphate in isolated hearts of these animals were observed [49]. These changes in cardiodynamics and cardiac metabolism were observed even at the low dose of ramipril which did not affect blood pressure and LVH. The beneficial changes could be prevented by chronic kinin receptor blockade with icatibant (Fig. 2). Thus, the observed cardiac effects of the ACE inhibitor were independent of blood pressure reduction and due to its kinin potentiating action.

Chronic NO synthase inhibition in rats

Endothelium-derived NO is an important modulator of vascular tone [64], and inhibition of its generation may be achieved using arginine analogues such as N^G-nitro-L-arginine-methyl ester (L-NAME) [65]. In the rat, acute administration of L-NAME is associated with a dose-dependent increase in arterial pressure and total vascular resistance [66, 67]. Recently, it has been reported that long-term inhibition of NO synthase will produce a sustained hypertension in otherwise normotensive rats and dogs [68–71], thus providing a new experimental model of hypertension and hypertrophy.

To evaluate the cardiac effects of ramipril on L-NAME-induced hypertension in rats we focused our interest on myocardial hypertrophy, dynamics and metabolism.

Chronic treatment with L-NAME in a dose of 25 mg/kg per day over 6 weeks caused myocardial hypertrophy and a significant increase in systolic blood pressure as compared to controls. Animals receiving simultaneously L-NAME and ramipril were protected against blood pressure increase and partially against myocardial hypertrophy [72] (Fig. 3).

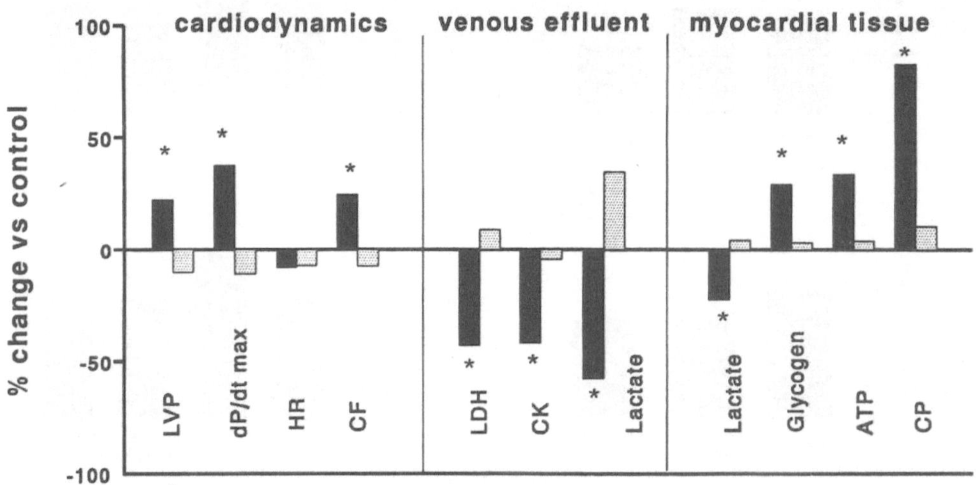

Fig. 2. Effect of chronic oral treatment, prenatally and subsequently up to the age of 20 weeks, with low dose ramipril (10 µg/kg/day) alone (black bars) and after cotreatment with the kinin receptor antagonist icatibant (500 µg/kg/day s.c.) (dotted bars) on myocardial function and metabolism in isolated perfused hearts from stroke prone spontaneously hypertensive rats (SHRSP). LVP indicates left ventricular pressure: dP/dt_{max}, differentiated left ventricular pressure; HR, heart rate; CF, coronary flow; LDH, lactate dehydrogenase; CK, creatine kinase; and CP, creatine phosphate. *$p < 0.05$ compared with vehicle control group.

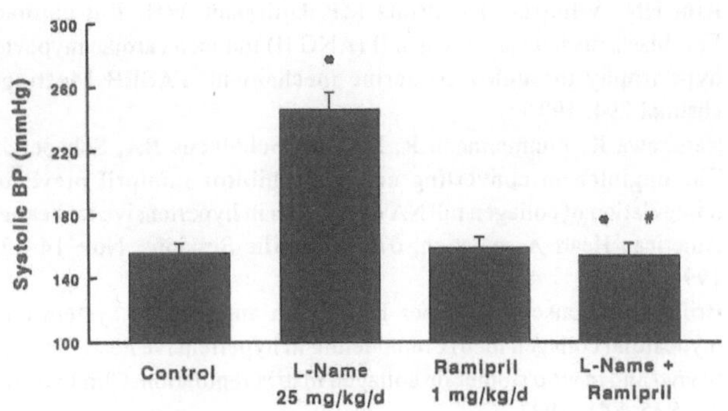

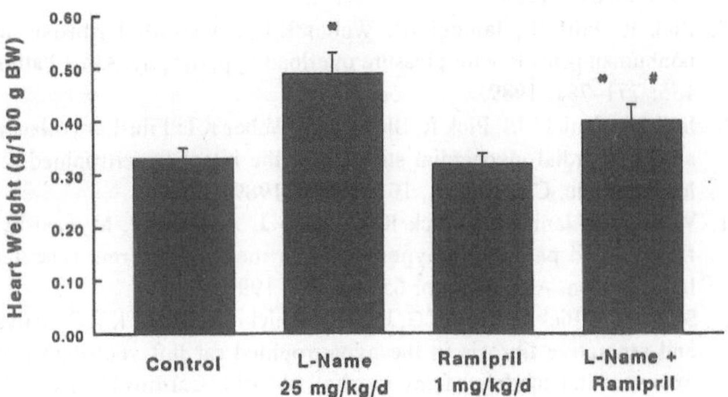

Fig. 3. Effect of oral L-NAME (25 mg/kg/day) and ramipril (1 mg/kg/day) treatment over 6 weeks as well as the combination of both on systolic blood pressure and heart weight in Wistar rats. *p < 0.05 vs. control, #p < 0.05 vs. L-NAME group.

Isolated hearts from these rats treated with L-NAME showed increased post-ischemic reperfusion injuries. Compared to controls duration and incidence of ventricular fibrillation was increased and coronary flow reduced. During ischemia the cytosolic enzymes lactate dehydrogenase and creatine kinase, as well as lactate in the venous effluent were increased. Myocardial tissue values of glycogen, ATP, and creatine phosphate were decreased, whereas lactate content was increased. Coadministration of ramipril reversed these effects.

Due to suppression of the modulating influence of NO by L-NAME, vasoconstrictor effects of ANG II may prevail. On the other hand, NO and PGI_2, when increased by inhibiting breakdown of BK and related kinins after ACE inhibition may contribute to the beneficial cardioprotective effects [73].

Conclusion

The cardiovascular actions of ACE inhibitors are not only mediated by reduction of ANG II but also by the inhibition of the degradation of endogenous BK and related kinins. This is evidenced by the comparable effects of ACE inhibitors and exogenously added BK in different physiological and pathophysiological situations and by the observation that the kinin receptor antagonist icatibant blocked the cardiovascular effects of ACE inhibitors as well as of BK in experimental models of LVH. The increase in local kinin concentrations by ACE inhibition exerts protective effects activating signal transduction pathways which generate second messengers such as cyclic GMP via an increase in NO or cyclic AMP via an increase in PGI_2 [73].

References

1. Messerli FH, Ketelhut R: Left ventricular hypertrophy: an independent risk factor. J Cardiovasc Pharmacol 17 (suppl 4): S59–S67, 1991
2. Schelling P, Fischer H, Ganten D: Angiotensin and cell growth: a link to cardiovascular hypertrophy? J Hypertension 9: 3–15, 1991
3. Giacomelli F, Anversa P, Wiener J: Effect of angiotensin-induced hypertension on rat coronary arteries and myocardium. Am J Pathol 84: 111–125, 1976
4. Kato H, Suzuki H, Tajima S, Ogata Y, Tominaga T, Sato A, Saruta T: Angiotensin II stimulates collagen synthesis in cultured vascular smooth muscle cells. J Hypertension 9: 7–22, 1991
5. Dzau VJ, Re RN: Evidence for the existence of renin in the heart. Circulation 75 (suppl I): I-34–I-36, 1987
6. Griendling KK, Murphy TJ, Wayne R, Alexander RW: Molecular biology of the renin-angiotensin system. Circulation 87 (6): 1816–1828, 1993
7. Kifor I, Dzau VJ: Endothelial renin-angiotensin pathway: Evidence for intracellular synthesis and secretion of angiotensins. Circ Res 60: 422–428, 1987
8. Lindpaintner K, Jin M, Wilhelm MJ, Suzuki F, Linz W, Schölkens BA, Ganten D: Intracardiac generation of angiotensin and its physiologic role. Circulation 77 (suppl I): 1–18, 1988
9. Linz W, Schölkens BA, Lindpaintner K, Ganten D: Cardiac renin-angiotensin system. Am J Hypertension 2: 307–310, 1989
10. Schunkert H, Dzau VJ, Tang SS, Hirsch AT, Apstein CS, Lorell BH: Increased rat cardiac angiotensin converting enzyme activity and mRNA expression in pressure overload left ventricular hypertrophy. J Clin Invest 36: 1913–1920, 1990
11. Yamada H, Fabris B, Allen AM, Jackson B, Johnston CI, Mendelsohn AO: Localization of angiotensin converting enzyme in rat heart. Circ Res 68: 141–149, 1991
12. Unger T, Gohlke P, Paul M, Rettig R: Tissue renin-angiotensin systems: fact or fiction? J Cardiovasc Pharmacol 18 (Suppl. 2): S20–S25, 1991
13. Unger T, Gohlke P, Gruber MG: Converting enzyme inhibitors. In: D. Ganten and P.J. Mulrow (eds). Handbook of Experimental Pharmacology, vol. 93. Springer Verlag Berlin Heidelberg, 1990
14. Unger T, Gohlke P: Converting enzyme inhibitors in cardiovascular therapy: current status and future potential. Cardiovasc Res 28: 146–158, 1994
15. Kramer HJ, Glänzer K, Meyer-Lehnert H, Mohaupt M, Predel HG: Kinin- and non-kinin-mediated interactions of converting-enzyme inhibitors with vasoactive hormones. J Cardiovasc Pharmacol 15 (Suppl 6): S91–S98, 1990
16. Scherf H, Pietsch R, Landsberg G, Kramer HJ, Düsing R: Converting-enzyme inhibitor ramipril stimulates prostacyclin synthesis by isolated rat aorta: Evidence for a kinin-dependent mechanism. Klin

Wochenschr 64: 742–745, 1986

17. Dahlöf B, Pennert K, Hansson L: Reversal of left ventricular hypertrophy in hypertensive patients. Am J Hypertens 5: 95–110, 1992
18. Schunkert H, Jackson B, Tang SS, Schoen FJ, Smits JFM, Apstein CS, Lorell B: Distribution and functional significance of cardiac angiotensin converting enzyme in hypertrophied rat hearts. Circulation 87: 1328–1339, 1993
19. Schunkert H, Tang SS, Litwin SE, Diamant D, Riegger G, Dzau VJ, Ingelfinger J: Regulation of intrarenal and circulating renin-angiotensin systems in severe heart failure in the rat. Cardiovasc Res 27: 731–735, 1993
20. Shirmada Y, Hearse DJ, Avkiran M: ACE inhibition and hypertrophy: short-term treatment with ramipril in the rat reduces vulnerability to arrhythmias without regression of hypertrophy. Circulation 90 (4/2), 1994 abstract
21. Gonzalez-Fernandez RA, Rivera M, Rodriguez PJ, Fernandez-Martinez J, Soltero LH, Diaz LM, Lugo JE: Prelevance of ectopic ventricular activity after left ventricular regression. Am J Hypertens 6: 308–313, 1993
22. Katz AM: Angiotensin II: hemodynamic regulator or growth factor? J Mol Cell Cardiol 22: 739–747, 1990
23. Re RN: The biology of angiotensin: paracrine, autocrine and intracrine actions in cardiovascular tissues. J Mol Cell Cardiol 21 (Suppl 5): 63–69, 1989
24. Linz W, Schölkens BA, Ganten D: Converting enzyme inhibition specifically prevents the development and induces regression of cardiac hypertrophy in rats. Clin and Exper Hypertens A11(7): 1325–1350, 1989
25. Linz W, Schaper J, Wiemer G, Albus U, Schölkens BA: Ramipril prevents left ventricular hypertrophy with myocardial fibrosis without blood pressure reduction: A one year study in rats. Br J Pharmacol 107: 970–975, 1992
26. Linz W, Henning R, Schölkens BA: Role of angiotensin II receptor antagonism and converting enzyme inhibition in the progression and regression of cardiac hypertrophy in rats. J Hypertension 9 (Suppl 6): S400–S401, 1991
27. Mohabir R, Young SD, Strosberg AM: Role of angiotensin in pressure overload-induced hypertrophy in rats: effects of angiotensin-converting enzyme inhibitors, an AT_1 receptor antagonist, and surgical reversal. J Cardiovasc Pharmacol 23: 291–299, 1994
28. Lièvre M, Guéret P, Gayet C, Roudaut R, Haugh MC, Delair S, Boissel J-P, on behalf of the HYCAR Study Group: Ramipril-induced regression of left ventricular hypertrophy in treated hypertensive individuals. Hypertension 25: 92–97, 1995
29. Linz W, Schölkens BA: A specific B_2 bradykinin receptor antagonist HOE 140 abolishes the antihyportrophic effect of ramipril. Br J Pharmacol 105: 771–772, 1992
30. Linz W, Wiemer G, Schölkens BA: Bradykinin prevents left ventricular hypertrophy in rats. J Hypertension 11 (Suppl 5): S96–S97, 1993
31. Garg UC, Hassid A: Nitric oxide-generating vasodilators and 8-bromo-cyclic guanosine monophosphate inhibit mitogenesis and proliferation of cultured rat vascular smooth muscle cells. J Clin Invest 83: 1774–1777, 1989
32. Thiemermann C: Biosynthesis and interaction of endothelium-derived vasoactive mediators. Eicosanoids 4: 187–202, 1991
33. Shirotani P, Yui Y, Hattori R, Kawai C: U-61, 431F, a stable prostacyclin analogue, inhibits the proliferation of bovine vascular smooth muscle cells with little antiproliferative effect on endothelial cells. Prostaglandins 41: 97–110, 1991
34. Wiemer G, Schölkens BA, Linz W: Endothelial protection by converting enzyme inhibitors. Cardiovasc Res 28: 166–172, 1994
35. Kim NN, Villarreal FJ, Printz MP, Dillmann WH: Rat cardiac fibroblasts mediate angiotensin II (ANG II) induced cardiac myocyte hypertrophy through a paracrine mechanism. FASEB Meeting, abstract 294, 1994
36. Nagasawa K, Zimmermann R, Linz W, Schölkens BA, Schaper J: The angiotensin converting enzyme inhibitor ramipril prevents upregulation of collagen mRNA expression in hypertensive rat hearts. American Heart Association, 67th Scientific Sessions, Nov. 14–17, 1994
37. Brilla CG, Maisch B, Weber KT: Renin-angiotensin system and myocardial collagen matrix remodeling in hypertensive heart disease: *in vivo* and *in vitro* studies on collagen matrix regulation. Clin Investig 71: S35–S41, 1993
38. Doering CW, Jalil JE, Janicki JS, Pick R, Aghili S, Abrahams C, Weber KT: Collagen network remodeling and diastolic stiffness of the rat left ventricle with pressure overload hypertrophy. Cardiovasc Res 22: 686–695, 1988
39. Pick R, Jalil JE, Janicki JS, Weber KT: Myocardial fibrosis in nonhuman primate with pressure overload hypertrophy. Am J Pathol 135: 771–781, 1989
40. Jalil JE, Janicki JS, Pick R, Shroff SG, Weber KT: Fibrillar collagen and myocardial myocardial stiffness in the intact hypertrophied rat left ventricle. Circ Res 64: 1041–1050, 1989
41. Weber KT, Janicki JS, Pick R, Capasso J, Anversa P: Myocardial fibrosis and pathologic hypertrophy in the rat with renovascular hypertension. Am J Cardiol 65: 1G–7G, 1990
42. Silver MA, Pick R, Brilla CG, Jalil JE, Janicki JS, Weber KT: Reactive and reparative fibrosis in the hypertrophied rat left ventricle: two experimental models of myocardial fibrosis. Cardiovasc Res 24: 741–747, 1990
43. Weber KT, Sun Y, Tyagi SC, Cleutjens JPM: Collagen network of the myocardium: function, structural remodeling and regulatory mechanisms. J Mol Cell Cardiol 26: 279–292, 1994
44. Unger T, Fleck T, Ganten D, Lang RE, Rettig R: 2-[N-[(S)]-1-ethoxycarbonyl-3-phenylpropyl-L-alanyl]-(1S.3S.5S)-2-azabicyclo [3.3.0]octane-3-carboxylic acid (HOE 498): antihypertensive action and persistent inhibition of tissue converting enzyme activity in spontaneously hypertensive rats. Drug Research 34: 1426–1430, 1984
45. Unger T, Ganten D, Lang RF, Schölkens BA: Is tissue converting enzyme inhibition a determinant of the antihypertensive efficacy of converting enzyme inhibitors? Studies with the two different compounds HOE 498 and MK 421 in spontaneously hypertensive rats. J Cardiovasc Pharmacol 6: 872, 1984
46. Becker RHA, Linz W, Schölkens BA: Pharmacological interference with the cardiac renin-angiotensin system. J Cardiovasc Pharmacol 14 (suppl 4): S10–S15, 1989
47. Gohlke P, Stoll M, Lamberty V, Mattfeldt T, Mall G, van Even P, Martorana PA, Unger T: Cardiac and vascular effects of chronic angiotensin converting enzyme inhibition at subantihypertensive doses. J Hypertension 10 (Suppl 6): S141–S144, 1992
48. Unger T, Mattfeld T, Lamberty V, Bock P, Mall G, Linz W Schölkens BA, Gohlke P: Effect of early onset ACE inhibition on myocardial capillaries in SHR. Hypertension 20: 478–482, 1992
49. Gohlke P, Linz W, Schölkens BA, Kuwer I, Bartenbach S, Schnell A, Unger T: Angiotensin converting enzyme inhibition improves cardiac function. Role of bradykinin. Hypertension 23: 411–418, 1994
50. Gohlke P, Linz W, Schölkens BA, Wiemer G, Martorana P, van Even P, Unger T: Effect of chronic high- and low-dose ACE inhibitor treatment on cardiac and vascular hypertrophy and vascular function in spontaneously hypertensive rats. Exp Nephrol 2: 93, 1994
51. Gohlke P, Lamberty V, Kuwer I, Bartenbach S, Schnell A, Unger T: Vascular remodelling in systemic hypertension. Am J Cardiol 71: 2E–7E, 1993

52. Linz W, Martorana PA, Schölkens BA: Local inhibition of bradykinin degradation in ischemic hearts. J Cardiovasc Pharmacol 15 (Suppl 6): S99–S109, 1990
53. Mall G, Zimmer G, Baden S, Mattfeld T: Capillary neoformation in the rat heart stereological studies on papillary muscles in hypertrophy and physiologic growth. Basic Res Cardiol 85: 531–540, 1990
54. Odori T, Paskins-Hurlburt A, Hollengberg N: Increase in collateral endothelial cell proliferation induced by captopril after renal artery stenosis in the rat. Hypertension 5: 307–311, 1993
55. Rösen P, Eckel J, Reinauer H: Influence of bradykinin on glucose uptake and metabolism studied in isolated cardiac myocytes and isolated perfused rat hearts. Hoppe-Seyler's Z Physiol Chem 364: 431–438, 1983
56. Ferder L, Inserra F, Romano L, Ercole L, Pszenny V: Effects of angiotensin-converting enzyme inhibition on mitochondrial number in the aging mouse. Am J Physiol 265: C15–C18, 1993
57. Kass RW, Kotler MN, Yazdanfar S: Stimulation of coronary collateral growth: current developments in angiogenesis and future clinical applications. Am Heart J 123(2): 486–496, 1992
58. Baker KM, Chernin MI, Wixson SK, Aceto JF: Renin-angiotensin system involvement in pressure-overload cardiac hypertrophy in rats. Am J Physiol 259: H324–H332, 1990
59. Kromer EP, Riegger AJ: Effects of long-term angiotensin converting enzyme inhibition on myocardial hypertrophy in experimental aortic stenosis in the rat. Am J Cardiol 62: 161–163, 1988
60. Nakamura F, Nagano M, Higaki J, Higashimori K, Morishita R, Mikami H, Ogihara T: The antiotensin-converting enzyme inhibitor, perindopril, prevents cardiac hypertrophy in low-renin hypertensive rats. Clin Exp Pharmacol Physiol 20: 135–140, 1993
61. Matsubara BB, Matsubara LS, Janicki JS: Low dose angiotensin-converting enzyme inhibitor (ACEI) prevents myocardial fibrosis but not hypertrophy in young rats with renovascular hypertension. FASEB Meeting, 1994
62. Rhaleb N-E, Yang X-P, Scicli AG, Carretero OA: Increase in blood pressure induced by aortic coarctation is a primary determinant of left ventricular hypertrophy in rats. Hypertension 22: 440, 1993 abstract
63. Gohlke P, Lamberty V, Kuwer I, Bartenbach S, Schnell A, Linz W, Schölkens BA, Wiemer G, Unger T: Long-term low-dose angiotensin converting enzyme inhibitor treatment increases vascular cyclic Guanosine 3′,5′-monophosphate. Hypertension 22: 682–687, 1993
64. Palmer RM, Ferrige AG, Moncada S: Nitric oxide release accounts for the biological activity of endothelium-derived relaxing factor. Nature 327: 524–526, 1987
65. Rees DD, Palmer RM, Schulz R, Hodson HF, Moncada S: Characterization of three inhibitors of endothelial nitric oxide synthase *in vitro* and *in vivo*. Br J Pharmacol 101: 746–752, 1990
66. Gardiner SM, Compton AM, Kemp PA, Bennett T: Regional and cardiac hemodynamic effects of NG-nitro-L-arginine methyl ester in conscious, Long Evans rats. Br J Pharmacol 101: 625–631, 1990
67. Wang YX, Gavras I, Wierzba T, Lammek B, Gavras H: Inhibition of nitric oxide, bradykinin and prostaglandins in normal rats. Hypertension 19 (Suppl II): II-225–II-261, 1992
68. Arnal JF, Warin L, Michel JB: Determinants of aortic cyclic guanosine monophosphate in hypertension induced by chronic inhibition of nitric oxyde synthase. J Clin Invest 90: 647–652, 1992
69. Baylis C, Mitruka B, Deng A: Chronic blockade of nitric oxide synthesis in the rat produces systemic hypertension and glomerular damage. J Clin Invest 90: 278–281, 1992
70. Ribeiro MO, Antunes E, de Nucci G, Lovisolo SM, Zatz R: Chronic inhibition of nitric oxide synthesis. A new model of arterial hypertension. Hypertension 20: 298–303, 1992
71. Salazar FJ, Pinilla JM, Lopez F, Romero JC, Quesade T: Renal effects of prolonged synthesis inhibition of endothelium-derived nitric oxide. Hypertension 20: 113–117, 1992
72. Hropot M, Grötsch H, Klaus E, Langer KH, Linz W, Wiemer G, Schölkens BA: Ramipril prevents the detrimental sequels of chronic NO synthase inhibition in rats: hypertension, cardiac hypertrophy and renal insufficiency. Naunyn Schmiedeberg's Arch 350: 646–652, 1994
73. Linz W, Wiemer G, Schölkens BA: ACE inhibition induces NO-formation in cultured bovine endothelial cells and protects isolated ischemic rat hearts. J Mol Cell Cardiol 24: 909–919, 1992

Molecular and Cellular Biochemistry **147**: 99–103, 1995.

Influence of global ischemia on the sarcolemmal ATPases in the rat heart

Norbert Vrbjar, Andrej Džurba and Attila Ziegelhöffer

Institute for Heart Research, Department of Biochemistry, Slovak Academy of Sciences, Dubravska cesta 9, 842 33 Bratislava, Slovak Republic

Abstract

To elucidate the effect of global ischemia on the energy utilizing processes, regarding the molecular principles, the kinetic and thermodynamic properties of the sarcolemmal ATPases were investigated in the rat heart. The activation energy for hydrolysis of ATP during ischemia was higher when the reaction was catalyzed by Ca-ATPase or Mg-ATPase. For the Na,K-ATPase reaction, no changes in the activation energy were observed. With respect to the enzyme kinetics, ischemia in a time-dependent manner induced important alterations in K_M and V_{max} values of Na,K-ATPase, Ca-ATPase and Mg-ATPase. The V_{max} value decreased significantly already after 15 min of ischemia, and it also remained low after 30, 45 and 60 min for all 3 enzymes. The significant diminution of K_M values occurred later in the 30th min for Ca-ATPase, in the 45th min for Na,K-ATPase. The observed drop in K_M indicates the increase in the affinity of the enzymes to substrate, suggesting thus the adaptation to ischemic conditions on the molecular level. This effect could be attributed to some conformational changes of the protein molecule in the vicinity of the ATP-binding site developing after longer duration of ischemia. (Mol Cell Biochem 147: 99–103, 1995)

Key words: cardiac sarcolemma, (Na,K)-ATPase, Mg-ATPase, Ca-ATPase, enzyme kinetics, activation energy, ischemia

Introduction

The plasma membrane of cells performs many functions, including the generation and maintenance of ion gradients, the transport of metabolites, and transduction of hormonal signals. More of these activities in cardiac sarcolemma (SL) require energy supply. One of the common energy sources is ATP which is utilized for maintaining the homeostasis of ions by sarcolemmal ATPases like the (Na,K)-ATPase, (Ca,Mg)-ATPase, Ca-ATPase with low affinity to calcium. Profound alterations of energy metabolism resulting in progressive reduction in ATP are induced by ischemia [1]. Investigation of the myocardial ATP content as a function of duration of total ischemia revealed that most critical are the first 60 minutes [2, 3]. In acute experiments using the model of global ischemia the lower ATP content in cardiac tissue induced diminution of the (Na,K)-ATPase activity [4]. To clarify the molecular basis of the inhibition induced by ischemia the present paper deals with investigation of the ATP-binding site by enzyme kinetics and the energy barrier of ATPase reaction in ischemic myocardium for the sarcolemmal (Na,K)-ATPase, Ca-ATPase with low affinity to calcium and the Mg-ATPase.

Material and methods

Quickly excised hearts from male rats (200–250 g) were incubated at 37°C for variable periods of global ischemia. Cardiac sarcolemma was prepared by the hypotonic shock-NaI treatment method as described previously [5]. The protein content was assayed according to [6] using bovine serum albumin as a standard.

Kinetic parameters of all three ATPases were estimated measuring the splitting of ATP by 30–50 μg sarcolemmal proteins at 37°C in the presence of increasing concentrations of ATP in the range of 0.08–6.0 mmol/l in a total volume of 0.5 ml of medium containing 50 mmol/l Imidazole (pH 7.4).

Address for offprints: N. Vrbjar, Institute for Heart Research, Department of Biochemistry, Slovak Academy of Sciences, Dubravska cesta 9, 842 33 Bratislava, Slovak Republic

For the respective ATPases specific ionic cofactors were present in the incubation medium. For the (Na,K)-ATPase 4 mmol/l $MgCl_2$, 10 mmol/l KCl and 100 mmol/l NaCl; for Mg-ATPase 4 mmol/l $MgCl_2$ and for the Ca-ATPase with low affinity to calcium 4 mmol/l $CaCl_2$ were used. Following 10 min of preincubation in the substrate free medium the reaction was started by addition of ATP and after 20 min was terminated by 1 ml of 12% solution of trichloroacetic acid. The inorganic phosphorus liberated was determined according to [7]. ATP hydrolysis that occurred in the presence of Mg only 'absence of Na and K' was subtracted from the activity measured in the presence of all three cofactors Mg^{2+}, Na^+ and K^+ in order to calculate the (Na,K)-ATPase activity. The activities of Mg-ATPase and Ca-ATPase were calculated as a difference of hydrolysis in the presence and absence of respective ionic cofactor.

The activation energy of ATPase reaction was estimated for all three enzymes analyzing the temperature dependence of ATP splitting in the range of 1–45°C at constant concentration of ATP (4 mmol/l).

All results were expressed as means±SEM. The significance of differences between the individual groups was determined with the use of the unpaired Student's *t*-test.

Results

Ischemia in all investigated cases from 15–60 min induced for all three studied enzymes a significant decrease of their activity as it is shown by significant diminution of maximum velocities (Vm) of ATP hydrolysis in ischemic hearts (Fig. 1). The most representative change was observed in the first 15 min of ischemia with no further change of Vm till the 60 min. Only for the Mg-ATPase an additional diminution of Vm value was observed after 60 min (Fig. 1). Statistical evaluation of this additional decrease revealed that there is no significant difference between the Vm after 45 or 60 min of ischemia but the change is statistically significant ($p < 0.05$) when comparing the value in 60th min to values in 15th or 30th min of ischemia.

Concerning the Km value the enzymes answered to ischemia on various ways. For the Mg-ATPase the Km value was not changed significantly (Fig. 2). For the Ca-ATPase a significant decrease occurred after 30 min and it remained on the same level after 45 and also 60 min of ischemia. For the (Na,K)-ATPase the decrease of Km value was significant after 45 and 60 min of ischemia.

Evaluation of temperature dependence of actual enzyme activities resulted in the value of activation energy (Ea) for the ATP hydrolysis catalyzed by sarcolemmal ATPases. Ischemia did not influence the Ea for the reaction when it was catalyzed by the (Na,K)-ATPase. When the reaction was catalyzed by the Ca-ATPase or Mg-ATPase Ea was in ischemic hearts markedly higher comparing to control hearts (Fig. 3).

Discussion

The presented decrease of Vm values for all three investigated sarcolemmal ATPases is in agreement with the time course of the decrease of ATP content in cardiac tissue [2, 3]. Al-

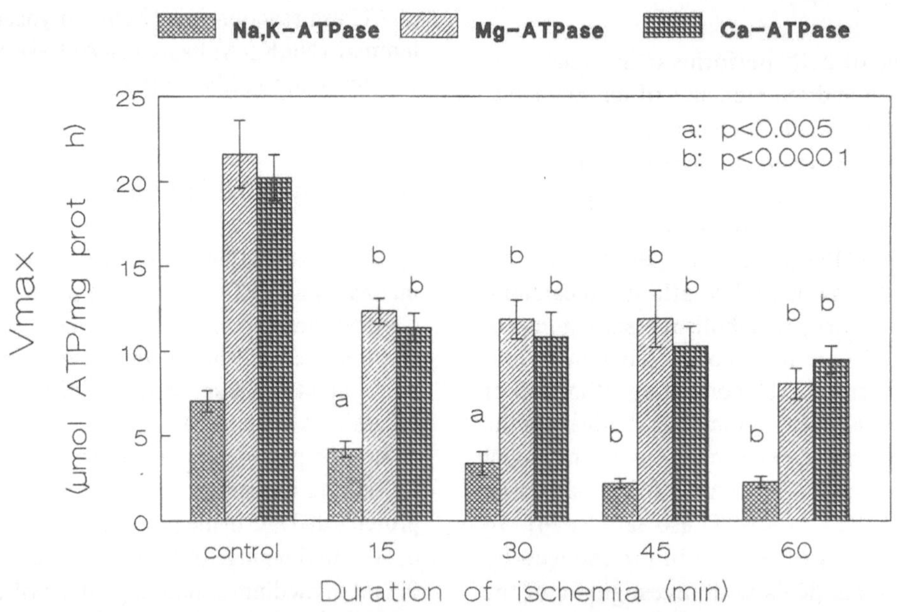

Fig. 1. Time course of Vm of ATPases in cardiac sarcolemma during ischemia. The values represent the means ± SEM (n=10–18).

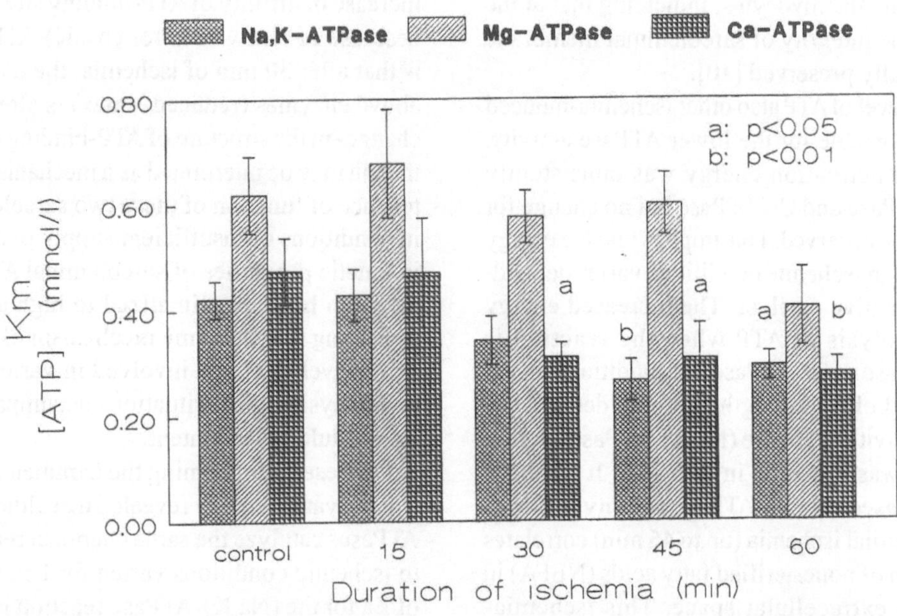

Fig. 2. Time course of Km of ATPases in cardiac sarcolemma during ischemia. The values represent the means ± SEM (n = 10–18).

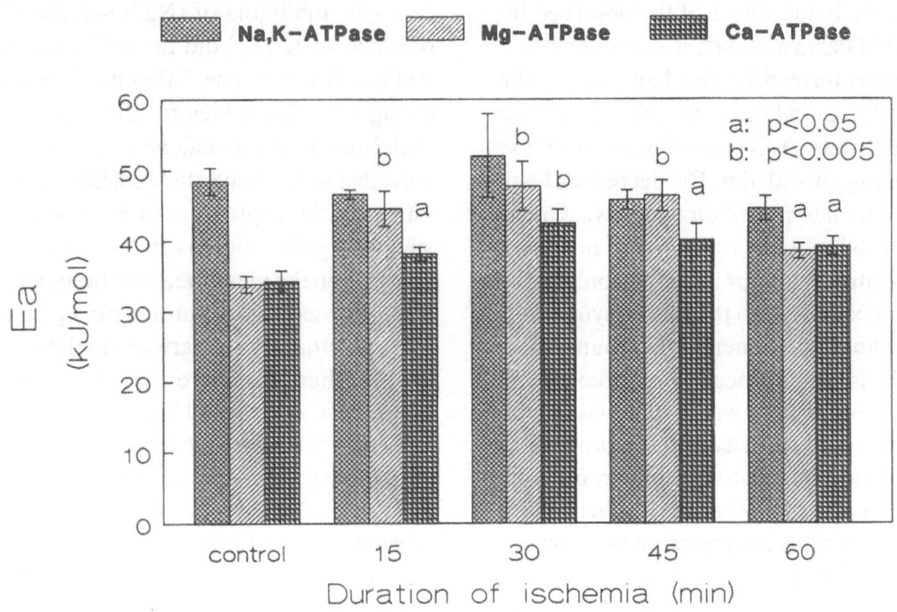

Fig. 3. Time course of Ea of ATPases in cardiac sarcolemma during ischemia. The values represent the means ± SEM (n = 10–18).

ready 15 min after the onset of ischemia a significant decrease in sarcolemmal ATPase activities was detected. At 30 min of ischemia the decrease of Vm value ended. The Vm was approximately the same as in the 15th min and it remained unchanged also after ischemia prolonged to 45 or 60 min. How far may these findings reflect real damage to myocardial sarcolemma is difficult to judge since considerable damage to the integrity of the glycocalyx should be followed by massive calcium entry into the myocardium [8], without, however, any detectable changes in the plasma membrane [9]. Cytochemical data revealed a considerable decrease in the intensity of specific precipitate of K-dependent, ouabain sensitive p-NPPase in cardiomyocytes after 30 min of ischemia. In spite of clearly evident cellular damage manifested by swollen mitochondria, reduced glycogen deposits and at least 50% loss of the (Na,K)-ATPase activity, the lanthanum deposits remained outside the myocytes, indicating that at the 30 min of ischemia the integrity of sarcolemmal membrane remains at least partially preserved [10].

Besides the lower level of ATP also other ischemia-induced effects might be responsible for the lower ATPase activity. During ischemia the activation energy was consistently higher for the Mg-ATPase and Ca-ATPase but no change for the (Na,K)-ATPase was observed. This implies that the energy barrier of the reaction in ischemic conditions varies depending on the nature of active catalyst. The increased energy barrier for the hydrolysis of ATP when the reaction is catalyzed by Mg-ATPase or Ca-ATPase is an additional negative ischemia-induced effect strengthening the decrease of respective ATPase activities. For the (Na,K)-ATPase another additional influence was observed in ischemia. It has been shown that the decreased (Na,K)-ATPase activity during a short period of myocardial ischemia (up to 45 min) correlates with the accumulation of nonesterified fatty acids (NEFA) in the cytosolic and/or extracellular space. This ischemia-induced inhibition of (Na,K)-ATPase is of extramembraneous origin because the content of NEFA in sarcolemma did not correlate with a decrease of the enzyme activity [4].

To describe more precisely the nature of the observed inhibition of sarcolemmal ATPases in ischemic hearts their ATP binding sites were characterized by the Km value. The decrease of Km values for Ca-ATPase and (Na,K)-ATPase indicates an increase in affinity of ATP-binding sites of these enzymes at ischemia lasting 30–60 min. The decrease of Km value in ischemia might be interpreted on two ways. One is that ischemia lasting 30–60 min in cardiac tissue induces a presence of an uncompetitive inhibitor that binds only to the enzyme-substrate complex and not to the free enzyme. This hypothetical inhibitor should be a membrane-bound compound because during the isolation procedure of sarcolemmal fraction all soluble compounds were washed out. Basing on the above assumption it would be reasonable to propose as this inhibitor a membrane-bound protein in sarcolemma which might be activated after 30–60 min of ischemia. An ischemia-induced synthesis of stress protein was shown in isolated and perfused rat hearts [11]. Influence of this stress protein on the ATPase reaction is at present only speculative and its verification requires detailed studies. Another protein with confirmed inhibitory effect on the (Na,K)-ATPase was documented in red cell membranes [12] and in myometrial sarcolemma [13]. This membrane-bound protein is calcium dependent and is responsible for the Ca^{2+} sensitivity of the (Na,K)-ATPase [12, 13]. In myocytes injured by longer periods of ischemia when the intracellular calcium concentration is higher [14] the increased activity of such inhibitory protein might be suggested. However, in cardiac sarcolemma such calcium dependent protein with inhibitory effect on the (Na,K)-ATPase was not described yet.

The second way for interpretations of ischemia-induced increase of affinity of ATP-binding site as appeared from the decrease of Km values for (Na,K)-ATPase and Ca-ATPase is that after 30 min of ischemia, the decrease in turnover of above enzymes (reduced Vmax) is already compensated by changes in the structure of ATP-binding site. The latter phenomenon may be interpreted as a mechanism securing the maintenance of function of these two sarcolemmal ATPases even in conditions of insufficient supply of ATP. Similar changes in kinetic properties of sarcolemmal ATPases were also detected in hearts acclimatized to high altitude hypoxia [15] indicating that the same mechanism of adaptation at the enzyme level might be involved in various physiological and pathophysiological situations accompanied with decreased intracellular ATP content.

Our results concerning the parameters of enzyme kinetics and activation energy revealed that although the sarcolemmal ATPases catalyze the same chemical reaction, their response to ischemic conditions varied for Km and Ea. The stability of Ea for the (Na,K)-ATPase reaction is contradictory to the elevation of Ea observed for the other two enzymes. This contrast may be explained by the specificity of (Na,K)-ATPase which differs in its nature from the other two ATPases. For example, inhibitors of (Na,K)-ATPase like ouabain [16] or orthovanadate [17] did not inhibit the Ca-ATPase and Mg-ATPase [18, 19]. The difference between ischemia-induced changes of Km values for the Mg-ATPase and Ca-ATPase with low affinity to calcium is intriguing from the point of view that several investigators have suggested that these activities are the expression of a single enzyme. The evidence supporting this view is that both activities were enriched similarly in the membrane fractions, activities were not additive, they exhibited similar pH optima and they were depressed similarly by various inhibitors [for review see 20]. On the other hand there are studies providing evidence that Ca-ATPase and Mg-ATPase may be two separate enzymes. The enzymes differed in sensitivity to ADP and P_i, in sensitivity to deoxycholate treatment, in sensitivity to Mn^{2+} [for review see 20]. Our finding that the affinity of enzyme to substrate is lower for the Ca-ATPase after 30–60 min of ischemia with no change for the Mg-ATPase indicates that the two enzymes are distinct at least in the vicinity of ATP-binding site.

References

1. Reimer KA, Jennings RB: Myocardial ischaemia, hypoxia and infarction. In: H.A. Fozzard, E. Haber, R.B. Jennings, A.M. Katz, H.E. Morgan (eds). The Heart and Cardiovascular System. Raven Press, New York, 1989, pp 1133–1201
2. Fedelesova M, Ziegelhöffer A, Styk J, Slezák J: Effect of myocardial ischemia on metabolic adaptability in the dog heart. I. In: J. Slezak (ed). Artificial circulation. VEDA, Publishing House of the Slovak Academy of Sciences, 1978, pp 189–199
3. Reimer KA, Jennings RB, Hill ML: Total myocardial ischemia, *in*

vitro. 2. High energy phosphate depletion and associated defects in energy metabolism, cell volume regulation, and sarcolemmal integrity. Circ Res 49: 901–911, 1981

4. Kim EA, Danilenko MP, Murzakhmetova MK, Vaschenko VI, Esyrev OV: Na,K-ATPase activity in the myocardium sarcolemma under ischemia. Ukr Biokhim J 61: 61–65, 1989
5. Vrbjar N, Soos J, Ziegelhöffer A: Secondary structure of heart sarcolemmal proteins during interaction with metallic cofactors of (Na,K)-ATPase. Gen Physiol Biophys 3: 317–325, 1984
6. Lowry OH, Rosebrough NJ, Farr AL, Randall RJ: Protein measurement with the folin phenol reagent. J Biol Chem 193: 265–275, 1951
7. Taussky HH, Shorr EE: A microcolorimetric method for the determination of inorganic phosphorus. J Biol Chem 202: 675–685, 1953
8. Langer GA: The ultrastructure and function of the myocardial cell surface. Am J Physiol 235: H461–H468, 1978
9. Langer GA, Frank JS, Philipson KD: Ultrastructure and calcium exchange of the sarcolemma, sarcoplasmic reticulum and mitochondria of the myocardium. Pharmac Theor 96: 331–376, 1982
10. Vrbjar N, Slezák J, Ziegelhöffer A, Tribulová N: Features of the (Na,K)-ATPase of cardiac sarcolemma with particular reference to myocardial ischaemia. Eur Heart J 12 (Suppl F): 149–152, 1991
11. Currie RW: Effects of ischaemia and perfusion temperature on the synthesis of stress-induced (heat shock) proteins in isolated and perfused rat hearts. J Mol Cell Cardiol 19: 795–1987, 1987
12. Yingst DR, Marcowitz MJ: Effect of hemolysate on calcium inhibition of the (Na^+ + K^+)-ATPase of human red blood cells. Biochem Biophys Res Commun 111: 970–979, 1983
13. Turi A, Torok K: Myometrial (Na^+ + K^+)-activated ATPase and its Ca^{2+} sensitivity. Biochim Biophys Acta 818: 123–131, 1985
14. Shen AC, Jennings RB: Myocardial calcium and magnesium in acute ischemic myocardial injury. Am J Path 46: 367–386, 1972
15. Ziegelhöffer A, Prochazka J, Pelouch V, Ošťádal V, Džurba A, Vrbjar N: Increased affinity to substrate in sarcolemmal ATPases from hearts acclimatized to high altitude hypoxia. Physiol Bohemoslov 36: 403–415, 1987
16. Adams RJ, Schwartz A, Grupp G, Grupp I, Lee S, Wallick ET: High affinity binding site and low-dose positive inotropic effect in rat myocardium. Nature 296: 167–169, 1982
17. Cantley LC Jr, Cantley LC, Josephson L: A characterization of vanadate interactions with the (Na,K)-ATPase. Mechanistic and regulatory implications. J Biol Chem 253: 7361–7368, 1978
18. Tuana BS, Dhalla NS: Purification and characterization of a Ca^{2+}/Mg^{2+} ecto-ATPase from rat heart sarcolemma. Mol Cell Biochem 81: 75–88, 1988
19. Zhao D, Dhalla NS: Characterization of heart plasma membrane Ca^{2+}/Mg^{2+} ATPase. Arch Biochem Biophys 263: 281–292, 1988
20. Dhalla NS, Zhao D: Cell membrane Ca^{2+}/Mg^{2+} ATPase. Prog Biophys Molec Bio 52: 1–37, 1988

Molecular and Cellular Biochemistry **147**: 105–114, 1995.

Response of the rat heart to catecholamines and thyroid hormones

Heinz-Gerd Zimmer, Michael Irlbeck and Claudia Kolbeck-Rühmkorff

Department of Physiology, University of Munich, Germany

Abstract

Catecholamines and thyroid hormones have a similar influence on heart function and metabolism, but this may occur in a differential manner and to a different extent. In this study, the effects of norepinephrine (NE) and of triiodothyronine (T_3) were studied in regard to the function of the left (LV) and right ventricle (RV) and to the oxidative pentose phosphate pathway (PPP). NE was applied in rats as continuous i.v. infusion (0.2 mg/kg/h) for three days. T_3 was given as daily s.c. injections (0.2 mg/kg) for the same period of time. LV and RV function was measured in the closed-chest trapanal-anesthetized animals using special Millar ultraminature catheter pressure transducers. NE induced an increase in heart rate, in mean arterial pressure, and in total peripheral resistance (TPR). The cardiac RNA/DNA and the left ventricular weight/body weight ratios were increased by about 40%. These effects were prevented by simultaneous α- and β-receptor blockade with prazosin and metoprolol, respectively, but not by verapamil which abolished the hemodynamic effects. RVSP was significantly elevated by NE in a dose-dependent manner. The functional effects of T_3 on the LV were not as pronounced as those induced by NE. Heart rate and LV dp/dt_{max} were increased by T_3, and this increase was prevented by concomitant β-receptor blockade with metoprolol. In contrast to NE, T_3 induced an increase in cardiac output and a concomitant decrease in TPR. The RNA/DNA ratio was elevated and cardiac hypertrophy had developed after treatment for three days with T_3. These changes were not affected by β-receptor blockade with metoprolol. RVSP was increased by T_3 to a lesser extent than with NE. In metabolic terms it turned out that only NE, but not T_3 had a stimulating effect on the cardiac PPP. NE increased the mRNA and activity of glucose-6-phosphate dehydrogenase (G-6-PD), the first and regulating enzyme of this pathway. However, there was no effect of T_3 on G-6-PD activity nor on 6-phosphogluconate dehydrogenase activity, one of the following enzymes in the pathway within the first 5 days of T_3 treatment. These results demonstrate that the functional effects of T_3 were not as pronounced as or even different from those of NE, and that T_3 lacked a stimulating effect on the cardiac PPP. (Mol Cell Biochem **147**: 105–114, 1995)

Key words: pentose phosphate pathways, heart function, heart metabolism, catecholamine effects, thyroid hormone effects

Introduction

Catecholamines and thyroid hormones have a similar influence on the cardiovascular system. This is not surprising, since thyroid hormones have a permissive effect on catecholamines. However, there may be differences in quantitative terms and in the mechanisms underlying the effects of these hormones.

The simplified scheme of Fig. 1 shows that NE increases total peripheral resistance, and this may be involved in triggering cardiac hypertrophy. As to the specific effects on the myocardium, β-receptor stimulation leads to the increase of adenylate cyclase activity and to the elevation of cAMP [1]. Activation of cAMP-dependent protein kinase A (PKA) induces the known metabolic effects such as the increase in lipolysis and glycogenolysis. In addition, phosphorylation of several proteins is initiated. For instance, a channel protein which is involved in transsarcolemmal Ca^{++}-transport becomes phosphorylated. In this way, the positive inotropic effect is brought about. Also phospholamban, a component of the sarcoplasmic reticulum, is phosphorylated subsequent to β-adrenergic stimulation. As a consequence, the re-uptake of Ca^{++} into the sarcoplasmic reticulum is facilitated thus resulting in the increased relaxation velocity. This is the lusitropic effect [2]. Besides the positive chronotropic, inotropic and lusitropic effect, stimulation of β-adrenergic

Address for offprints: H-G. Zimmer, Physiologisches Institut, Universität München, Pettenkoferstr. 12, 80336 München, Germany

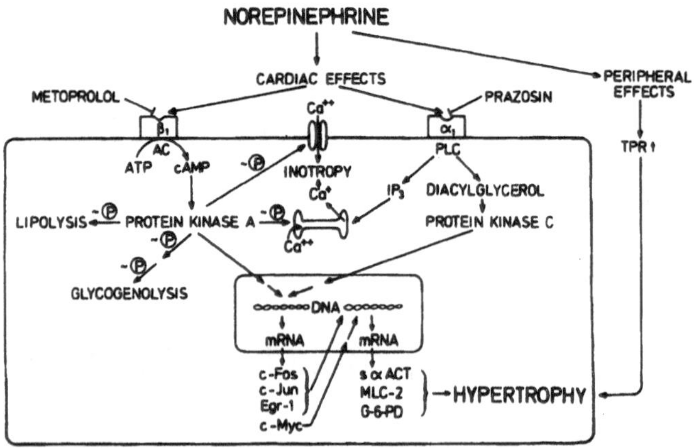

Fig. 1. Schematic presentation of the effects of norepinephrine on the peripheral circulation and on the heart. For details see Introduction.

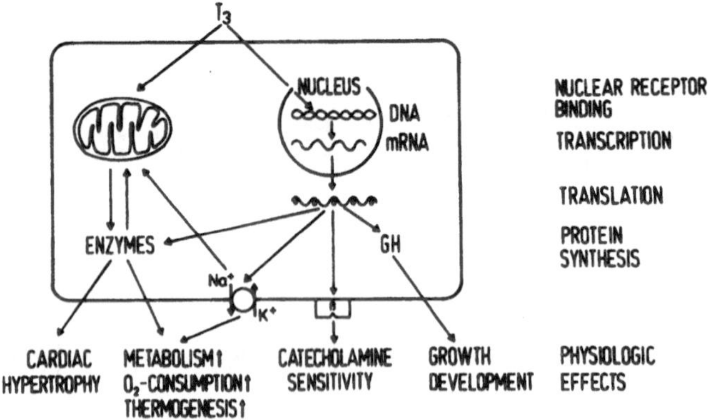

Fig. 2. Schematic presentation of the action of triiodothyronine (T_3) at the cellular level, and the physiological effects. For details see Introduction.

receptors is also known to shift the isomyosin from V_3- to the V_1-form [3] and to induce cardiac hypertrophy [4].

Stimulation of α-adrenergic receptors is associated with the elevation of the second messengers inositoltrisphosphat (IP_3) and diacylglycerol [5, 6]. IP_3 liberates Ca^{++}-ions from the sarcoplasmic reticulum and thus induces a positive inotropic effect [7, 8]. Diacylglycerol stimulates protein kinase C (PKC).

Both PKA and PKC influence the DNA in the cell nucleus either directly or indirectly thus leading to the expression of proto-oncogenes such as c-*fos*, c-*jun*, c-*myc* [9] and the early growth response gene 1, *egr*-1. Proto-oncogenes may play an important role in the regulation of cardiac growth and of the transcription of cardiac-specific genes. Components of this immediate early gene program may be involved in promoting directly or indirectly the reexpression of the fetal program of genes coding for the main contractile proteins. These are the skeletal isoform of alpha actin [10], and the myosin-light chain-2 [11]. In addition, G-6-PD is induced [12].

In contrast to NE, T_3 binds not to a receptor located at the cell membrane, but to receptors in the nucleus [13] and in the mitochondria. After receptor binding there is an increase in the transcription and translation process that leads to the new synthesis of several proteins and enzymes (Fig. 2). The receptor is a chromatin-associated protein with a molecular weight of 50000–55000. Unlike steroid hormones, a cytoplasmic form of a T_3-receptor has not been identified or characterized. The thyroid hormone receptor protein is similar to that encoded by the c-erb-A-oncogene which also binds T_3 specifically [14]. T_3 rapidly increases the transcription of the growth hormone gene in the pituitary [15] which is then involved in growth and development. The number of β-adrenergic receptors in cardiac membranes of hyperthyroid animals is increased thus leading to an enhanced sensitivity of the heart to catecholamines [16]. T_3 also augments the Na^+/K^+-transport via activation of the Na^+/K^+ ATPase which functions as a pump [17]. Consequent changes in the ATP/ADP ratio stimulate mitochondrial respiratory rates resulting in increased oxygen consumption, stimulation of metabolism, and enhanced thermogenesis. The T_3-induced increase in protein synthesis also concerns several enzymes which remain in the cytoplasm or will be transported into the mitochondria such as α-glycerophosphate dehydrogenase. Increase of the activity of many enzymes contributes to stimulation of metabolism and is in some cases associated with the development of cardiac hypertrophy, although it may not always be essential for the hypertrophic response [18, 19]. T_3 may also increase transcription and translation of several mitochondrial proteins [20, 21] such as cytochrome oxidase. In the heart, glycolysis is enhanced, in particular, the activity of phosphofructokinase, a key enzyme in this pathway [22]. Likewise, the activity of some enzymes of the Krebs cycle such as citrate synthase and malate dehydrogenase is enhanced [23]. In this way, ATP can be produced both via glycolysis and via oxidative phosphorylation that is necessary for the increased activity of the Na^+/K^+-ATPase. Furthermore, the distribution of the three different isoenzymic forms of myosin which exist in a number of animal species is controlled by thyroid hormones [24, 25]. Under the influence of T_3, the myocardial content of V_1-myosin which is associated with the highest ATPase activity is elevated and the synthesis of V_3-myosin is depressed [26, 27].

In this contribution, results of our experimental studies are presented in which the effects of norepinephrine (NE) and triiodothyronine (T_3) on the function and metabolism of the rat heart will be compared. As to the functional effects, particular emphasis will be placed on the changes in right heart function. The right heart has not been studied extensively, particularly in small laboratory animals, since a simple and reliable method was not available. Only recently it has become feasible to manufacture ultraminiature catheter pressure transducers that can be applied not only for left heart catheterization [28], but also for right heart catheterization in rats [29].

In metabolic terms it is a particular concern of this comparative study to examine the oxidative pentose phosphate pathway (PPP). This pathway is the link between carbohydrate and nucleotide metabolism. Glucose-6-phosphate originating from glycogenolysis or from glucose taken up by the myocardial cell is metabolized predominantly by glycolysis. A small portion of G-6-P, however, enters the oxydative PPP of which glucose-6-phosphate dehydrogenase (G-6-PD) is the first and rate-limiting enzyme. This pathway serves mainly two functions: 1. It provides reducing equivalents in the form of NADPH which can be used for the synthesis of free fatty acids and for the reduction of oxidized glutathione (GSSG). This is important for detoxification processes. 2. In this pathway ribose-5-phosphate is generated which can be transformed to 5-phosphoribosyl-1-pyrophosphate (PRPP), and this is an essential precursor substance for the synthesis of both pyrimidine and purine nucleotides. There are connections between this pathway and glycolysis on two levels via the transaldolase and transketolase reactions [30].

A characteristic feature of this pathway is that its capacity in the heart is very low, as a consequence, the available pool of PRPP and the rate of purine nucleotide biosynthesis are also very limited [30]. Basically, there are two possibilities to affect the capacity of the oxidative PPP in the heart. The first is to bypass the first and rate-limiting step in the pathway, that is the reaction which is catalyzed by G-6-PD. This can be done with ribose. Ribose is taken up by the myocardial cell and becomes phosphorylated to ribose-5-phosphate. This is the immediate precursor of PRPP which is essential not only for the biosynthesis of adenine nucleotides from small molecular precursor substances, but also for the salvage of hypoxanthine to IMP and of adenine to AMP. In addition, orotic acid is converted to OMP. It has been shown in previous studies that ribose is capable to attenuate or even to prevent any experimentally induced decrease in the cardiac ATP pool [31–33]. Interestingly, the normalization of the high-energy phosphates is accompanied by an improvement of global heart function in intact rats [34, 35] and by an effect on the structural integrity of the heart [32].

The second possibility for intervention is the stimulation of G-6PD, the first and rate-limiting enzyme of the oxidative PPP, so that more PRPP is provided for nucleotide synthesis. This can be done by all catecholamines that have been tested so far [9, 36]. It was therefore interesting to examine whether thyroid hormones may also have a stimulating effect on the first two enzymes G-6-PD and 6-phosphogluconate dehydrogenase (6-PGD).

Material and methods

The experiments were done on female Sprague-Dawley rats (200–250 g body weight) obtained from Savo GmbH (Kisslegg, Germany) fed a control rat chow diet (Altromin C 100 from Altromin GmbH, Lage, Germany) with free access to tap water. Norepinephrine (NE, 0.2 mg/kg/h) was administered as continuous intravenous infusion via a catheter (Vygon, Aachen, Germany) which was positioned in the left jugular vein in ether anesthesia. The catheter was tunneled under the skin and let out at the neck of the animals. It was connected to a 20 ml syringe placed in an infusion pump (Infors AG, Basel, Switzerland). The infusion rate was 4 ml/kg/h. The animals could move around freely in their cages during the infusion periods.

In previous studies this NE dose had induced a marked degree of cardiac hypertrophy within 3 days [4]. Sodium chloride-infused animals served as controls. NE was combined with the α_1-blocker prazosin (0.1 mg/kg/h), with the β_1-blocker metoprolol (1 mg/kg/h), and with both drugs. Isoproterenol was subcutaneously injected once at a dose of 25 mg/kg. The catecholamines were dissolved in 0.9% NaCl. To prevent oxidation of the catecholamines, 100 mg/l ascorbic acid was added to the solutions. It was also included in 0.9% NaCl that was used for the control experiments. The syringes with the substances were protected against light. 3,3′,5-Trioodothyronine (T_3) was administered in daily subcutaneous injections (0.2 mg/kg) for 3 days in rats that received continuous i.v. infusion of 0.9% NaCl or metoprolol (1 mg/kg/h).

Norepinephrine HCl (NE), isoproterenol, and 3,3′5-triiodothyronine (T_3) were purchased from Sigma Chemie GmbH, München, Germany. L-(+)-Ascorbic acid was obtained from Merck, Darmstadt, Germany. Prazosin was donated by Pfizer, Karlsruhe, Germany, metoprolol-tartrate was obtained from Ciba-Geigy, Wehr, Germany.

The following sequence specific cDNA clones were used for RNA hybridization: A 2400 bp cDNA coding for rat G-6-PD (gift from Dr. Ye-Shih Ho, Laboratory of Molecular Biology, Division of Allergy, Critical Care and Respiratory Medicine, Department of Medicine, Duke University Medical Center, Durham, NC 27710 USA), a 880 bp cDNA coding for rat 6-PGD (gift from Dr. Howard C. Towle, Department of Biochemistry, University of Minnesota, Minneapolis, MN 5545–0347, USA), and a 350 bp DNA coding for murine 18S rRNA subcloned from a 1900 bp DNA (gift from Dr. Ilse Oberbäumer, Max-Planck-Institut für Biochemie Martinsried bei München, Germany).

Measurement of left and right heart function

When the function of the left heart was measured, the animals were anesthetized with thiopental sodium (Trapanal[R] Byk Gulden, Konstanz, Germany, 80 mg/kg, i.p.). The depth of anesthesia was tested by eliciting reflexes in the legs by a forceps. After tracheotomy a catheter was placed in the trachea to maintain airway patency, to allow suction of secretions, and to institute artificial respiration, if necessary. The ultraminiature catheter pressure transducers (models PR-249 and PR-291, Millar Instruments, Inc., Houston, Texas) were used for left [28] and right heart catheterization [29], respectively. The left heart catheter was inserted into the right carotid artery and advanced upstream the aorta into the left ventricle. The right heart catheter was inserted into the right jugular vein, advanced into the right atrium and then rotated counterclockwise and thereby placed in the right ventricle. The catheters were attached to a Millar control unit (model TC-100) which was connected to a HSE electromanometer (Hugo Sachs Elektronik, March-Hugstetten, Germany) for the measurement of systolic pressure (LVSP, RVSP). The maximal rate of rise in ventricular pressure (LV and RV dp/dt_{max}) was obtained with an electronic differentiation system (Physio-Differentiation, Hugo Sachs Elektronik).

Calibration of pressure and dp/dt_{max} was done using a mercury manometer (type 36, Hugo Sachs Elektronik) and the calibration device that is built into the Physio-Differentiator. Heart rate was measured with the HSE Digi-Puls rate meter (Hugo Sachs Elektronik) and continuously recorded together with ventricular pressure and dp/dt_{max} on a Brush 2600 recorder (Gould Inc., Cleveland, Ohio). Cardiac output was determined by using the thermodilution technique [4].

Measurement of metabolic parameters

The activities of G-6-PD (EC 1.1.1.49) and 6-PGD (EC 1.1.1.44) were measured according to the methods of Glock and McLean [37, 38]. After various periods of *in vivo* exposure to the different substances, the rats were anesthetized with ether, a cannula was placed in the ascending aorta after thoracotomy and tightly fixed there. The hearts were quickly excised, and the coronary arteries were perfused via the cannula with an ice-cold KCl solution (0.15 M/l containing 8 ml of 0.02 M $KHCO_3$) to remove blood and to stop beating of the heart. After homogenization of the hearts in the perfusion medium, pH control (7.0) and centrifugation (Beckman ultracentrifuge model L5–65 at 20000 RPM for 30 min), the supernatants were dialyzed overnight. The enzyme activities were then measured spectrophotometrically. Protein concentration in the dialysate was determined using the modified biuret reaction [39]. The mean specific activity of both enzymes was expressed as units/g protein.

RNA isolation and Norther blot analysis

Total RNA was isolated from rapidly frozen rat hearts using a modified method of Birnboim [40]. Briefly, a rat heart was homogenized in 0.3 M sodium acetate pH 5.2 and 10 mM EDTA with an Ultra-Turrax (Janke and Kunkel KG, Ika-Werk, Staufen i. Breisgau, Germany). Sodium dodecyl sulfate (SDS) was added to 1%, and the sample was vortexed. Two phenol/chloroform extractions were followed by ethanol precipitation. The precipitate was resolved in RES-buffer (0.5 M LiCl, 1 M urea, 0.25% SDS, 0.02 M sodium citrate, 2.5 mM cyclohexanediamine tetraacetate (CDTA), pH 6.8), followed by proteinase K treatment (final concentration 50 μg/ml, 30 min at 50°C), phenol extraction and precipitation at 0°C overnight with LiCl/ethanol (3 vol 5 M LiCl, 2 vol ethanol). The precipitate was resolved in CCS-buffer (1 mM sodium citrate, 1 mM CDTA, 0.1% SDS, pH 6.8) and precipitated twice with ethanol. The RNA was quantified by measuring optical density at 260 nm and directly used for Northern blot analysis.

For Northern analysis, total RNA was electrophoresed on a 1% agarose gel containing formaldehyde [41], transferred to a Hybond N membrane (Amersham-Buchler, Braunschweig, Germany) and UV crosslinked (Stratalinker 1800, Stratagene, Heidelberg, Germany). Prehybridization (3–4 h) and hybridization (16–20 h) with ^{32}P-labeled cDNA were performed in 50% formamide, 5 × SSC (1 × SSC: 0.15 M NaCl, 0.015 sodium citrate), 5 × Denhardt's (1 × Denhardt's: 0.02% bovine serum albumin, 0.02% polyvinylpyrrolidone, 0.02% ficoll) and 100 μg/ml herring sperm DNA at 42°C. cDNAs were labeled using a multiprime labeling kit (Amersham-Buchler) and ^{32}P-dATP (specific activity: 3000 Ci/mMol, Amersham-Buchler) according to the suppliers protocol.

After hybridization, filters were washed twice in 2 × SSC at room temperature, then 20 min in 2 × SSC, 1% SDS at 42°C, 20 min in 0.1 × SSC at room temperature and finally 20 min in 0.1 × SSC at 58°C. Filters were then exposed (Kodak, X-Omat AR) at –80°C using intensifying screens. Subsequent to development, intensity of bands was measured by densitometry (Elscript 400, Hirschmann Gerätebau, Unterhaching, Germany) normalized to the rehybridization signal obtained with 18S RNA probe to correct for difference in RNA amounts and calculated as percentage of values derived from RNA of control rat hearts.

Statistical analysis

All values are expressed as mean ± SEM. For calculating significance, the Student's *t*-test for unpaired data was used. Differences were considered significant at a value of $p < 0.05$ [42].

Results

Functional studies

After 3 days of continuous i.v. infusion of NE, heart rate, mean arterial pressure, and total peripheral resistance were increased. Cardiac output was slightly lower. To assess the development of cardiac hypertrophy, a metabolic parameter, the RNA/DNA ratio, and a morphological parameter, the left ventricular weight/body weight (LVW/BW) ratio, were measured. They were both elevated by about 40% (Fig. 3, left hand side). The NE-induced cardiac hypertrophy may be triggered by the increase of total peripheral resistance or may be induced by direct stimulation of adrenergic receptors (Fig. 1). To examine the first possibility, the functional effects of NE were eliminated by the calcium antagonist verapamil.

Verapamil prevented the NE-induced increase in mean arterial pressure and total peripheral resistance. Only heart rate was still significantly elevated. The RNA/DNA and the LVW/BW ratios were elevated to about the same extent as with NE alone (Fig. 3, right hand side). Thus, the increase in total peripheral resistance seems not to be the trigger for the development of NE-induced cardiac hypertrophy. However, combined application of prazosin and metoprolol reversed the NE-induced functional changes. The LVW/BW ratio was entirely normalized, the RNA/DNA ratio was only slightly elevated (Fig. 4, right hand side).

In view of these changes it was of interest to examine what the T_3-induced hemodynamic alterations may be and whether they can be influenced by adrenergic blockade. After 3 days of daily s.c. injections of T_3 heart rate was increased by 27%. Also LV dp/dt_{max} was elevated to a similar extent as with NE. To examine whether and to what extent these T_3-induced changes are due to catecholamines, β-adrenergic receptors were blocked with metoprolol. This blocker induced a negative chronotropic and inotropic effect and prevented entirely

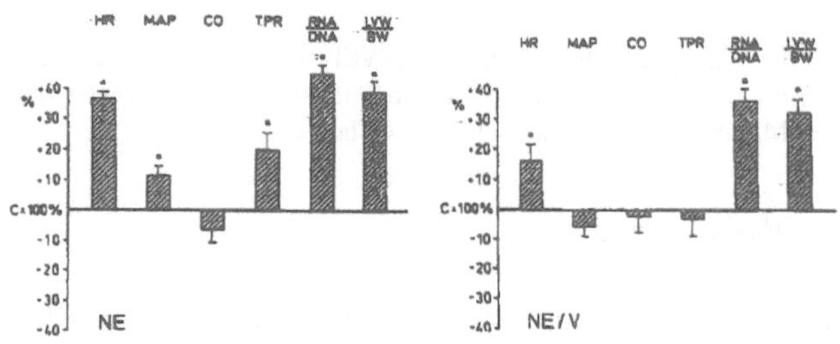

Fig. 3. Percentage changes in heart rate (HR), mean arterial pressure (MAP), cardiac output (CO), total peripheral resistance (TPR), the RNA/DNA ratio, and the left ventricular weight/body weight (LVW/BW) ratio in norepinephrine (NE)-treated rats (0.2 mg/kg/h continuous i.v. infusion for 3 days, left hand side), and the effects of verapamil (V, 1 mg/kg/h, i.v. infusion for 3 days, right hand side). Number of experiments: 8–13. Significant changes are indicated by asterisks ($p < 0.05$ vs. control).

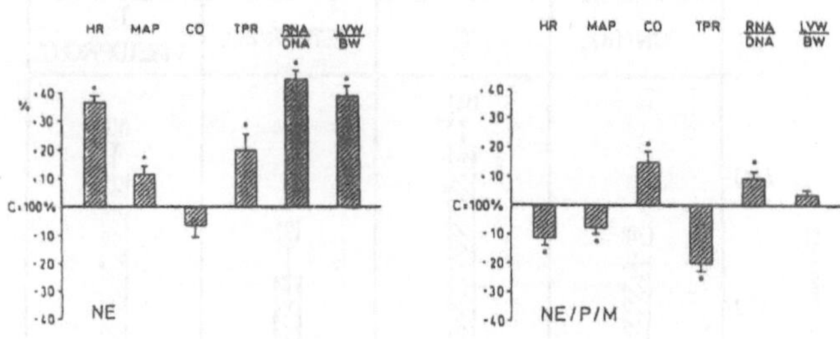

Fig. 4. Percentage changes in functional (HR, MAP, CO, TPR), metabolic (RNA/DNA ratio), and morphological parameters (LVW/BW ratio) in rats treated with NE (0.2 mg/kg/h, continuous i.v. infusion for 3 days, left hand side) and the effects of combined treatment with prazosin (0.1 mg/kg/h) and metoprolol (1 mg/kg/h, continuous i.v. infusion for 3 days, right hand side). Number of experiments: 8–13. Statistically significant changes are indicated by asterisks ($p < 0.05$ vs. control).

the T_3-induced increase in heart rate and LV dp/dt_{max} (Fig. 5). In contrast to NE, T_3 induced a marked increase in cardiac output. Metoprolol had no effect of its own and did not affect the T_3-induced increase (Fig. 6). Likewise, the T_3-induced increase in the RNA content and in the RNA/DNA ratio (Fig. 7) as well as in the heart weight and in the heart weight/body weight ratio (Fig. 8) was not influenced by metoprolol which by itself had no effect.

A characteristic feature of NE and T_3 concerns their effects on the right ventricle. After 3 days of NE infusion, RVSP was elevated in a dose-dependent fashion. In contrast, LVSP was first elevated, but with the highest dose of NE it was even lower than the control. Thus, the pressure increase was more pronounced in the RV (Fig. 9). T_3 had also a more marked effect on the RVSP than on the LVSP. After 3 days, the T_3-induced increase in LVSP was only marginal. The elevation of RVSP was more pronounced (Fig. 10).

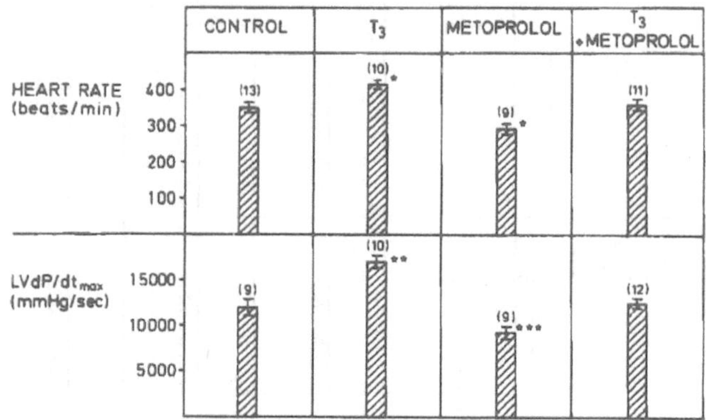

Fig. 5. Changes in heart rate and LV dp/dt_{max} in rats after 3 days of treatment with triiodothyronine (T_3, 0.2 mg/kg, s.c.) and with metoprolol (1 mg/kg/h, continuous i.v. infusion) alone and in combination. Number of experiments in parentheses. Mean values ± SEM. *$p < 0.005$, **$p < 0.0005$, **$p < 0.025$ vs. control.

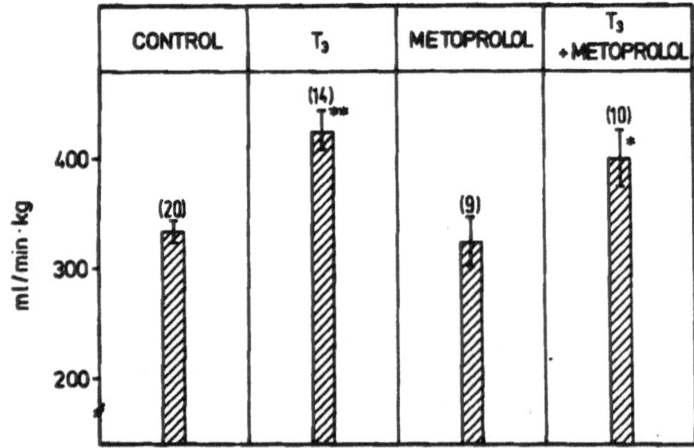

Fig. 6. Effects of T_3 and metoprolol alone and in combination on cardiac output. Doses and time of application as in Fig. 5. Mean values ± SEM, number of experiments in parentheses. *p < 0.001, **p < 0.0005 vs. control.

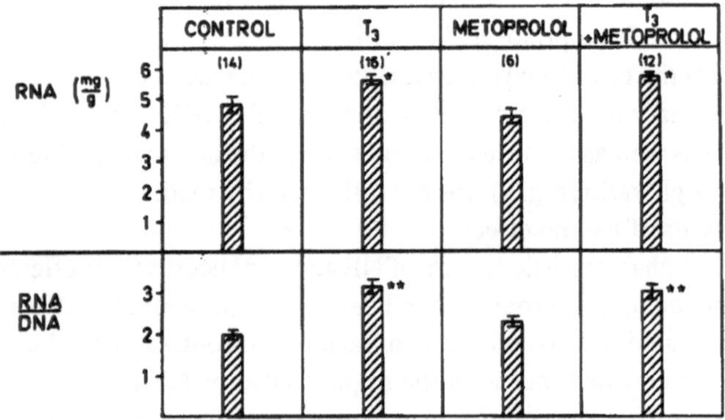

Fig. 7. RNA concentration and the RNA/DNA ratio in hearts of rats under the influence of T_3 and metoprolol alone or in combination. Mean values ± SEM, number of experiments in parentheses. *p < 0.01, *p < 0.0005 vs. control.

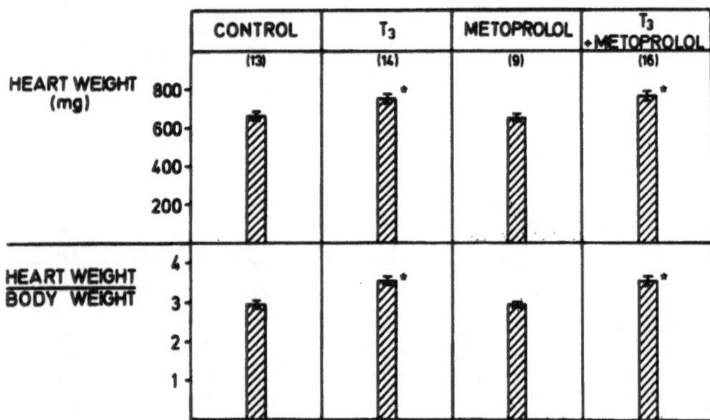

Fig. 8. Heart weight and the heart weight/body weight ratio in rats treated with T_3 and with metoprolol alone and in combination for 3 days. Data are mean values ± SEM, number of experiments in parentheses. *p < 0.0005 vs. control.

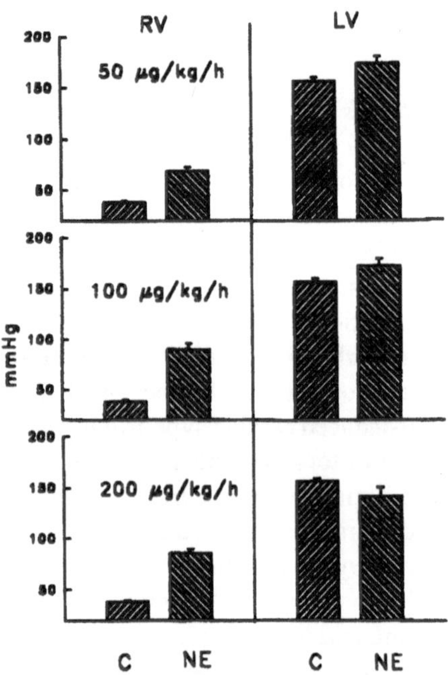

Fig. 9. Effect of norepinephrine (NE) in increasing doses on right ventricular (RV) and left ventricular (LV) systolic pressure after 3 days of continuous i.v. infusion. Data are mean values of 6 experiments. Mean values ± SEM, C: control.

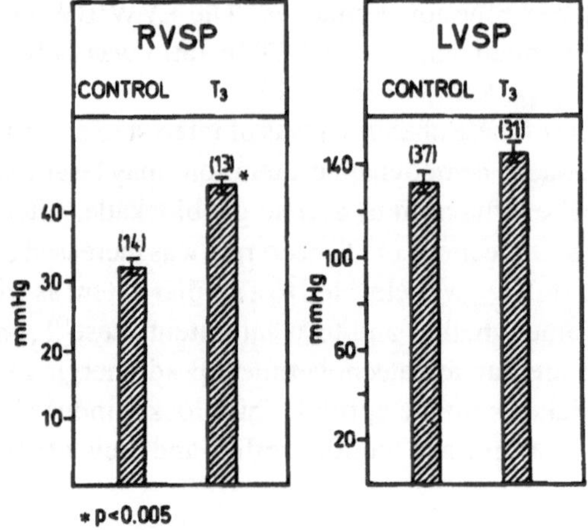

Fig. 10. Changes in right ventricular (RVSP) and left ventricular systolic pressure (LVSP) after 3 days of daily administrations of T_3. Mean values ± SEM, number of experiments in parentheses.

Metabolic studies

A typical metabolic effect of catecholamines which has been discovered recently [12, 36] concerns the stimulation of the PPP. Both isoproterenol and NE increased the activity of cardiac G-6-PD. The NE-induced increase was partially antagonized by the β-receptor blocker metoprolol and by the α-receptor blocker prazosin. When α- and β-blockers were combined, the NE-elicited G-6-PD stimulation was abolished (Fig. 11). Thus, both α- and β-receptor stimulation increases

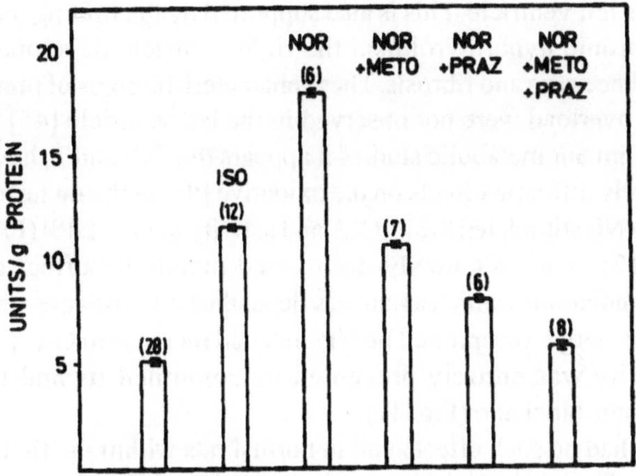

Fig. 11. Effect of isoproterenol (ISO, 25 mg/kg, s.c., measurement after 24 h) and of norepinephrine (NOR, 0.2 mg/kg/h continuous i.v. infusion for 48 h) alone and in combination with metoprolol (METO, 1 mg/kg/h), with prazosin (PRAZ, 0.1 mg/kg/h), and with both on the activity of cardiac glucose-6-phosphate dehydrogenase (units/g protein). Mean values ± SEM; number of experiments in parentheses.

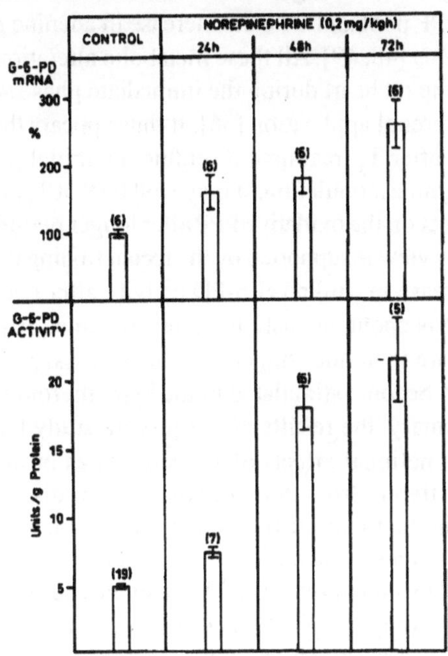

Fig. 12. Densitometric measurement of glucose-6-phosphate dehydrogenase (G-6-PD) mRNA levels and enzyme activity in hearts of control rats and after continuous i.v. infusion of NE. Mean values ± SEM, number of experiments in parentheses.

cardiac G-6-PD activity. When the changes in the mRNA content and in the activity of G-6-PD are plotted over the first 3 days, there was a steady increase in the mRNA which was followed by the elevation in enzyme activity (Fig. 12). In contrast to NE, T_3 had no effect on the activity of cardiac G-6-PD and of 6-PGD within the first 5 days of daily injections (Fig. 13).

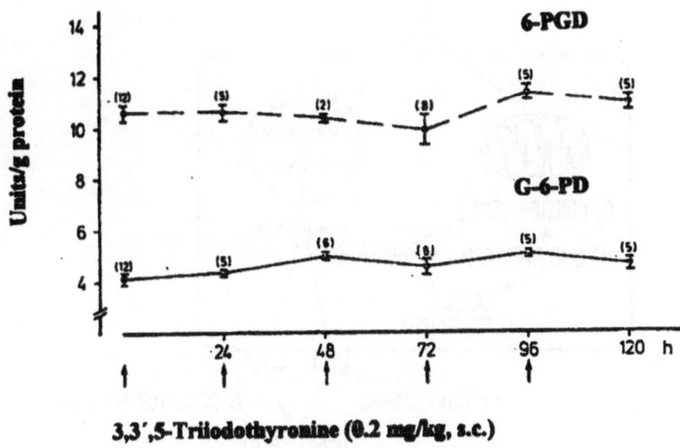

Fig. 13. Activity of glucose-6-phosphate dehydrogenase (G-6-PD) and of 6-phosphogluconate dehydrogenase (6-PGD) in rat hearts treated with daily injections of T_3. Mean values ± SEM; number of experiments in parentheses.

Discussion

In this study the short-term effects of NE and T_3 were examined in the intact rat. Our results indicate that NE has marked effects on basal functional parameters of the heart and the vascular system. NE proved to be positive chronotropic and inotropic, it increased total peripheral resistance and induced cardiac hypertrophy. The latter seems not to be due to the increase in total peripheral resistance, since it was present when the NE-induced functional alterations were almost entirely abolished by the calcium antagonist verapamil (Fig. 3). Thus, after 3 days, the development of NE-induced cardiac hypertrophy occurred independently of the concomitant increase in total peripheral resistance, but was dependent on the stimulation of cardiac α- and β-adrenergic receptors, since the combination of metoprolol and prazosin prevented cardiac hypertrophy in terms of LVW/BW increase (Fig. 4).

Similar to NE, T_3 had also positive chronotropic and inotropic effects (Fig. 5), but in contrast to NE it increased cardiac output (Fig. 6). This was due to the decrease in total peripheral resistance [43] which is opposed to the increase observed with NE treatment (Fig. 3). Thus, the effects of NE and T_3 on the systemic circulation are different. The T_3-induced positive chronotropic and inotropic effects were mediated by catecholamines, since they were entirely prevented by β-adrenergic receptor blockade with metoprolol (Fig. 5). However, the increase in cardiac output (Fig. 6) and in the cardiac RNA/DNA ratio (Fig. 7) as well as the development of cardiac hypertrophy (Fig. 8) were all not affected by metoprolol. This is in contrast to NE, since both the functional and morphological effects of NE were abolished with combined α- and β-receptor blockade. It thus appears that T_3 induces cardiac hypertrophy by an entirely different mechanism than NE. NE acts clearly via stimulation of α- and β-adrenergic receptors and thereby induces cardiac hyper-

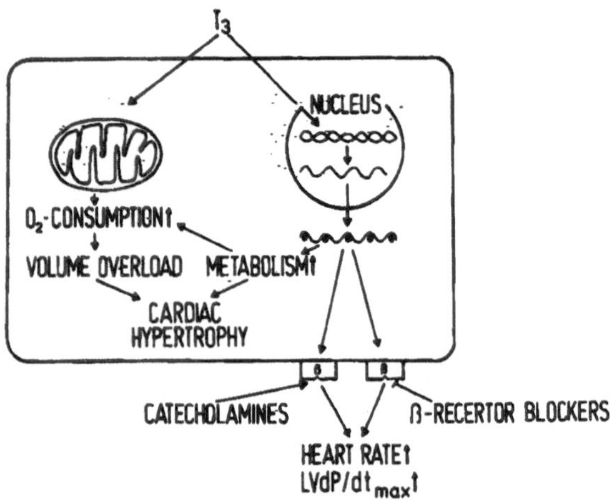

Fig. 14. Schematic presentation of the action of T_3 at the cellular level, and the physiological effects on heart function, the development of cardiac hypertrophy, and the permissive effect on catecholamines. Also shown is the effect of β-receptor blockers.

trophy. However, T_3 has an effect on β-adrenergic receptors only via increasing the number ([16], Fig. 14). When the β-adrenergic receptors were blocked, all the T_3-mediated metabolic and morphologic changes leading to the development of cardiac hypertrophy were undisturbed.

Not only the trigger mechanism for induction of cardiac hypertrophy seems to be different for NE and T_3, but also the consequences that are associated with hypertrophy. It is well known that isoproterenol induces focal myocardial cell lesions [44]. In addition, it was shown previously that cardiac DNA content increased linearly with the elevation in total peripheral resistance induced by NE [4]. However, the DNA tissue concentration in the left ventricle was lower in T_3-treated rats than in controls [43]. As a consequence, morphological cell lesions did not occur in the left ventricle of rats even in chronic hyperthyroidism [45]. These have only been observed and documented in the right ventricle.

In this regard it is interesting that both NE (Fig. 9) and T_3 (Fig. 10) induced an increase in RVSP. The effect was more pronounced with NE than with T_3. Although the NE-induced percent increase in RVSP was higher than that in LVSP, left ventricular hypertrophy was much more pronounced than right ventricular hypertrophy (unpublished observations). On the other hand, the T_3-induced right ventricular hypertrophy was more pronounced than left ventricular hypertrophy [43]. This corresponds well to data published earlier [46]. Also at the cellular level, hypertrophy of isolated cardiac myocytes was more extensive in the right ventricle [47]. This seems to be due to the fact that the right ventricle has to manage a pressure overload in addition to the volume overload which has been documented as an increase in cardiac output (Fig. 6). As a consequence, there was a disproportionate greater degree of cardiac hypertrophy in the right ventricle compared to the left ventricle. This is also supported by the finding that in chronic hyperthyroidism the right ventricle developed focal necrosis and fibrosis. These characteristic signs of pressure overload were not observed in the left ventricle [45].

From our metabolic studies it appears that NE and T_3 have entirely different effects on the oxidative PPP in the rat heart. Only NE stimulated the mRNA and activity of G-6-DPP (Fig. 11, 12). This is a newly discovered metabolic effect of catecholamines which again was dependent on cardiac α- and β-adrenergic receptors. The NE-induced increase in G-6-PD activity was entirely prevented by combined α- and β-receptor blockade (Fig. 11).

T_3 had no such effect at all in normal rats within the first 5 days of treatment. This seems to be surprising, since the number of cardiac β-adrenergic receptors is increased under the influence of thyroid hormones [16]. Furthermore, the cardiac content of cAMP is increased [49] and thus glycogenolysis is stimulated. Also the glycolytic capacity of the heart was shown to be enhanced within the first days of thyroid hormone treatment [22]. These changes were associated with the elevation of the PRPP pool and with the increase in adenine nucleotide biosynthesis [48, 49]. All these metabolic alterations did also occur in the rat heart during the immediate phase subsequent to isoproterenol application [36]. It thus appears that the first few days after T_3 treatment resemble the initial period after catecholamine stimulation. It may well be that T_3 has a stimulating effect on the oxidative PPP after longer periods of treatment. This view is supported by the recent finding that T_3 leads to an increase in cardiac G-6-PD activity after two weeks of treatment is spontaneously hypertensive rats [50]. Alternatively, more than one trigger may be necessary for G-6-PD activity to become stimulated in the hyperthyroid heart.

In summary, the results of our present study have shown that the functional effects of T_3 were not as pronounced as or even different from those of NE. A major discrepancy relates to the fact that all functional, metabolic and morphological changes induced by NE were greatly attenuated or prevented by concomitant blockade of adrenergic receptors. In contrast, only the positive chronotropic and inotropic effects of T_3 were abolished by β-receptor blockade. The T_3-induced increase in cardiac output and the development of cardiac hypertrophy occurred independently of adrenergic receptors. On the other hand, NE and T_3 had similar effects on the RVSP which was elevated though to a different degree. In metabolic terms, T_3 lacked the stimulating effect on the cardiac PPP which was very pronounced with NE.

Acknowledgements

This work was supported by grants from the Deutsche Forschungsgemeinschaft (Zi 199/8–2,3). The technical assistance of Sabine D'Avis and Sabine Weiland is gratefully acknowledged. Dr. C. Kolbeck-Rühmkorff is a fellow of the Deutsche Forschungsgemeinschaft.

References

1. Sutherland EW, Robison GA, Butcher RW: Some aspects of the biological role of adenosine 3',5'-monophosphate (cyclic AMP). Circulation 37: 279–306, 1968
2. Tada M, Katz AM: Phosphorylation of the sarcoplasmatic reticulum and sarcolemma. Ann Rev Physiol 44: 401–423, 1982
3. Rupp H, Berger H-J, Pfeifer A, Werdan K: Effect of positive inotropic agents on myosin isozyme population and mechanical activity of cultured rat heart myocytes. Circ Res 68: 1164–1173, 1991
4. Zierhut W, Zimmer H-G: Significance of myocardial α- and β-adrenoceptors in catecholamine-induced cardiac hypertrophy. Circ Res 65: 1417–1425, 1989
5. Berridge MJ, Irvine RF: Inositol trisphosphate, a novel second messenger in cellular signal transduction. Nature 312: 315–321, 1984
6. Nishizuka Y: The molecular heterogeneity of protein kinase C and its implications for cellular regulation. Nature 334: 661–665, 1988
7. Schmitz W, Scholz H, Scholz J, Steinfath M: Increase in IP_3 precedes α-adrenoceptor-induced increase in force of contraction in cardiac muscle. Eur J Pharmacol 140: 109–111, 1987
8. Kohl C, Schmitz W, Scholz H, Scholz J, Toth M, Döring V, Kalmar P: Evidence for α_1-adrenoceptor-mediated increase of inositol trisphosphate in the human heart. J Cardiovasc Pharmacol 13: 324–327, 1989
9. Kolbeck-Rühmkorff C, Horban A, Zimmer H-G: Effect of pressure and volume overload on proto-oncogene expression in the isolated working rat heart. Cardiovasc Res 27: 1998–2004, 1993
10. Long CS, Ordahl CP, Simpson PC. α_1-Adrenergic receptor stimulation of sarcomeric actin isogene transcription in hypertrophy of cultured rat heart muscle cells. J Clin Invest 83: 1078–1082, 1989
11. Lee HR, Henderson SA, Reynolds R, Dunnmon P, Yuan D, Chien KR: α_1-Adrenergic stimulation of cardiac gene transcription in neonatal rat myocardial cells. Effects on myosin light chain-2-gene expression. J Biol Chem 263: 7352–7358, 1988
12. Zimmer H-G, Lankat-Buttgereit B, Kolbeck-Rühmkorff C, Nagano T, Zierhut W: Effects of norepinephrine on the oxidative pentose phosphate pathway in the rat heart. Circ Res 71: 451–459, 1992
13. Oppenheimer JH: Thyroid hormone action at the cellular level. Science 203: 971–979, 1979
14. Weinberger C, Thompson CC, Ong ES, Lebo R, Gruol DJ, Evans RM: The c-erb-A gene encodes a thyroid hormone receptor. Nature 324: 641–646, 1986
15. Martial JA, Baxter JD, Goodman HM, Seeburg PH: Regulation of growth hormone messenger RNA by thyroid and glucocorticoid hormones. Proc Natl Acad Sci USA 74: 1816–1820, 1977
16. Williams LT, Lefkowitz RJ, Watanabe AM, Hathaway DR, Besch HR: Thyroid hormone regulation of β-adrenergic receptor number. J Biol Chem 252: 2787–2789, 1977
17. Ismail-Beigi F, Bissell DM, Edelman IS: Thyroid thermogenesis in adult rat hepatocytes in primary monolayer culture. Direct action of thyroid hormone *in vitro*. J Gen Physiol 73: 369–383, 1979
18. Bartolome J, Huguenard J, Slotkin TA: Role of ornithine decarboxylase in cardiac growth and hypertrophy. Science 210: 793–794, 1980
19. Pegg AE, Hibasami H: Polyamine metabolism during cardiac hypertrophy. Am J Physiol 239: E372–E378, 1980
20. Schimmelpfennig K, Sauerberg M, Neubert D: Stimulation of mitochondrial RNA synthesis by thyroid hormone. FEBS Letters 10: 269–272, 1970
21. Nelson BD, Joste V, Wielburski A, Rosenqvist U: The effects of triiodothyronine on the synthesis of mitochondrial proteins in isolated rat hepatocytes. Biochim Biophys Acta 608: 422–426, 1980
22. Seymour A-ML, Eldar H, Radda GK: Hyperthyroidism results in increased glycolytic capacity in the rat heart. A ^{31}P-NMR study. Biochim Biophys Acta 1055: 107–116, 1990
23. Tanaka T, Morita H, Koide H, Kawamura K, Takatsu T: Biochemical and morphological study of cardiac hypertrophy. Effects of thyroxine on enzyme activities in the rat. Basic Res Cardiol 80: 165–174, 1985
24. Hoh JFY, McGrath PA, Hale PT: Electrophoretic analysis of multiple forms of rat cardiac myosin: effects of hypophysectomy and thyroid replacement. J Mol Cell Cardiol 10: 1053–1076, 1977
25. Hoh JFY, Egerton LJ: Action of triiodothyronine on the synthesis of rat ventricular myosin isoenzymes. FEBS Letters 101: 143–148, 1979
26. Chizzonite RA, Zak R: Regulation of myosin isoenzyme composition in fetal and neonatal rat ventricle by endogenous thyroid hormones. J Biol Chem 259: 12628–12632, 1984
27. Gustafson TA, Markham BE, Morkin E: Effects of thyroid hormone on α-actin and myosin heavy chain gene expression in cardiac and skeletal muscle of the rat: measurement of mRNA content using synthetic oligonucleotide probes. Circ Res 59: 194–201, 1986
28. Zimmer H-G: Measurement of left ventricular hemodynamic parameters in closed-chest rats under control and various pathophysiological conditions. Basic Res Cardiol 78: 77–84, 1983
29. Zimmer H-G, Zierhut W, Seesko RC, Varekamp AE: Right heart catheterization in rats with pulmonary hypertension and right ventricular hypertrophy. Basic Res Cardiol 83: 48–57, 1988
30. Zimmer H-G: The oxidative pentose phosphate pathway in the heart: Regulation, physiological significance, and clinical implications. Basic Res Cardiol 87: 303–316, 1992
31. Zimmer H-G, Ibel H: Effects of ribose on cardiac metabolism and function in isoproterenol-treated rats. Am J Physiol 245: H880–H886, 1983
32. Zimmer H-G, Ibel H, Steinkopff G, Korb G: Reduction of the isoproterenol-induced alterations in cardiac adenine nucleotides and morphology by ribose. Science 207: 319–321, 1980
33. Zimmer H-G, Ibel H: Ribose accelerates the repletion of the ATP pool during recovery from reversible ischemia of the rat myocardium. J Mol Cell Cardiol 16: 863–866, 1984
34. Zimmer H-G: Normalization of depressed heart function in rats by ribose. Science 220: 81–82, 1983
35. Zimmer H-G, Martius PA, Marschner G: Myocardial infarction in rats: effects of metabolic and pharmacologic interventions. Basic Res Cardiol 84: 332–343, 1989
36. Zimmer H-G, Ibel H, Suchner U: β-Adrenergic agonists stimulate the oxidative pentose phosphate pathway in the rat heart. Circ Res 67: 1525–1534, 1990
37. Glock GE, McLean P: Further studies on the properties and assay of glucose 6-phosphate dehydrogenase and 6-phosphogluconate dehydrogenase of rat liver. Biochem J 55: 400–408, 1953
38. Glock GE, McLean P: Levels of enzymes of the direct oxidative pathway of carbohydrate metabolism in mammalian tissues and tumours. Biochem J 56: 171–175, 1954
39. Robinson HW, Hodgen CG: The biuret reaction in the determination of serum proteins. J Biol Chem 135: 727–731, 1940
40. Birnboim HC: Rapid extraction of high molecular weight RNA from cultured cells and granulocytes for Northern analysis. Nucleic Acids Res 16: 1487–1497, 1988
41. Maniatis T, Fritsch EF, Sambrook J: Molecular Cloning. Cold Spring Harbor, NY. Cold Spring Harbor Laboratory, 1982

42. Wallenstein S, Zucker CL, Fleiss JL: Some statistical methods useful in circulation research. Circ Res 47: 1–9, 1980
43. Zierhut W, Zimmer H-G: Differential effects of triiodothyronine on rat left and right ventricular function and the influence of metoprolol. J Mol Cell Cardiol 21: 617–624, 1989
44. Rona G, Chappell CI, Kahn DS: The significance of factors modifying the development of isoproterenol-induced myocardial necrosis. Am Heart J 66: 389–395, 1963
45. Gerdes AM, Moore JA, Bishop SP: Failure of propranolol to prevent chronic hyperthyroid induces hypertrophy and multifocal cellular necrosis in the rat. Can J Cardiol 1: 340–345, 1985
46. Van Liere EJ, Sizemore DA, Hunnel J: Size of cardiac ventricles in experimental hyperthyroidism in the rat. Proc Soc Exp Biol Med 132: 663–665, 1969
47. Gerdes AM, Moore JA, Hines JM: Regional changes in myocyte size and number in propranolol-treated hyperthyroid rats. Lab Invest 57: 708–713, 1987
48. Zimmer H-G, Ibel H, Gerlach E: Significance of the hexose monophosphate shunt in experimentally induced cardiac hypertrophy. Basic Res Cardiol 75: 207–213, 1980
49. Zimmer H-G, Peffer H: Metabolic aspects of the development of experimental cardiac hypertrophy. Basic Res Cardiol 81 (Suppl 1): 127–137, 1986
50. Heckmann M, Zimmer H-G: Effects of triiodothyronine in spontaneously hypertensive rats. Studies on cardiac metabolism, function and heart weight. Basic Res Cardiol 87: 333–343, 1992

Molecular and Cellular Biochemistry **147**: 115–122, 1995.

On the mechanism and possible therapeutic application of delayed adaptation of the heart to stress situations

Ernst-Georg Krause[1] and László Szekeres[2]

[1]*Max Delbrück Centre for Molecular Medicine, Department of Molecular Cardiology, Berlin-Buch, Germany;* [2]*Institute of Pharmacology, Albert Szent-Györgyi Medical University, Szeged, Hungary*

Abstract

Mild (not harmful) stress may initiate an *adaptive mechanism*, protecting the heart from harmful consequences of a more severe stress. There are at least three known types of cardiac adaptation to stress, such as:

a) the gradually developing, long lasting adaptation to chronic mechanical overload, leading to cardiac hypertrophy, later to cardiomyopathy and heart failure,
b) the rapidly developing adaptation to moderate stress initiated by 'preconditioning' brief coronary occlusion(s) or brief periods of rapid cardiac pacing, protecting for less than 1 h against consequences of a subsequent, severe stress,
c) the later appearing, more prolonged cardio-protective adaptation, described by us in 1983, induced by various forms of more severe but not injurious stimuli, such as an optimal dose of prostacyclin or its stable analogues; or a series of brief periods of rapid pacings.

This form of cardiac adaptation to stress protects for 24–48 h against consequences of a more severe stress such as:

1. myocardial ischaemia;
2. early and late postocclusion and reperfusion arrhythmias;
3. early morphologic changes secondary to ischaemia and reperfusion;
4. ischaemia induced myocardial loss of K^+ and accumulation of Na^+ and Ca^{++};
5. it may increase the tolerance to the toxic effects of cardiac glycosides.

A reduced response to beta-adrenergic stimuli and a concomitant increase in activity and amount of PDE I and IV was shown by us earlier. The hypothesis that these factors may play a role in the mechanism of delayed protection was confirmed by our present findings according to which 7-oxo-PgI_2-treatment greatly attenuated the dose dependent isoprenaline-induced increase in contractility, relaxation and myocardial cAMP level in rat hearts isolated 48 h after 7-oxo-PgI_2. In addition all these values are in close correlation with each other.

The endogenous 'self-defence' of the heart, based on adaptation represents a *new therapeutic concept*, different from the classical drug-receptor interaction produced protection. Its possible exploitation to therapeutic use requires that the adaptation inducing stress should be *applicable* to *patients*, furthermore prolongation of duration of protection should be possible. As a first step in testing applicability to therapy we had to show that drug induced adaptive protection is existing in the conscious animal. In our present study we found an attenuation of rapid pacing induced elevation of the ST-segment in the endocardial electrogram and in the left ventricular end diastolic pressure in conscious rabbits 24–48 h after treatment with Iloprost. Besides we found an attenuation of tachycardia and arrhythmias due to two stage coronary artery ligation in conscious dogs 48 h after pretreatment with 7-oxo-PgI_2. Finally we were able to demonstrate that protection against coronary artery occlusion-induced ST segment elevation and arrhythmias can be prolonged at will by periodically repeated maintenance doses. (Mol Cell Biochem **147**: 115–122, 1995)

Key words: cardioprotection, delayed adaptation, cAMP, PDE-isoenzymes, prolongation of protection

Address for offprints: E. -G. Krause, Max Delbrück Centre for Molecular Medicine, Department of Molecular Cardiology, Robert-Rössle-Strasse 10, D-13125 Berlin-Buch, Germany

Introduction

Benign (not injurious) stress might initiate an *adaptation* process in the heart (and the whole organism), leading to changes, which may protect against harmful consequences of a subsequent, more severe stress. This endogenous 'self-defence' of the heart represents a *new therapeutic concept*, different from the protection produced by the classical drug-receptor interaction.

Apart from the well known and widely studied *chronic cardiac adaptation* to long lasting mechanical overload, resulting in cardiac hypertrophy, at least two forms of cardiac adaptation to stress are known at present.

1. *Rapid adaptation*, evoked by a brief ischaemic 'preconditioning' stress, like short-term coronary artery occlusion or rapid cardiac pacing, or a brief heat stress may afford a short-lived (1–3 h) protection against harmful effects of a subsequent, more severe stress, such as a prolonged coronary artery occlusion. (See last review: 1.)
2. *Delayed adaption* provides a longer protection, lasting for a few days. It was described by us first in 1983 [2]. We could induce this phenomenon by drugs [3], such as prostacyclin (PgI_2) and its stable analogs: 7-oxo-PgI_2 and Iloprost (Carbacyclin); as well as by a series of brief periods of rapid cardiac pacing [4, 5].

The *protective effects* of drug induced delayed adaptation to stress appeared as a:

1. *moderation of myocardial ischaemia* due to occlusion of the left anterior descending (LAD) coronary artery [6, 7];
2. prevention of ischaemia and reperfusion induced changes in *transmembrane ion-transport*, such as myocardial loss of K^+ and accumulation of Na^+ and Ca^{++} [8, 9];
3. prevention of ischaemia and reperfusion induced early changes in the *myocardial ultrastructure* [8–10, 13];
4. *moderation of arrhythmias* appearing early [10–13] and late [14] after LAD coronary artery occlusion and reperfusion, as well as after toxic doses of cardiac glycosides [15–17];
5. *Electrophysiological changes*, such as prolongation of the refractory period and duration of the action potential [18–21].

Objective of our present investigations was:

1. to further elucidate the mechanism of this delayed form of adaptation, providing a longer protection against consequences of severe stress. We have shown earlier that delayed adaptation is associated with an increased activity and induction of key enzymes, such as Na/K-ATPase [22] and PDE I and IV [23]. The question was whether diminished response to adrenergic stimuli, an essential factor in the mechanism of cardioprotection due to delayed adaptation correlates with similar changes in myocardial cAMP-level, further how are these changes related to the above quoted data on increased activity and synthesis of PDE I and IV due to delayed adaptation?
2. The second question investigated was whether cardioprotection due to delayed adaptation to stress, could be *exploited to use in therapy*?

To answer this question certain requirements should be fulfilled:

1. The stress, needed to induce adaptation should be *applicable* to clinical therapy.
2. *Duration* of protection should be long enough to justify the use of an adaptation initiating stress, which like the use of drugs may involve the risk of side effects.

What concerns the first requirement, it is obvious that even brief coronary occlusions (which represent the most frequently used form of preconditioning stresses in animal experiments) cannot be applied under clinical conditions. The use of global ischemia, evoked by rapid pacing or exercise-induced tachycardia or thermal stress is limited by the state of health of the patient, involving the risk of rhythm disturbances or heart failure, if the heart is also diseased. In the light of these considerations the use of drugs described by us first with prostacyclin in 1983 [2] and later with its stable derivatives [3] seems to be the most appropriate procedure to initiate delayed adaptation for possible therapeutic application. In connection with this latter problem we wanted to investigate:

1. whether drug induced delayed adaptation to stress can be induced in the conscious animal, protecting the heart against a moderate and a more severe form of stress;
2. whether the protective action can be prolonged by periodical administration of the drug initiating delayed adaptation to stress.

Materials and methods

Materials

DL-Isoprenaline was purchased from Sigma Chemicals Company, Germany; cAMP and cGMP from Boehringer Mannheim Gmbh, Germany; 7-oxo-PgI_2, a stable analogue of prostacyclin was provided by the courtesy of Chinoin, Hungary and Iloprost (carbacycline derivative) another stable PgI_2-analogue was donated from Schering AG, Germany.

Animals, preparations and experimental protocols

All animal experiments were performed in accordance with the Declaration of Helsinki and internationally accepted principles concerning the care and the use of laboratory animals. Male albino Wistar rats (250–300 g) kept on standard pellet diet with free access to water were used in all experiments.

The rats were injected with physiological NaCl buffer (control) or with 50 μg/kg body weight 7-oxo-PgI_2 i.m. 48 h prior to excision of the heart. This was performed in anesthetized (30 mg/kg i.m. pento-barbiturate), heparinized animals and the hearts isolated according to Langendorff were perfused at a constant flow of 10 ml/min at 37°C with a modified Krebs-Henseleit bicarbonate-buffered medium containing 1.5 mM $CaCl_2$ and 11 mM glucose and gassed with a 19:1 mixture of oxygen + carbon-dioxide. In the isolated heart diastolic tension was pre-calibrated by a constant load (8 g). The beta adrenergic responsiveness was tested by switching to a perfusate which contained 1.0, 10.0 or 100.0 nM isoprenaline and the hearts were freeze-clamped 30 sec later. Tissues were stored in liquid nitrogen until use. Contractile parameters were monitored using an isometric force-displacement transducer attached to the apex of the heart. The developed contractile force and maximal velocity of tension development (+df/dt) and the rate of relaxation (–df/dt) were monitored before and during drug intervention.

To test the validity of drug induced delayed cardioprotection in the conscious animal male white New Zealand rabbits of 2.5–3.0 kg body weight were used. Details of surgical procedure for implantation of catheters were described earlier [5]. In the conscious rabbit with a catheter implanted in the left ventricle and a pacing electrode in the right ventricle a *mild* form of stress-test was used. A brief global ischemia evoked by rapid pacing (500/min for 5 min) resulted in a temporary elevation in the ST segment of the endocardial electrogram and in the left ventricular end-diastolic pressure (LVEDP). A week after surgery 6 instrumented control animals were given NaCl buffer, and other 6 instrumented animals 10 ug/kg Iloprost i.m.LVEDP and ST segment elevation were estimated before and immediately after rapid pacing in nontreated controls as well as Iloprost-treated animals 20 min, 2, 24, 48, 72 and 96 h after drug treatment.

In mongrel dogs of 15–20 kg body weight a *more severe* stress-test was used, namely late postocclusion arrhythmias, appearing 24 h after two-stage ligation of the LAD coronary artery. This was performed under aseptic conditions in pentobarbitone anaesthesia (30 mg/kg). 7-oxo-PgI_2 (50 μg/kg i.m.) was given 24 h before ligation to avoid interference of this latter with the process of adaptation. Thus 48 h elapsed when late dysrhythmias were first tested with the aid of telemetric ECG recording from chest leads. Since the second test-day coincided with the decline of protection (72 h), additional 25 μg/kg 7-oxo-PgI_2 was given to prolong protection.

To test whether drug-induced protective action of delayed adaptation can be prolonged beyond 48 h, groups of mongrel dogs (6 animals each) were anesthetized, thoracotomized and subjected to 25 min LAD coronary artery occlusion 6, 24, 48 and 72 h after 50 μg/kg 7-oxo-PgI_2. A fifth group served as untreated control (but similarly subjected to coronary occlusion). The sixth group received the same inducing dose and every third day a 25 μg/kg maintenance-dose was given; altogether four times. Protection of pretreatment was tested by:

1. ST segment elevation in the epicardial ECG;
2. Number of extrasystoles (ES) during occlusion and reperfusion;
3. Inhomogeneity of conduction (maximal delay).

Assay of cAMP and protein concentration

Tissue levels of cAMP were estimated in neutralized trichloracetic acid extracts prepurified by column chromatography. Protein concentration was measured according to Lowry *et al.* with egg albumin as standard. For further details see [23].

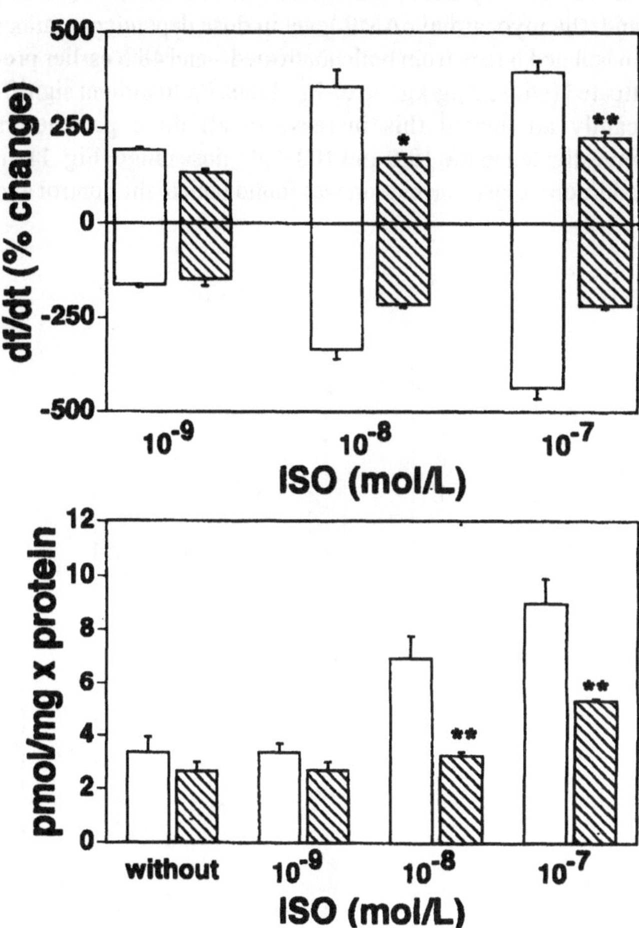

Fig. 1. Effect of 1.0, 10.0 and 100.0 nM isoprenaline (ISO) on per cent changes (vs control = 100 per cent) on: a) maximal velocity of tension development (+df/dt); b) rate of relaxation (–df/dt); and c) on the cAMP-level in isolated rat hearts prepared from animals 48 h after treatment with saline (controls: empty column) and with 50 μg/kg 7-oxo-PgI_2 (treated: shaded column). Data are expressed as mean ± SEM of at least 5 experiments. Significant difference from control group is given at level *$p < 0.05$; **$p < 0.005$.

Statistics

The data are given as means ± SEM of at least 5 experiments. The significant of differences was calculated with Students paired *t*-test preceded by analysis of variance for repeated measurements, and with the u test. Significance was assumed when p was less than or equal to 0.05.

Results

Beta adrenergic stimulation on contractile force, relaxation and myocardial cAMP-level

Beta adrenergic stimulation by 1.0, 10.0 and 100.0 nM Isoprenaline increased contractility as maximal velocity of tension development (+df/dt), the rate of relaxation (–df/dt) and the myocardial cAMP level in dose dependent manner in isolated hearts from both nontreated – and 48 h earlier pretreated (with 50 µg/kg 7-oxo-PgI_2) rats. Pretreatment significantly attenuated this increase in all three parameters investigated in the 10.0 and 100.0 nM dose range (Fig. 1). In addition a close correlation was found both in the control and pretreated group between Isoprenaline produced increase in contractile response and tissue cAMP-level (r = 0.982 for controls; r = 0.977 for pretreated); and relaxation and cAMP-level (control r = 0.998; pretreated r = 0.690).

Effect of Iloprost on ST-segment and LVEDP elevation due to rapid cardiac pacing in the conscious rabbit

In the conscious, chronically instrumented rabbit rapid pacing induced elevation of ST-segment in the endocardial electrogram and the rise of LVEDP were not affected 20 min, however markedly reduced 24 and 48 h but not 72 h after i.v. injection of 10 µg/kg Iloprost. Thus there was no early protection (within 1 h after drug administration). (Fig. 2.)

Effect of pretreatment on extrasystolic activity and tachycardia in the conscious dog 24 and 48 h after two-stage ligation of the LAD

The changes are most expressed 24 h after two-stage ligation, showing a marked increase in heart rate and in the number of

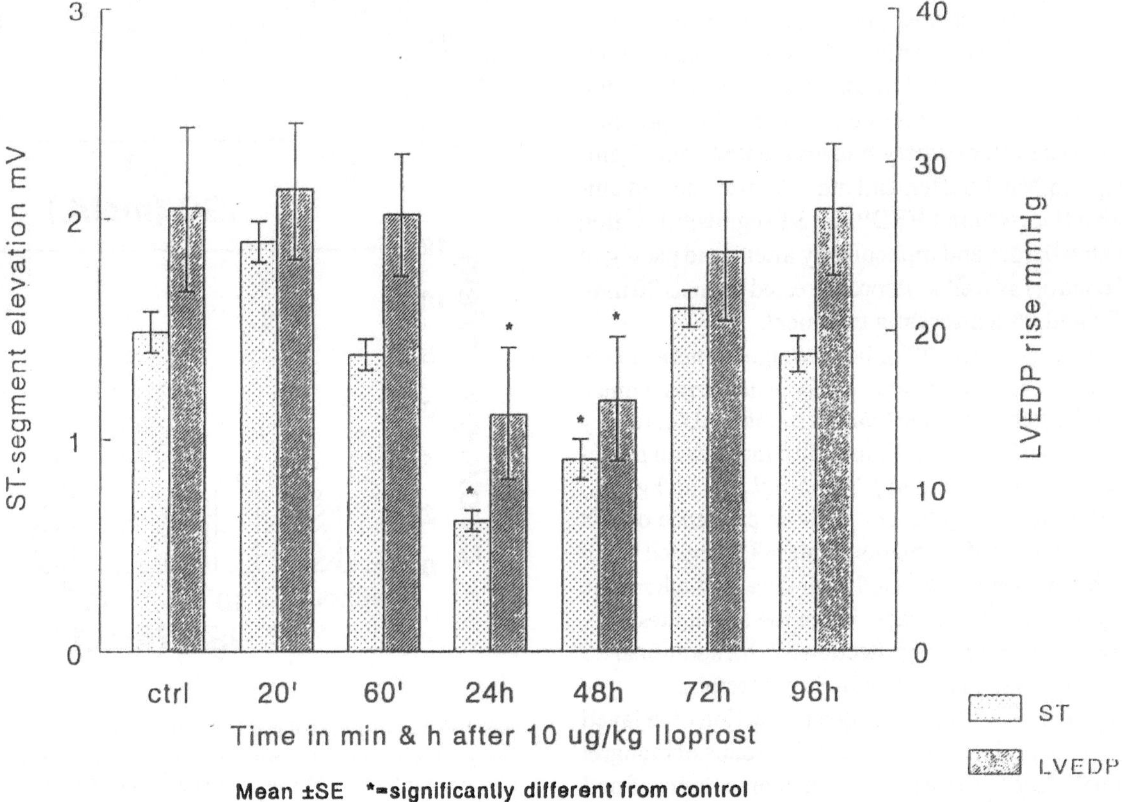

Fig. 2. Effect of a single dose of 10 µg/kg Iloprost on ST-segment elevation in the endocardial electrogram and on the rise in LVEDP produced by rapid cardiac pacing (500/min for 5 min) in the conscious, chronically instrumented rabbit at different times after treatment. Light column = ST segment elevation; dark column = LVEDP; Ordinate left: ST-segment elevation in mV; Ordinate right: LVEDP rise in mmHg. Abscissa = ctrl: control; minutes and hours after Iloprost administration. Data are expressed as mean ± SEM of 6 animals. * = significant difference from control values.

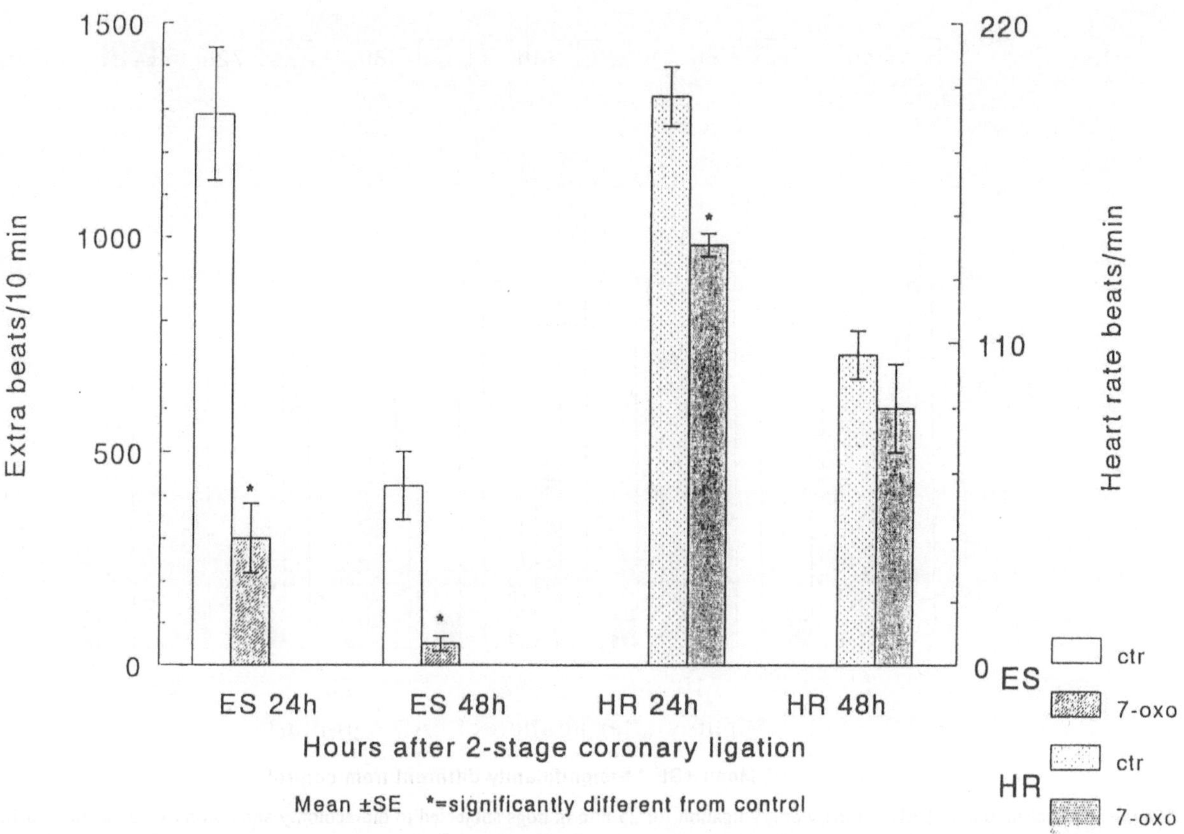

Fig. 3. Effect of 50 μg/kg 7-oxo-prostacyclin given 24 h before two-stage ligation of the LAD coronary artery on extrasystolic activity and heart rate, 24 and 48 h after coronary occlusion in the conscious dog. Light column = control; dark column = 7-OXO-pretreated. Ordinate left: Extra beats/10 min; Ordinate right: Heart rate beats/min; Abscissa: hours after two-stage coronary artery ligation. Other signs and symbols as in Fig. 2.

extra-beats/10 min. Both changes were considerably attenuated in the 7-oxo-prostacyclin pretreated group. (Fig. 3)

Prolongation of 7-oxo-PgI$_2$ induced cardioprotection in dogs by periodical administration of maintenance doses

In dogs the time interval between administration of the inducing 50 μg/kg 7-oxo-PgI$_2$ dose and ligation of the LAD coronary artery for 25 min markedly influenced the response to coronary occlusion. It significantly attenuated ST segment elevation, inhomogeneity of intraventricular conduction (expressed as maximal conduction delay in msec) and the number of extra beats at time intervals of 6, and particularly 24 and 48 h but not 72 h after 7-oxo-PgI$_2$ administration. (Figs 4–6) This drug induced protection lasting for 2 days could be extended by administration every second day of 25 μg/kg 7-oxo-PgI$_2$ as a maintenance dose, which given alone failed to initiate delayed protective effect. In such a way protection could be prolonged for 14 days in our experiments, showing the most expressed reduction of changes due to coronary occlusion among all time intervals investigated (see black column in Figs 4–6).

Discussion

The strong correlation between dose dependent beta adrenergic increase of contractility and relaxation on one hand and the concomitant increase of myocardial cAMP-level on the other hand suggests that these factors could be closely interrelated. This supports our hypothesis mentioned in our recent paper [23] that an essential factor in the mechanism of prostacyclin-induced delayed cardioprotection, namely the anti beta adrenergic effect is due to the adaptative attenuation of stress induced rise in myocardial cAMP-level. In this study [23] a significant increase of PDE isoforms I and IV was described 48 h after treatment with 7-oxo-PgI$_2$. Diminution of tissue cAMP-level could result from decreased generation, or from an increased breakdown of the second messenger. According to findings of the same paper, the first possibility could be excluded, since beta-adrenoceptor densities and affinities, further G protein patterns and adenylyl cyclase activity were not significantly affected by 7-oxo-PgI$_2$-pretreatment. On the other hand the enhanced capacity of PDE isoforms IV directly contributes to breakdown of severe stress produced excess

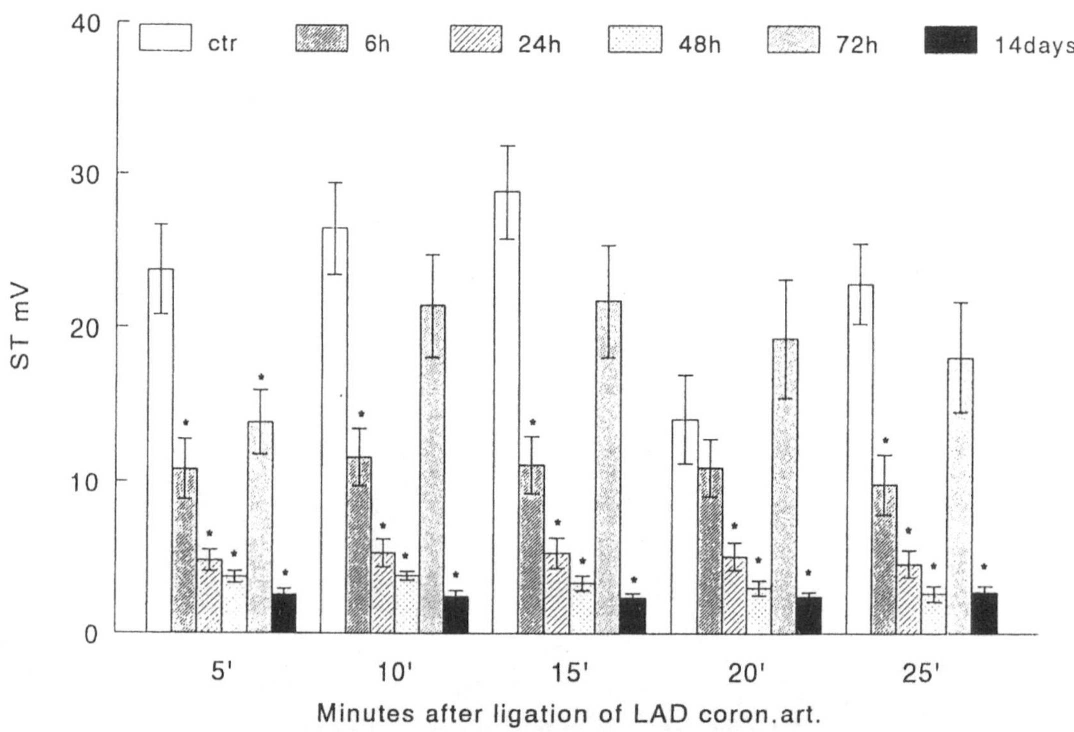

Fig. 4. ST-segment elevation due to LAD coronary artery ligation for 25 min in dogs subjected to thoracotomy and coronary occlusion at different times after treatment with 50 μg/kg 7-oxo-PgI_2. Ordinate: ST mV; Abscissa: min after ligation of the LAD coronary artery. Other signs and symbols as in Fig. 2.

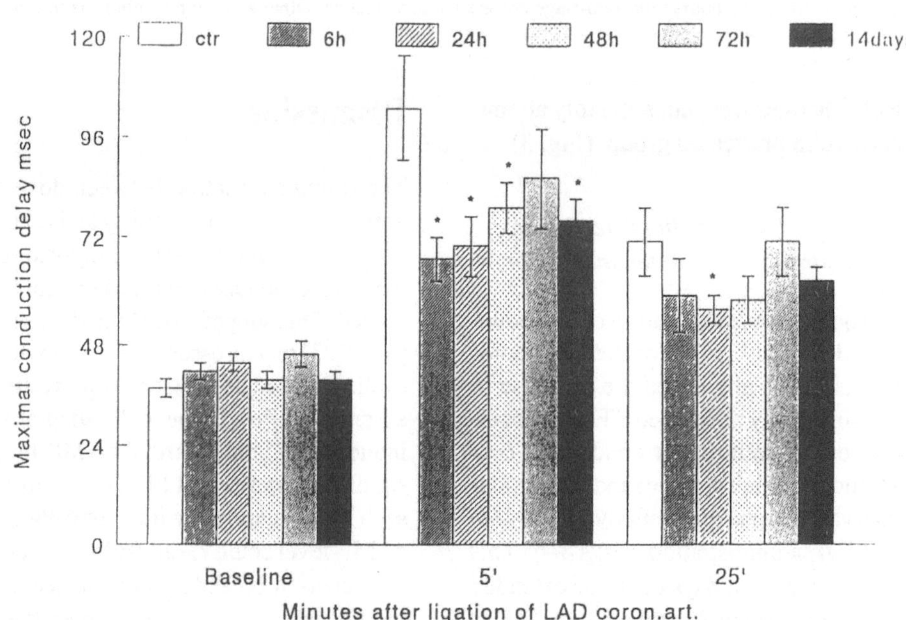

Fig. 5. Inhomogeneity of conduction due to LAD coronary artery ligation for 25 min in dogs subjected to thoracotomy and coronary occlusion at different times after treatment with 50 μg/kg 7-oxo-PgI_2. Ordinate: Maximal conduction delay msec; Abscissa: min after ligation of the LAD coronary artery. Other signs and symbols as in Fig. 2.

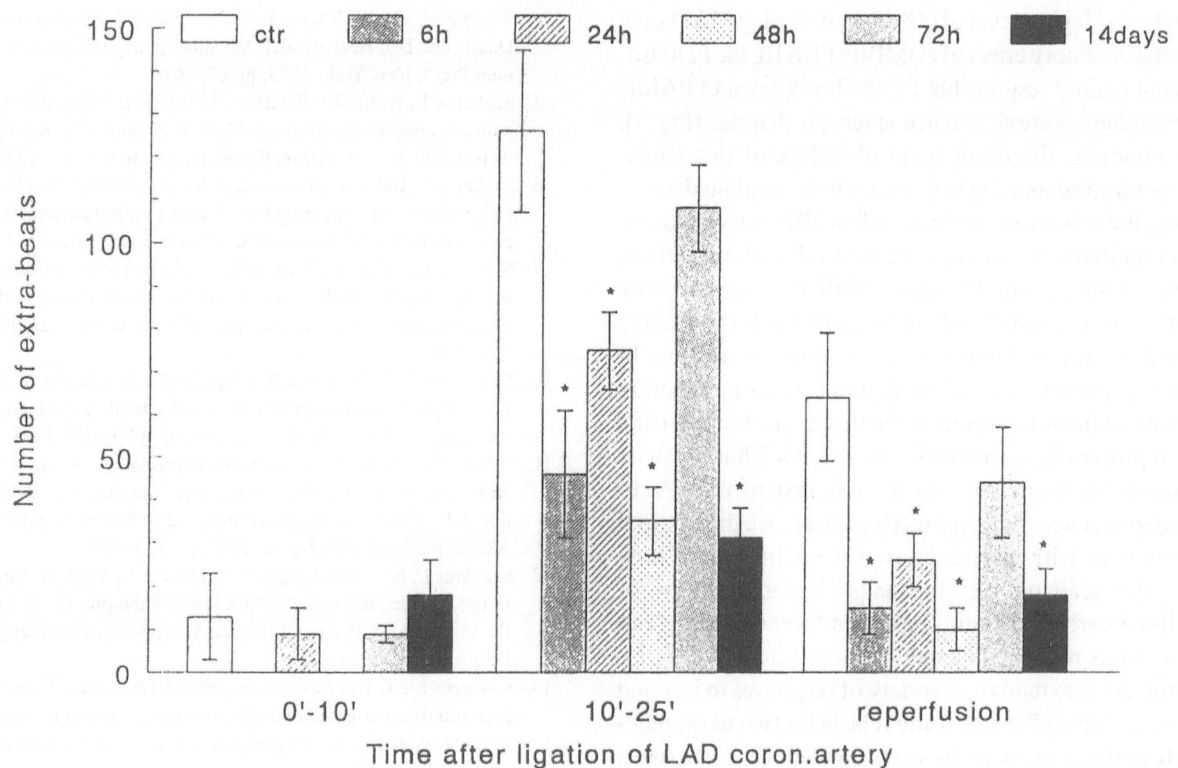

Fig. 6. Extrasystolic activity due to LAD coronary artery ligation for 25 min in dogs subjected to thoracotomy and coronary occlusion at different times after treatment with 50 μg/kg 7-oxo-PgI_2. Ordinate: Number of extra beats; Abscissa: min after ligation of the LAD coronary artery. Other signs and symbols as in Fig. 2.

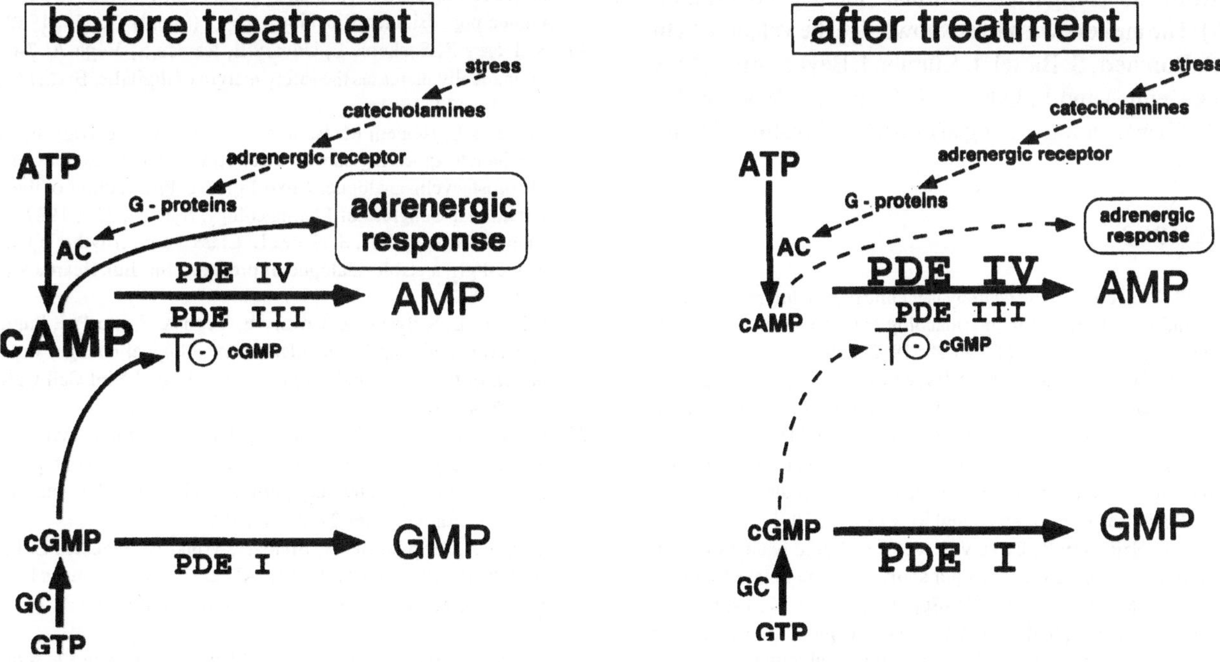

Fig. 7. Schematic diagram of ischaemia and/or adrenergic beta stimulation induced changes in PDE isoforms I and IV and their action on cAMP before and after treatment with 7-oxo-PgI_2.

cAMP, whereas higher level of cGMP-hydrolysing PDE I may alleviate the inhibitory effect of cGMP on PDE III, the PDE isoform, being mainly responsible for the breakdown of cAMP. These interrelations are shown in a schematic diagram (Fig. 7).

What concerns the other main objective of this study, whether delayed adaptation to stress could be exploited to use in therapy, the results are in favour of an affirmative answer. The first requirement, namely a not harmful, initiating stress, applicable to the patient, has been fulfilled by us earlier by showing the successful use of drugs (prostacyclin and stable analogues) to initiate delayed adaptive cardioprotection. In the present paper we could show that this type of protection can be induced in the conscious animal, such as the rabbit and the dog, representing rather different species. Thus it can be hoped that the mechanism is working in man as well. As to the time dependence of the protective action, we did not find an early (within 1 h) protection, so it is not likely that drug induced delayed adaptive protection is a 'second window' of a short-lived early protection based on 'preconditiong'. In these conscious models we could demonstrate that drug induced protection extends to a variety of responses to less and more severe forms of stresses and it lasts for two days, independently of the severity of the stress applied. These two days could be prolonged by periodic stimulation of induction of key enzymes, using low, maintenance doses of the drug.

Acknowledgements

A part of this work was supported by a grant-in-aid (No. 1353) from the Hungarian National Science Foundation (OTKA). The authors wish to acknowledge the valuable help of Drs G. Borchert, S. Bartel, I. Küttner, I. Beyerdörfer, (Max Delbrück Centre) and É. Udvary, Z. Szilvássy, Å. Végh (Institute of Pharmacology, Szeged) for their invaluable help.

References

1. Parratt JR: Protection of the heart by ischemic preconditioning: mechanisms and possibilities for pharmacological evaluation. Trends in Pharmacological Sciences (TIPS) 15: 19–25, 1994
2. Szekeres L, Krassói I, Udvary É: Delayed antischemic effect of PgI_2 and of a new stable PgI_2 analogue 7-oxo-prostacyclin-Na in experimental model angina in dogs. J Mol Cell Cardiol 15 (Suppl I): 394, 1983
3. Szekeres L, Pataricza J, Szilvássy Z, Udvary É, Végh Å: Cardioprotection: endogenous protective mechanisms promoted by prostacyclin. Basic Res Cardiol 86 (Suppl 3): 215–221, 1991
4. Szekeres L, Szilvássy Z, Udvary É, Végh Å: Rapid pacing evoking myocardial ischaemia induces both short-term and delayed cardioprotection. J Mol Cell Cardiol 23 (Suppl V): S72, p 47, 1991
5. Szekeres L, Papp JGy, Szilvássy Z, Udvary É, Végh Å: Moderate stress by cardiac pacing may induce both short term and long term cardioprotection. Cardiovasc Res 27: 593–596, 1993
6. Szekeres L, Krassói I, Pataricza J, Udvary É: Delayed antiischaemic effect of prostaglandin I_2 and of a new stable prostaglandin I_2 analogue, 7-oxo-prostacyclin-Na, in experimental model angina in dogs. In: N.S. Dhalla and D.J. Hearse (eds). Advances in Myocardiology, Vol 6. Plenum Press, New York, 1985, pp 607–618
7. Szekeres L, Koltai M, Pataricza J, Takáts I, Udvary É: On the late antiischaemic action of the stable PgI_2 analogue: 7-oxo-PgI_2-Na and its possible mode of action. Biomed Biochim Acta 43: 135–142, 1984
8. Szekeres L, Bálint Zs, Karcsú S, Tósaki Å, Udvary É: On the 7-oxo-PgI_2 induced late appearing long-lasting cytoprotective effect. Prostaglandins in Clinical Research: Cardiovascular System 143–147, 1989
9. Szekeres L, Bálint Zs, Karcsú S, Tósaki Å: Delayed protection by 7-oxo-PgI_2 against cardiac transmembrane ion shifts and early morphological changes due to ischaemia and reperfusion. Cardiosci 1: 280–286, 1990
10. Ravingerová T, Tribulova N, Ziegelhöffer A, Džurba A, Szekeres L: 7-oxo-PgI_2 prevents partially the postischemic reperfusion injury of the rat heart. J Mol Cell Cardiol 23 (Suppl V): 104, 1991
11. Udvary É, Szekeres L: Prostacyclin: antiischaemic or cardioprotective? Advances in pharmacological research and practice. Volume 3, Section 7. In: V. Kecskeméti, K. Gyires, and G. Kovács (eds). Prostanoids. Akadémiai Kiadó, Budapest, 1986, pp 333–338
12. Szekeres L, Németh M, Papp JGy, Udvary É, Végh Å, Virág L: Neue Entwicklungen der antiarrhythmischen Therapie. In: B. Lüderitz and B. Antoni (eds). Perspektiven der Arrhythmiebehandlung. Springer, Berlin, 1988, pp 24–34
13. Ravingerová T, Tribulová N, Ziegelhöffer A, Styk J, Szekercs L: Suppression of reperfusion induced arrhythmias in the isolated rat heart: pretreatment with 7-oxo-prostacyclin *in vivo*. Cardiovasc Res 27: 1052–1055, 1993
14. Végh Å: 7-oxo-PgI_2 induced delayed protective action from late post-occlusion arrhythmias in conscious dogs. East European Subsection Meeting, International Society for Heart Research, May 13–18. Smolenice CSR: 93, 1990
15. Szekeres L, Szilvássy Z, Udvary É, Végh Å: 7-oxo-PgI_2 induced late appearing protection against ouabain induced cardiac arrhythmias in anesthetized guinea pigs. Pharmacol Res Commun 20: 77–78, 1988
16. Szilvássy Z, Szekeres L, Udvary É, Végh Å: On the 7-oxo-PgI_2 induced lasting protection against ouabain arrhythmias in anesthetized guinea pigs. Biomed Biochim Acta 47 (Suppl): 35–38, 1988
17. Szilvássy Z, Szekeres L, Udvary É, Karcsú S, Végh Å: 7-oxo-PgI_2 dramatically increases the safety margin of digitalis. Bratisl Lek Listy 92: 134–137, 1991
18. Szekeres L, Németh M, Papp JGy, Szilvássy Z, Udvary É, Végh Å: On the late appearing cardiac electrophysiological actions of the stable prostacyclin analogue: 7-oxo-PgI_2-Na. Proceedings of the 10th International Congress of Pharmacology, Sydney, 356, 1987
19. Németh M, Papp JGy, Szekeres L: Class 3 antiarrhythmic features of 7-oxo-PgI_2, a stable analogue of prostacyclin. Eur Heart J 9 (Suppl): 232, 1988
20. Szekeres L, Szilvássy Z, Udvary É, Végh Å: 7-oxo-PgI_2 induced late appearing and long-lasting electrophysiological changes in the heart *in situ* of the rabbit, guinea pig, dog and cat. J Mol Cell Cardiol 21: 545–554, 1989
21. Szekeres L, Németh M, Papp JGy, Udvary É: Short incubation with 7-oxo-prostacyclin induces long lasting prolongation of repolarisation time and effective refractory period in rabbit papillary muscle preparation. Cardiovasc Res 24: 37–41, 1990
22. Džurba A, Ziegelhöffer A, Breier A, Vrbjar N, Szekeres L: Increased activity of sarcolemmal ($Na^+ K^+$) -ATPase is involved in the late cardioprotective action of 7-oxo-prostacyclin. Cardioscience 2: 105–108, 1991
23. Borchert G, Bartel S, Beyerdörfer I, Küttner I, Szekeres L, Krause EG: Long lasting anti-adrenergic effect of 7-oxo-prostacyclin in the heart: a cycloheximide sensitive increase of phosphodiesterase isoform I and IV activities. Mol Cell Biochem 132: 57–67, 1994

Molecular and Cellular Biochemistry **147**: 123–128, 1995.

Ischaemic preconditioning in the rat heart: The role of G-proteins and adrenergic stimulation

Tanya Ravingerová, Nigel J. Pyne and James R. Parratt
Department of Physiology and Pharmacology, University of Strathclyde, Glasgow, Scotland, UK

Abstract

Since recent findings indicate the involvement of G-proteins in the mechanisms of ischaemic preconditioning (PC), the present study was aimed to investigate the role of adrenergic mechanisms, such as G-proteins and stimulation of adrenergic receptors, in this phenomenon. For this purpose, isolated Langendorff-perfused rat hearts were subjected to regional ischaemia (30 min occlusion of LAD) followed by reperfusion. The effect of PC (a single 5 min occlusion/reperfusion before a long occlusion) on ischaemia- and reperfusion-induced arrhythmias was studied in conjunction with an assessment of G-proteins in the myocardial tissue by means of Western blotting and ADP-ribosylation with bacterial toxins. To follow the link between G-proteins and adrenergic receptors, their stimulation by exogenous norepinephrine (NE) was applied to test whether it can mimic the effect of PC on arrhythmias. Thirty min ischaemia and subsequent reperfusion induced high incidence of ventricular tachycardia (VT) and fibrillation (VF). PC significantly reduced a total number of extrasystoles, incidence of VT and abolished VF. It was, however, insufficient to suppress reperfusion-induced sustained VF. Measurement of G-proteins revealed that PC led to a reduction of stimulatory Gs proteins, whereas inhibitory Gi proteins were increased. NE (50 nmol) introduced in a manner similar to PC (5 min infusion, 10 min normal perfusion) reduced ischaemic arrhythmias in the same way, as PC. In addition, in NE-pretreated hearts reperfusion induced mostly transient VF, which was spontaneously reverted to a normal sinus rhythm. A transient increase in heart rate and perfusion pressure during NE infusion completely waned before the onset of ischaemia, indicating that antiarrhythmic effect was not related to haemodynamic changes and to conditions of myocardial perfusion. Conclusion: Antiarrhythmic effect of PC may be mediated by a stimulation of adrenergic receptors coupled to appropriate G-proteins. Consequently, the inhibition of adenylate cyclase activity and reduction in cAMP level, as well as the activation of protein kinase C may be considered as two possible pathways leading to a final response. (Mol Cell Biochem **147**: 123–128, 1995)

Key words: ischaemic preconditioning, adrenergic mechanisms, G-proteins, catecholamines

Introduction

Brief episodes of myocardial ischaemia have been found to enhance resistance of the heart to a subsequent prolonged period of the same stress [1], an effect called preconditioning (PC). Protection afforded by PC includes reduction in cellular damage and in life-threatening arrhythmias, and was observed in various species [2, 3]. Although a vast number of endogenous substances (e.g., adenosine, prostanoids, bradykinin, nitric oxide) has been suggested as mediators of this protective effect [4], and the involvement of K^+atp-channels [5], A_1-receptors [2] or O_2 radicals [6] has been strongly implicated as well, the precise mechanisms of this phenomenon still remain not fully elucidated. Moreover, in species like rats none of the above mentioned factors seems to mediate protection. Recent findings indicate that ischaemic PC can be abolished by pertussis toxin, at least in some models [7, 8], suggesting the possible involvement of G-proteins in this mechanism. However, up till now G-proteins have not been directly assessed in the myocardium under conditions of brief ischaemia and/or preconditioning. Since G-proteins are important mediators of signal transduction in the cells [9], their participation in the mechanism of PC suggests the involvement of other components of the cell signalling system including adrenergic receptors. Their stimulation by catecholamines under normal conditions is known to exert an unequivocally deleterious effect on the heart [10]. On the contrary, in pathological situations, like heart failure or

Address for offprints: T. Ravingerová, Institute for Heart Research, Slovak Academy of Sciences, 9 Dubravska cesta, 842 33 Bratislava, Slovak Republic ischaemia, adrenergic stimulation has been shown to attenuate electrophysiological changes in the dog heart [11, 12]. Although the levels of endogenous catecholamines released from the sympathetic nerve endings in the myocardium increase during brief ischaemia [13], their role in the mechanisms of PC is still unclear. Therefore, the present study was undertaken to investigate whether adrenergic mechanisms participate in the ischaemic PC in rats. This was achieved by studying the effect of PC on severe ventricular arrhythmias in conjunction with a direct estimation of G-proteins in the myocardial tissue of the preconditioned hearts, as main transducers of adrenergic signals in the cells. Secondly, the link between G-proteins and adrenergic receptors was followed. For this purpose, their stimulation by exogenous norepinephrine was applied to test whether it can mimic the effect of stimulation by endogenous catecholamines caused by PC, on ischaemia/reperfusion-induced arrhythmias in the isolated rat heart.

Materials and methods

All studies were performed in accordance with the Guide for the care and use of laboratory animals published by the US National Institutes of Health (NIH publication No 85–23, revised 1985). Hearts were excised from the anaesthetised (sodium barbitone) male Wistar rats (250–300 g) and perfused (Langendorff) at a constant flow of 10 ml/min. Krebs-Henseleit solution contained (in mM): NaCl 118.0; KCl 3.2; $MgSO_4$ 1.66; $NaHCO_3$ 25.0; KH_2PO_4 1.18; $CaCl_2$ 2.5; glucose 5.55; Na pyruvate 2.0; it was gassed (95% O_2 + 5% CO_2) and maintained at 37°C; pH 7.4. All hearts were allowed to stabilise for 20 min before the experimental protocol began. Regional ischaemia was induced by occlusion of LAD coronary artery and lasted 30 min followed by a 10 min reperfusion. Efficacy of occlusion/reperfusion was confirmed by a 45% increase/decrease in perfusion pressure corresponding to a size of the occluded area. Epicardial ECG was continuously recorded (GOULD Polygraph), and arrhythmias were characterized in accordance with Lambeth Conventions [14]. Arrhythmias were analysed by counting the total number of ventricular premature beats (VPB), by determining the incidences of ventricular tachycardia (VT) and fibrillation (VF) and their duration. VF lasting more than 2 min was considered as sustained, and in such hearts normal sinus rhythm has never been restored. In addition, severity of arrhythmias was characterised using a 5 point arrhythmia score ranging from 1 (for VPBs) to 5 (for sustained VF). PC was induced by a single brief occlusion of LAD (5 min) and a consequent 10 min reperfusion before a long occlusion. Exogenous norepinephrine (Arterenol SIGMA, 10^{-6} M) was introduced in a manner similar to PC: 5 min infusion, 10 min normal perfusion before the onset of ischaemia. The samples for immunochemical processing were taken from the ischaemic area of left ventricles of the hearts subjected to ischaemia (at the time of the maximal incidence of ventricular arrhythmias), to PC, as well as to PC followed by ischaemia, and from the corresponding area of the time-matched control hearts. Tissue was fresh-frozen and stored at –70°C.

Preparation of membranes

Frozen tissue was homogenized in medium containing 0.25 M sucrose, 1 mM EDTA, 10 mM Tris, 2 mM benzamidine and 0.1 mM PMSF and centrifuged at 48 000 × g for 10 min at 4°C. After resuspension of the pellet in the same buffer, it was recentrifuged for 20 min, and the resulting pellet was again resuspended and stored frozen at –20°C.

Measurement of G-proteins was performed by means of Western blotting and by ADP-ribosylation with bacterial toxins: pertussis toxin for Gi- and cholera toxin for Gs-proteins [15]. Proteins concentrations were determined spectrophotometrically.

Western blotting

Membranes were subjected to SDS-PAGE and transferred to nitrocellulose. After transfer, the sheets were blocked in 3% gelatin, washed and incubated with an anti-Gs or anti-Gi antibody for 12 h at 30°C. After washing in buffer containing Tween-20, the sheets were then incubated with a horseradish-linked anti-rabbit antibody for 2 h at room temperature. After washing, the sheets were reacted with dianisidine, hydrogen peroxide and Tris for the development of the reaction and for the visualization of the immunoreactive bands.

Toxin-catalyzed ADP-ribosylation

Membranes were combined with ADP-ribosylation cocktail containing potassium phosphate, thymidine, $MgCl_2$, [^{32}P] NAD$^+$; (ATP and arginine for pertussis toxin); (DTT and $CaCl_2$ for cholera toxin) and incubated for 2 h at 37°C with pre-activated either pertussis toxin, or cholera toxin. After incubations, the samples were subjected to SDS-PAGE, fixation of gels and autoradiography.

Statistical evaluation was performed using a one-way analysis of variance (ANOVA) and unpaired Student's *t*-test for Gaussian distributed data, as well as a χ^2 test for binomially distributed variables. $p < 0.05$ was considered as significant.

Results

During 30 min ischaemia, the highest incidence of severe ventricular arrhythmias including VPBs, VT and VF occured between 15 and 20 min of LAD occlusion (Fig. 1), which was not observed in the preconditioned hearts neither in the same time interval, nor later on. Accordingly, the samples for the determination of G-proteins were taken in the 18 min of ischaemia, or time-matched control perfusion. Figure 2A, B demonstrates the effect of PC and NE infusion on the incidence of severe ischaemic arrhythmias. In the ischaemic group (n = 6), 30 min occlusion of LAD induced 100% and 67% incidence of VT and VF, respectively, as well as 542 ± 84 VBPs. In the group with PC protocol before a long occlusion (n = 6), incidences of VT and VF were significantly reduced to 17% and 0%, respectively ($p < 0.01$), and the total number of VPBs to 30 ± 11 ($p < 0.001$). Infusion of NE (n = 6) suppressed ischaemic arrhythmias in the same way: incidences of VT and VF to 17% and 0%, respectively, the number of VPBs to 15 ± 6 ($p < 0.001$). Duration of VT and VF (Fig. 3) and the severity of ischaemic arrhythmias characterized by arrhythmia score (Fig. 4) were also significantly decreased after both, PC and infusion of NE. However, PC was not sufficient to suppress reperfusion-induced sustained VF, which occured in all the hearts. In these hearts sinus rhythm has not been restored (Fig. 5). On the contrary, in the NE-pretreated hearts, reperfusion-induced sustained VF was significantly suppressed, and transient VF was spontaneously reverted to a normal sinus rhythm. Infusion of NE caused a

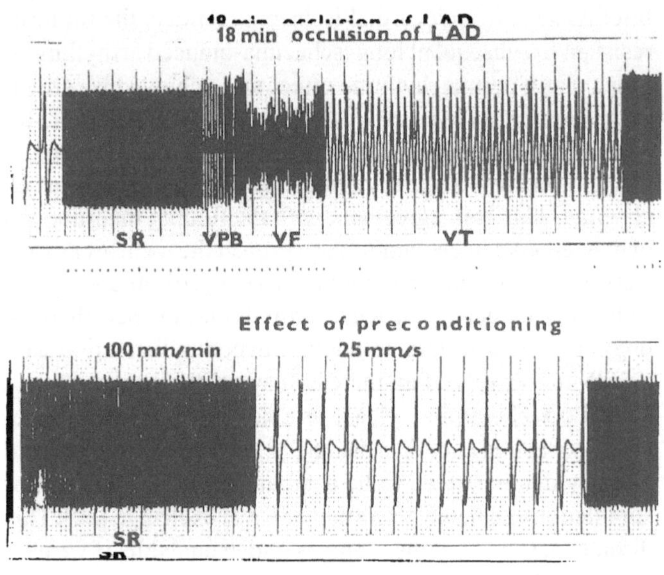

Fig. 1. Original recording demonstrating ischaemia-induced arrhythmias in the 18 min of LAD occlusion in non-preconditioned and in the preconditioned heart. VT – ventricular tachycardia, VF – ventricular fibrillation, VPB – ventricular premature beats.

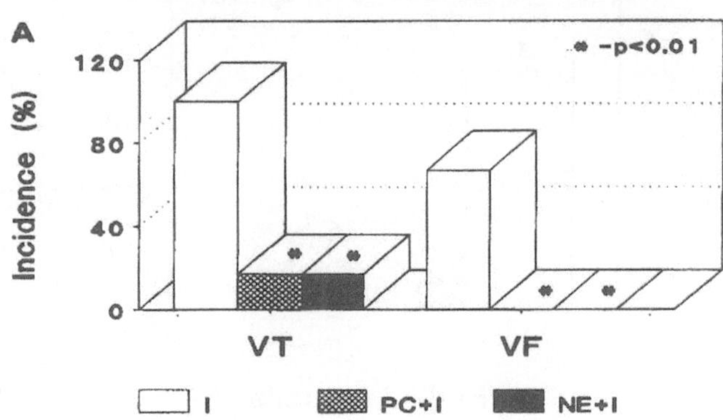

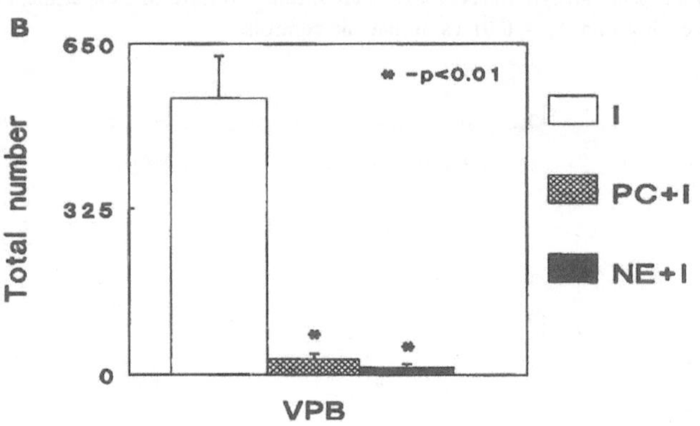

Fig. 2A,B. Effect of preconditioning and norepinephrine infusion on ischaemic arrhythmias. Abbreviations as in Fig. 1.

Fig. 3. Duration of ventricular tachycardia and fibrillation. Data are means ± S.E.M. of 6 experiments in each group. *$p < 0.01$ vs. ischaemic controls.

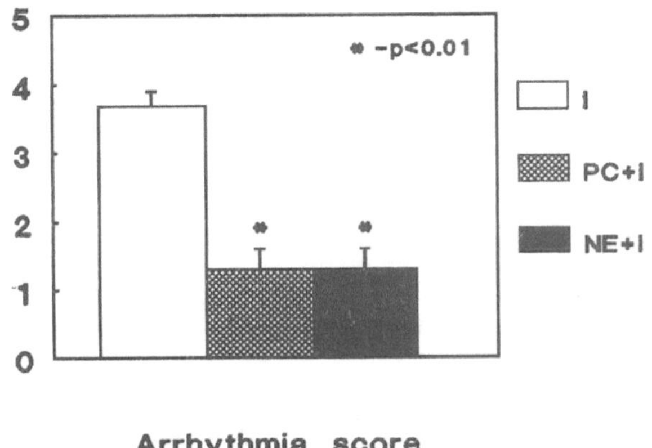

Fig. 4. Severity of ischaemic arrhythmias. Arrhythmias were assessed by a 5-point arrhythmia score. Data are means ± S.E.M. of 6 experiments in each group. *p < 0.01 vs. ischaemic controls.

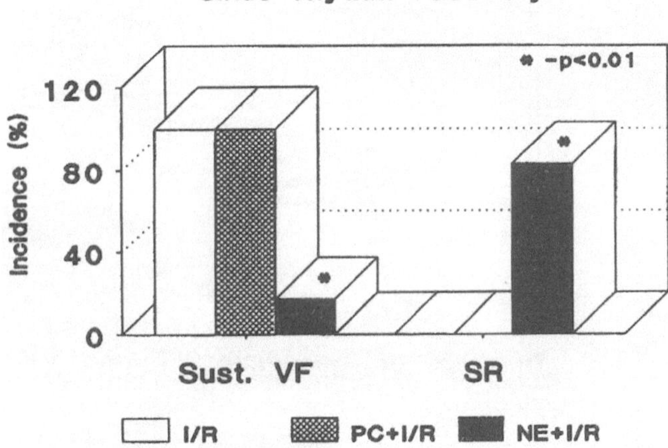

Fig. 5. Effect of preconditioning and norepinephrine infusion on reperfusion-induced sustained ventricular fibrillation and sinus rhythm (SR) recovery.

transient increase in heart rate and in perfusion pressure before the onset of ischaemia (Fig. 6). These hemodynamic effects, however, completely waned by the end of a 10 min recovery period.

Western blot analysis using anti-Gsα and anti-Giα antibodies revealed the bands of 45 and 41 kDa, corresponding to Gs- and Gi-proteins, respectively. ADP-ribosylation with both, pertussis and cholera toxins confirmed these findings. Qualitative analysis of G-proteins in the myocardial tissue by both methods revealed the presence of Gs- and Gi-proteins in the control hearts and in the ischaemic area of the hearts subjected to LAD occlusion. PC induced a reduction in Gs-

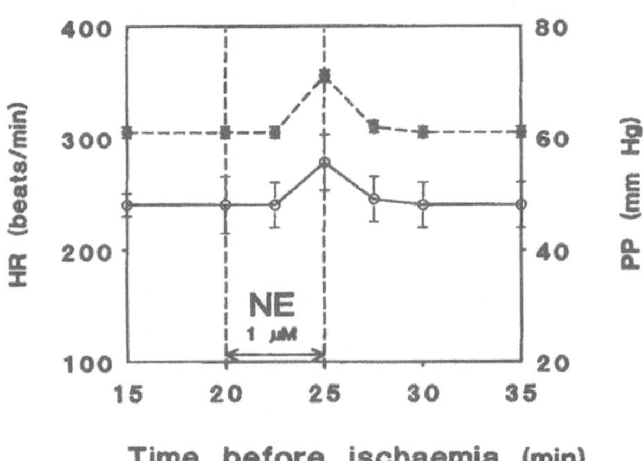

Fig. 6. Effect of norepinephrine infusion on heart rate (dashed line) and perfusion pressure (solid line) in isolated rat heart. Norepinephrine (NE, 10^{-6} M) was infused for 5 min at a rate of 10 ml/min.

proteins, which persisted during subsequent long-lasting ischaemia. On the contrary, Gi-proteins were increased after PC, and this was maintained during subsequent ischaemia (Table 1).

Discussion

The present study demonstrated that PC induced by a single brief episode of myocardial ischaemia renders the rat heart resistant to subsequent long ischaemia-induced arrhythmias. Since similar antiarrhythmic effect was achieved by adrenergic stimulation induced by exogenous norepinephrine, we can suggest that catecholamines participate in the induction of the protective antiarrhythmic effect caused by a brief ischaemia. This is in agreement with the finding that infusion of exogenous catecholamines can protect the rat heart against ischaemia-induced arrhythmias *in vivo* [16], and suppress ischaemia-induced ventricular arrhythmias in anaesthetized dogs [17]. However, this finding is in contradistinction with a well-known role of an increased sympathetic activity and a local catecholamine release as a key arrhythmogenic factor in ischaemia and reperfusion [18, 19]. The latter has been recently demonstrated to be attributed to α_{1A}-adrenoceptor stimulation-mediated activation of Na/H exchange [20]. This discrepancy, on the other hand, is understandable, if we take into consideration the conditions of the experiment. The effects of catecholamines have been shown to depend on the concentration, time, protocol and experimental setup (*in vivo* or *ex vivo* experiments). Proarrhythmic effect of enhanced sympathetic activity in non-ischaemic myocardium is changed in partially depolarized tissue, where noradrenaline produces

Table 1. G-proteins in the myocardial tissue.

	C	I	PC	PC+I
Gs	+	+	±	±
Gi	+	+	++	+++

The samples were taken from the left ventricle of the hearts subjected to control perfusion (C), regional long-lasting ischaemia (I), preconditioning (PC) induced by 5 min of ischaemia followed by 10 min of reperfusion and preconditioning followed by long ischaemia (PC + I). A summarization of results obtained by Western blotting and ADP-ribosylation.

a transient hyperpolarization, an increase in action potential amplitude and duration, which may be expected to counteract the initiation of arrhythmias [19, 21]. In other words, what is deleterious in ischaemia, may be different and even beneficial in preconditioning. Indeed, it has been recently demonstrated that, like a brief ischaemia, noradrenaline can also precondition a canine heart against postischaemic myocardial dysfunction [22] and even to protect the rabbit heart against infarction [23], the effect being mediated via enhanced adenosine production and A_1 receptors stimulation, accompanied by a coronary vasodilatation. In our experiments, the PC-like effect of norepinephrine administration cannot be ascribed to its influencing the haemodynamics and thus, to the conditions of myocardial perfusion. Possible mechanisms of protection may be triggering of adaptation processes in the heart, similar to ischaemic PC. Those can be mediated by a stimulation of α1-adrenergic receptors, as it has been shown by Banerjee *et al.* [24], who have studied the recovery of heart function in rats under the same conditions as ours. In the present study the type of adrenergic receptors involved has not been investigated. However, since α-receptors are coupled to Gq- and Gi-proteins [9], our finding of enhanced Gi-proteins in the hearts preconditioned with a brief ischaemia indirectly points to the possible involvement of α-adrenergic stimulation in the mechanisms of protection afforded by both, ischaemic PC and PC-like effect of exogenous catecholamines. Gi-proteins may be also activated during brief ischaemia via A_1 receptors stimulation [25]. An increase in Gi- and a reduction in Gs-proteins can both result in a depressed stimulation of their effector – adenylate cyclase [9]. Accordingly, we can speculate about two possible G-proteins-mediated pathways of further signal transduction leading to a final protective response: 1. the inhibition of the activity of adenylate cyclase due to the changes in Gs- and Gi-proteins, with a consequent reduction in cAMP production, and 2. Gq-proteins-mediated translocation and activation of protein kinase C. Its important role in preconditioning has been recently reported by Liu *et al.* [26]. Further investigation of α signalling mechanisms in ischaemic PC may bring a new approach towards pharmacological induction of the protective effect.

Acknowledgement

This study was supported by the European Communities's grant for Cooperation in Sciences and Technology with Central and Eastern European Countries.

References

1. Murry CE, Jennings RB, Reimer KA: Preconditioning with ischemia; a delay of lethal cell injury in ischemic myocardium. Circulation 74: 1124–1126, 1986
2. Liu GS, Thornton J, Van Winkle DM, Stanley AWH, Olsson RA, Downey JM: Protection against infarction afforded by preconditioning is mediated by A_1 adenosine receptors in rabbit heart. Circulation 84: 350–356, 1991
3. Vegh A, Komori S, Szekeres L, Parratt JR: Antiarrhythmic effects of preconditioning in anaesthetised dogs and rats. Cardiovasc Res 26: 487–495, 1992
4. Parratt JR: Endogenous myocardial protective (antiarrhythmic) substances. Cardiovasc Res 27: 693–702, 1993
5. Parratt JR, Kane KA: K_{atp} channels in ischaemic preconditioning. Cardiovasc Res 28: 783–787, 1994
6. Tanaka M, Fujiwara H, Yamasaki K: N-2-Mercaptopropionyl glycine attenuates infarct size limiting effect of ischemic preconditioning. Circulation 86 (suppl 1): 1–30, 1992
7. Piacentini L, Wainwright CH, Parratt JR: The antiarrhythmic effect of ischaemic preconditioning in isolated rat heart involves a pertussis toxin sensitive mechanism. Cardiovasc Res 27: 674–680, 1993
8. Lasley RD, Mentzer RM: Pertussis toxin blocks adenosine A_1 receptor mediated protection of the ischemic heart. J Mol Cell Cardiol 25: 815–821, 1993
9. Fleming JW, Wisler PL, Watanabe AM: Signal transduction by G proteins in cardiac tissues. Circulation 85: 419–433, 1992
10. Dukes ID, Vaughan-Williams EM: Effects of selective α1-, α2-, β1- and β2-adrenoceptor stimulation on potentials and contractions in the rabbit heart. J Physiol (Lond) 355: 523–546, 1984
11. Li HG, Jones DL, Yee R, Klein G: Arrhythmogenic effects of catecholamines are decreased in heart failure induced by rapid pacing in dogs. Am J Physiol 265: H1654–H16662, 1993
12. Janse MJ, Schwartz PJ, Wilms-Schopman FJG, Petrs RJG, Durrer D: Effects of unilateral stellate ganglion stimulation and ablation on electrophysiologic changes induced by acute myocardial ischemia in dogs. Circulation 72: 585–595, 1985
13. Schömig A: Increase of cardiac and systemic catecholamines in myocardial ischemia. In: J. Brachman and A. Schömig (eds). Adrenergic system and ventricular arrhythmias in myocardial infarction. Springer-Verlag, Berlin, 1989, pp 61–77
14. Walker MJA, Curtis MJ, Hearse DJ *et al.*: The Lambeth conventions: guidelines for the study of arrhythmias in ischemia, infarction and reperfusion. Cardiovasc Res 22: 447–455, 1988
15. Grady M, Stevens PA, Pyne S, Pyne N: Adenylyl cyclase in lung from hypersensitive guinea pig displays increased responsiveness to guanine nucleotides and isoprenaline: the role of the G proteins Gs and Gi. Biochim Biophys Acta 1176: 313–320, 1993
16. Parratt JR, Campbell C, Fagbemi O: Catecholamines and early post-infarction arrhythmias: the effects of α- and β-adrenoceptor blockade. In: Catecholamines and the Heart. Springer-Verlag, Berlin/Heidelberg/New York, 1981, pp 269–284
17. Vegh A, Papp J Gy, Parratt JR: Intracoronary noradrenaline suppresses ischaemia-induced ventricular arrhythmias in anaesthetized dogs. J

Mol Cell Cardiol 26: LXXXVII, 1994 (abstract)
18. Benfey BG: Antifibrillatory effects of α1-adrenoceptor blocking drugs in experimental coronary artery occlusion and reperfusion. Can J Physiol Pharmacol 71: 103–111, 1993
19. Dietz R, Offner B, Dart AM, Schömig A: Ischaemia-induced noradrenaline release mediates ventricular arrhythmias. In: J. Brachman and A. Schömig (eds). Adrenergic system and ventricular arrhythmias in myocardial infarction. Springer-Verlag, Berlin, 1989, pp 313–321
20. Yasutake M, Avkiran M: Exacerbation of reperfusion arrhythmias by phenylephrine: a role for α_{1A}-adrenoceptor-mediated activation of Na^+/H^+ exchange? J Mol Cell Cardiol 26: CIX, 1994 (abstract)
21. Janse MJ: Why is increased adrenergic activity arrhythmogenic? In: J. Brachman and A. Schömig (eds). Adrenergic system and ventricular arrhythmias in myocardial infarction. Springer-Verlag, Berlin, 1989, pp 353–363
22. Kitakaze M, Hori M, Kamada T: Role of adenosine and its interaction with α adrenoceptor activity in ischaemic and reperfusion injury of the myocardium. Cardiovasc Res 27: 18–27, 1993
23. Thornton JD, Daly JF, Cohen MV, Yang X-M, Downey JM: Catecholamines can induce adenosine receptor-mediated protection of the myocardium but do not participate in ischemic preconditioning in the rabbit. Circ Res 73: 649–655, 1993a
24. Banerjee A, Locke-Winter C, Rogers KB, Mitchell MB, Brew EC, Cairns ChB, Bensard DD, Harken AH: Preconditioning against myocardial dysfunction after ischemia and reperfusion by an α1-adrenergic mechanism. Circ Res 73: 656–670, 1993
25. Thornton JD, Liu GS, Downey JM: Pretreatment with pertussis toxin blocks the protective effects of preconditioning: evidence for a G-protein mechanism. J Mol Cell Cardiol 25: 311–320, 1993b
26. Liu Y, Ytrehus K, Downey JM: Evidence that translocation of protein kinase C is a key event during ischemic preconditioning of rabbit myocardium. J Mol Cell Cardiol 26: 661–668, 1994

Molecular and Cellular Biochemistry **147**: 129–137, 1995.

Adaptation of the heart to ischemia by preconditioning: Effects on energy equilibrium, properties of sarcolemmal ATPases and release of cardioprotective proteins

Attila Ziegelhöffer, Norbert Vrbjar, Ján Styk, Albert Breier[1], Andrej Džurba and Tatjana Ravingerová

Institute for Heart Research, Slovak Academy of Sciences, 842 33 Bratislava; [1]Laboratory of Protein Chemistry, Institute of Molecular Physiology and Genetics and Institute for Heart Research, Bratislava, Slovak Republic

Abstract

Ischemic preconditioning of the heart is referred as a manifest increase in tolerance of the myocardium to otherwise damaging ischemic insult, achieved by one or few consequent initial short exposures to ischemia, each followed by reperfusion of the ischemic area. Several mechanisms such as opening of collateral vessels, the action of catecholamines, inositol phosphates, G-proteins and/or adenosine; inhibition of mitochondrial ATPase, the effects of different endogenous protective substances like heat stress or shock proteins, etc., are believed to cooperate in the mechanism of induction of preconditioning or in maintaining its effect. The present study is an attempt to extend the present knowledge about preconditioning from two aspects: i.) the peculiarities of energy equilibrium in preconditioned myocardium including adaptation of cardiac sarcolemmal ATPases to ischemia and/or hypoxia, and ii) participation of a new endogenous cardioprotective substance in the mechanism of preconditioning. The energy equilibrium in preconditioning is characterized by adaptation of cardiac energy demands to the capacity of energy production and delivery decreased by anaerobiosis and is manifested by constant ratios between ATP, ADP, AMP and the sum of ADN. Principles are proposed that may enable a prediction and mathematical modelling of the balanced energetic state in the preconditioned myocardium. These principles are based on thermodynamics and involve besides others a more economic handling of ATP by sarcolemmal ATPases. The latter enzymes adapt themselves to lowered availability of ATP by decreasing besides their Vmax also their values of K_m (increase in the affinity) for ATP and some of them even adjust their activation energy (the anaerobiosis-induced elevation of $E_{a.t.}$ is missing). It was also revealed that during preconditioning several up to now not described shock proteins unlike proteins (also glycoproteins) are released from the myocardium into the coronary blood. When these proteins indicated as a HS fraction were isolated, partially purified and in concentrated form applied into the coronary circulation, they were capable to induce in preliminary experiments a cardioprotective effect resembling that of the ischemic preconditioning. (Mol Cell Biochem **147**: 129–137, 1995)

Key words: ischemic preconditioning of the myocardium, equilibrium energetics of the myocardium, sarcolemmal ATPases, adaptation of ATPases to ischemia or hypoxia, glycoprotein release

Introduction

Since reported first by Murry *et al.* in 1986 [1], ischemic preconditioning of the myocardium became an issue with increasing importance from the point of view of both, it's early [2–4] and its late effects [5–9]. The latter are colloquially termed as the second window. According to its classical definition preconditioning represents a manifest increase in tolerance of the myocardium to otherwise damaging ischemic insult that is achieved by one or few consequent initial short exposures to ischemia, each followed by reperfusion of the ischemic area [2]. Numerous mechanisms have been reported to be involved in the phenomenon of preconditioning. They include: opening of collateral vessels [10, 11]; inhibition of

Address for offprints: A. Ziegelhöffer, Institute for Heart Research, Slovak Academy of Sciences, Dubravska cesta 9, 842 33 Bratislava, Slovak Republic mitochondrial ATPase [12]; release of endogenous mediators such as adenosine [13, 14] – the latter coupled with activation of 5'-nucleotidase [15], release of inositol phosphates [12]; release of catecholamines coupled with changes in G-proteins [16]; formation of nitric oxide [17] and cyclooxygenase products, particularly of prostaglandin [18]; induction of endogenous myocardial protective substances [19] and among them of bradykinin [20], heat stress proteins [6] and proto-oncogenes [21]. Other mechanisms suggested to be also involved in the process of preconditioning are: activation of ATP-sensitive K^+-channels [22] and release of oxygen free radicals [12]. However, radicals seem to exert also a positive, protecting mechanisms-triggering effect involving also an increase in Ca^{2+}/phospholipid-independent form of the protein kinase C [23]. All this points to large complexity of mechanisms that may hand in hand operate in preconditioning and finally unite in a new dynamic equilibrium state characterised by temporary adaptation of cardiac energy consuming systems to tolerate decreased energy production and/or delivery.

The aim of present communication is to contribute to the above knowledge from the following points of view: i) Decrease in energy demands of the ischemic myocardium. This process involves adaptation of sarcolemmal ATPases to work efficiently also in conditions with decreased availability of ATP. This may contribute to establish a new equilibrium state in cardiac energetics, characterised by more economic utilization of the available ATP, i.e., by lowered enthalpy of the system; ii) production and release of endogenous proteins that may participate in protection of the heart against ischemia.

Materials and methods

Treatment of animals

All studies have been performed in accordance with the Guide for the Care and Use of Laboratory Animals published by the US National Institute of Health (NIH publication No 85-23, revised 1985).

Experiments in rats

Hearts from male Wistar rats (a specific pathogen free strain) with body weight of 200–250 g kept on standard pellet diet and tap water ad libitum were used in all experiments. The animals were anaesthetized with sodium pentobarbitone (60 $mg.kg^{-1}$). Ten minutes after intraperitoneal administration of heparin (500 IU) the hearts were rapidly excised and arrested in ice cold buffer that should be used in further procedure. Up to this step the protocol was similar in majority of experiments. Deviations, if any, are always indicated in the respective place.

Animals acclimatised to high altitude hypoxia (preconditioning of the whole body with hypoxia)

A total of 80 male Wistar rats (70 days old) were exposed to a simulated altitude in a barochamber, 4 h per day, 5 days a week. Acclimatization was performed stepwise. The altitude of 7000 m (barometric pressure 40.9 kPa, PO_2 8.5 kPa, pCO_2 0.01 kPa) was reached after 13 exposures. The total number of exposures was 32. Experimental animals had all time free excess to water and were fed with standard laboratory pellet diet. Control animals were kept under normoxic conditions for an identical period. Immediately after the last hypoxic exposure the hearts were excised quickly by means of the method of Fulton *et al.* [24]. The barochamber in the Physiological Institute of the Czech Academy of Sciences in Prague was used for these experiments. The experimental procedure was essentially similar to that published in our earlier papers [8, 25]. Tissue samples were preserved in liquid nitrogen until isolation of the sarcolemmal fraction.

Ischemic preconditioning of rat hearts

Hearts from male Wistar rats were cannulated through the aorta and perfused in a non-recirculating mode according to Langendorff at a constant perfusion pressure equivalent to 75 mm Hg at 37°C. Perfusion medium was Krebs-Henseleit buffer (pH 7.4) gassed with 95% O_2 and 5% CO_2 and containing (in $mmol.l^{-1}$): NaCl (118.5); $NaHCO_3$ (25.0); Mg SO_4 (1.2); NaH_2PO_4 (1.2); $CaCl_2$ (1.4); KCI (3.0) and glucose (11.1). Precautions were taken to avoid the precipitation of calcium; the solution was filtered through a 5 µm porosity filter to remove contaminants. Ischemic preconditioning of the heart was applied in a mode: four repeated 4 min periods of global ischemia alternated with 5 min reperfusion periods. For further technical details see [16].

Rat hearts subjected to ischemia

Hearts were after excision subsequently subjected to global ischemia at 37°C for 45 min. This procedure was immediately followed by preparation of the sarcolemmal fraction. In the aim to increase the yield of final membrane fraction to amounts requested for study of enzyme kinetics, always two hearts were pooled for each isolation. More types of controls were used: i) absolute controls - hearts taken immediately after excision, and ii) preservation controls - hearts taken after 45 min of ischemia as well as iii) non-ischemic control hearts preserved for 24 h in liquid nitrogen. A comparison of ATPase activities in fresh hearts with those preserved in liquid nitrogen revealed that preservation of tissue samples induced an approximately 40% loss in ATPase activities. Nevertheless, preservation neither influenced the trends of changes induced by adaptation nor the values of main kinetic parameters of the sarcolemmal ATPases.

Superfusion of rat papillary muscles, estimation of their contractility and tolerance to anoxia

Papillary muscles from control hearts and hearts adapted to high altitude hypoxia were quickly dissected, mounted in a perfusion apparatus and superfused at 37°C with Krebs-Henseleit solution pH 7.4, gassed with 95% O_2 and 5% of CO_2. For a brief period of transient anoxia O_2 was replaced by nitrogen. Further technical details concerning perfusion technique and monitoring the contractility were described in our earlier communication [8].

Membrane preparation

Sarcolemmal fraction from ischemic and control rat hearts was isolated by the method of hypotonic shock combined with treatment with 0.6 mol.l^{-1}NaI [26].

Estimation of adenine nucleotides

Adenosine triphosphate, adenosine diphosphate and adenosine monophosphate contents in hearts were estimated in neutralized tissue extracts enzymatically by means of the Warburg's optical test. Tissue samples were taken with the aid of Wollenberger's tongs. For further details see [10].

Estimation of the activity and kinetic parameters of ATPases

For evaluation of K_m and V_{max} values the activities of ATPases were measured at 37°C in the presence of increasing concentrations of ATP (0.08–6.0 mmol.l^{-1}) by incubating 30–50 µg of membrane proteins in a total volume of 0.5 ml of medium containing 50 mmol.l^{-1} imidazole (pH 7.4) and the following metallic cofactors (in mmoles.l^{-1}): for (Na,K)-ATPase 4 $MgCl_2$; 100 NaCl, 10 KCl; for Mg-ATPase 4 $MgCl_2$; for Ca-ATPase with low affinity to calcium 4 $CaCl_2$. Following 10 min of preincubation in substrate free medium, the reaction was started by addition of ATP and after a reaction period of 20 min it was terminated by 1 ml of 12% ice cold solution of trichloroacetic acid. The inorganic phosphate liberated by ATP splitting was determined according to Taussky and Shorr [27]. Protein content was assayed by the procedure of Lowry *et al.* [28] using bovine serum albumin as standard. The activation energy for ATP hydrolysis by ATPases was estimated by the analysis of temperature dependence of ATP splitting in the range of 0–45°C. For these measurements the reaction medium was similar to that used in investigation of their reaction kinetics except that the concentration of ATP was kept constant at non-limiting value of 4 mmol.l^{-1}. Kinetic constants were established by means of non-linear curve fitting. Activation energies were evaluated according to the Arrhenius equation.

Ischemic preconditioning of the dog heart in situ

Ten adult mongrel dogs with an average weight of 12.5 kg were anaesthetized intravenously with pentobarbitone (SPOFA) 30 mg.kg^{-1} of body weight, heparinized (3.0 IU.kg^{-1}) and artificially ventilated (by means of a respirator Chirolog, Chirana) with a mixture of 95% O_2 and 5% CO_2. After left side thoracotomy in the 5th intercostal space, the pericard was opened and the heart exposed. Ischemic preconditioning was achieved by 3 repeated 4 min occlusions of the left anterior descending coronary artery (LAD). The first two occlusions were followed by 5 min of reperfusion of the occluded area. The efficacy of ischemic preconditioning was verified during 20 or 40 min ischemic episodes following a 60 min period of reperfusion after the last 4 min ischemic interval. Changes in size of the ischemic region were estimated by epicardial electrocardiography [10]; the frequencies of occurrence of left ventricular tachyarrhythmias and ventricular fibrillation as well as the adenine nucleotide content in non ischemic areas of the left ventricle were used as measures of efficacy of the preconditioning. Shame operated animals (n=10) were used as controls. During experiment, samples from the coronary sinus and peripheral venous blood were taken in regular intervals to monitor acid base equilibrium and to check the possible release of myocardial endogenous protective substances.

Isolation, characterisation and partial purification of proteins released from the heart into the blood during ischemic preconditioning

Samples from coronary sinus blood were taken at the beginning of experiment before the first ligation of the LAD (these served as controls), further after each of the three 4 min occlusions and the 5 min reperfusion intervals following the occlusions as well as also after the 120 min reperfusion following the last occlusion and finally 140 min after the last occlusion. Sera obtained by centrifugation (3000 × g) were precipitated with 50% ammonium sulfate and the precipitate spunned down at (3000 × g). Supernatants were repeatedly precipitated with 80% ammonium sulfate; again centrifuged at 3000 × g. Than the pellets were dissolved in isotonic saline-phosphate buffer solution (PBS) and dialyzed against

PBS overnight. Dialyzed samples of the highly soluble protein fraction (HS) were analyzed for the presence of Concanavaline A (Con A)-interacting glycoprotein by means of the ELISA and Western blot methods using Con A, mouse anti-Con A antibody and pig anti-mouse antibody connected with horseradish peroxidase for detection. Peroxidase reaction was visualised using o-phenylendiamine and was monitored spectrophotometrically at 595 nm on ELISA reader Labsystem multiscan MCC/340. The molecular weight of the isolated glycoprotein was determined by means of SDS-PAGE electrophoresis.

Testing the protective effect of the isolated and partially purified HS protein fraction

In this preliminary experiment one dog was exanguinated after preconditioning with three 4 min occlusions and the subsequent 5 min reperfusions. The HS protein fraction released by preconditioning into the circulating blood was isolated as it was described above. The final fraction was concentrated by freeze-drying. Than it was dissolved in 10 ml of saline and applied to an experimental animal intracoronarily 60 min prior to a 40 min occlusion of the LAD. A different animal pretreated with saline only served as the control. Similarly as in the group of animals with ischemic preconditioning, changes in size of the ischemic region estimated by epicardial electrocardiography [10] and the frequency of occurrence of left ventricular tachyarrhythmias as well as ventricular fibrillations were used as measures of efficacy of the endogenous cardioprotective HS proteins.

Results

Contents of adenine nucleotides and their mutual ratio in isolated perfused rat hearts subjected four times to 4 min of global ischemia, each followed by 5 min of reperfusion are presented in Fig. 1. Left hand panel shows that when plotted in a semilogarrhythmic way, the columns representing the contents of ATP, ADP and AMP in hearts belonging to the same group may be connected with a straight line. This indicates that independently on changes in actual levels of ATP, ADP and AMP their mutual interrelationship remains practically constant in both, the control hearts (group I) as well as in the hearts preconditioned with four repeated 4 min ischemic insults (group 11), i.e., their energy metabolism is in a near-to equilibrium state (right hand panel). On the other hand, hearts after only one ischemic episode with the same total duration (16 min. group III) exhibit disequilibrium. This means that it is synthetized more ATP from ADP than ADP from AMP.

The activities of heart sarcolemmal (Na,K)-ATPase, Mg-ATPase and Ca-ATPase with low affinity to calcium were

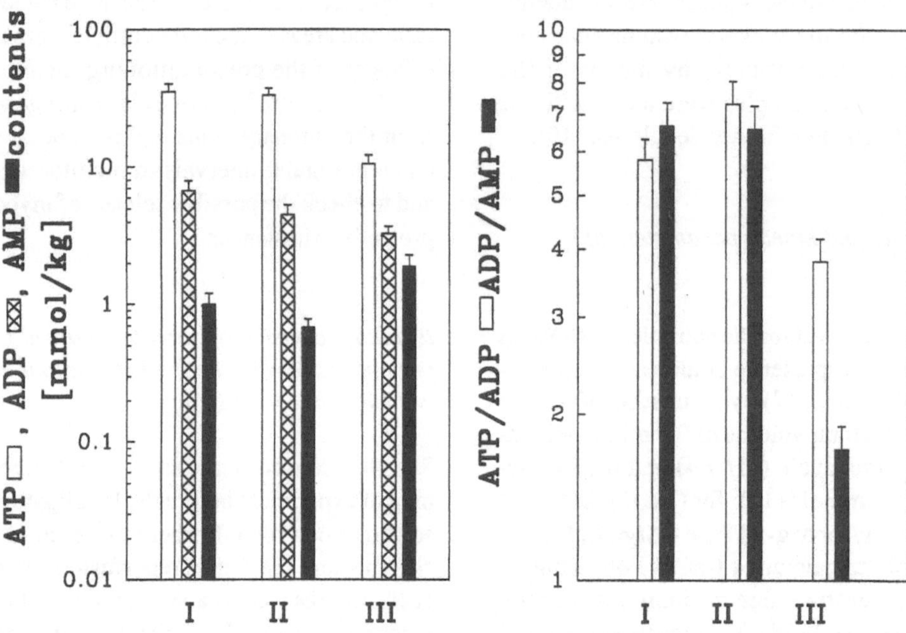

Fig. 1. Contents and mutual ratios of ATP, ADP and AMP in isolated perfused rat hearts submitted to preconditioning. The effect of repeated short ischemic insults versus one ischemic event of the same duration. Left hand panel – semilogarrhythmic plot of tissue contents of adenine nucleotides; Right hand panel – semilogarrhythmic plot of ATP/ADP and ADP/AMP ratios; I – normoxic control hearts; II – hearts preconditioned with four periods of 4 min ischemia followed by 5 min reperfusions; III – hearts preconditioned with one episode of 16 min lasting global ischemia; The total duration of perfusion was similar for in all experimental groups, differences were replaced with normoxic perfusion at the beginning of the experiment. Results are means ± S.E.M. from 10 hearts. Statistical significances were established by means of the Student's *t*-test. For other details see the Materials and methods.

determined at various concentrations of ATP in the range of 0.08–6.0 mmol.l^{-1}, in the presence of optimal pH and concentrations of cationic cofactors The control values of V_{max} amounted (in μmol P_i.mg^{-1}.h^{-1}): 13.30 ± 1.17, 21.61 ± 1.99 and 20.23 ± 1.35 for the (Na,K)-ATPase, MgATPase and Ca-ATPase, respectively. The control values for K_m given in mmol.l^{-1} of ATP were: 0.42 ± 0.04, 0.62 ± 0.07 and 0.48 ± 0.06 for the (Na,K)-ATPase, Mg-ATPase and Ca-ATPase, respectively and the control values for activation energy (E_a) expressed in kJ.mol.l^{-1} amounted 48.45 ± 1.92 for the (Na,K)-ATPase, 33.89 ± 1.16 for the Mg-ATPase and 34.28 ± 1.51 for the Ca-ATPase with low affinity to calcium. A comparison of V_{max}, K_m and E_a values for (Na,K)-ATPase, Mg-ATPase and Ca-ATPase in control hearts, hearts of rats subjected to high altitude hypoxia and in hearts subjected to 45 min ischemia at 37 oC are given in Fig. 2. Results indicate that intermittent high altitude hypoxia decreased significantly the maximal reaction velocity of (Na,K)-ATPase ($p < 0.001$) and Ca-ATPase ($p < 0.01$) exerting the smallest effect on V_{max} value of the Mg-ATPase ($p < 0.05$) in cardiac sarcolemma (Fig. 2. Upper panel). In hearts subjected to ischemia lasting 45 min all investigated ATPases showed significantly depressed V_{max} values ($p < 0.001$). In respect to decrease of K_m, values most sensitive to ischemia or hypoxia proved to be again the (Na,K)-ATPase ($p < 0.01$) followed by the Ca-ATPase ($p < 0.05$) but the Mg-ATPase failed to exhibit any considerable modulation in affinity of its active site towards ATP ($p > 0.05$; Fig. 2. Middle panel). Differences in individual response of investigated ATPases to ischemia were also seen in activation energy (Fig 2. Lower panel). For technical reason, changes in activation energy were investigated in ischemia only. Results revealed that after 45 min of ischemia no significant change in activation energy could be detected in the case of (Na,K)-ATPase. This points to high degree of adaptation of this enzyme to ischemia [8, 25, 35, 36]. In contrast, significant elevation in E_a required for Ca-ATPase ($p < 0{,}05$) and particularly for Mg-ATPase ($p > 0{,}005$) indicates that from the point of view of their activation energy these enzymes are less adapted to work in ischemic conditions.

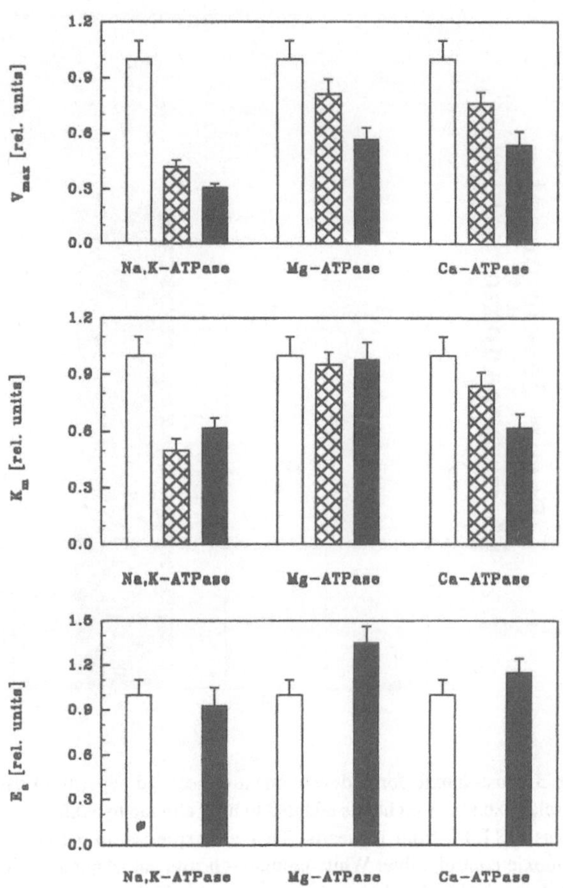

Fig. 2. Kinetic charactenstics of sarcolemmal (Na,K)-ATPase, Mg-ATPase and Ca-ATPase with low affinity to calcium in hearts of rats adapted to intermittent high altitude hypoxia and in hearts subjected to 45 min ischemia. White columns – control values from normal hearts; cross-lined columns – hearts adapted to high altitude hypoxia; black columns hearts after 45 min of ischemia at 370C. Results are means ± S.E.M. from 10–15 hearts and they are expressed in relative units referring to the respective control values taken as 1. Statistical significances were established by means of the Student's *t*-test. For other details see the Materials and methods.

In spite of diverse experimental models that we have applied in evaluating the effects of ischemia and hypoxia, the changes in kinetic parameters of investigated enzymes exhibited essentially similar trends. This indicated that they may be triggered by similar mechanisms, i.e., that the adaptation changes termed as preconditioning are not confined to ischemia only. As a particular proof of this may be considered the significantly ($p < 0.05$) better post-anoxic force development of isolated papillary muscles from hearts preconditioned by adaptation to high altitude hypoxia in comparison to non-adapted controls (Fig. 3). Moreover, this finding also proves that the adaptation of cardiac energy utilising systems, including sarcolemmal ATPases, to lowered energy production and delivery in ischemia/hypoxia does not mean only a more economic handling with ATP on biochemical level, but it has also important functional consequences. On the other hand, the intentionally chosen diversity of experimental models applied for adaptation to ischemia/hypoxia in this study made little sense for a crossevaluation of their preconditioning-efficacy, at least from the point of view of modulation of the ATPase activities.

Investigation of ATP and ADN contents in adjacent, non-ischemic parts of the left ventricle after preconditioning with regional ischemia revealed a balanced energy status in the dog hearts (Fig. 4). Expectedly, after three subsequent 4 min lasting occlusions of the LAD coronary artery, each followed by 5 min reperfusion of the ischemic area, the actual tissue levels of ATP and ADN were found decreased significantly ($p < 0.05$). Nevertheless, the ratio of ADN to ATP in preconditioned hearts remained preserved on the control value. This

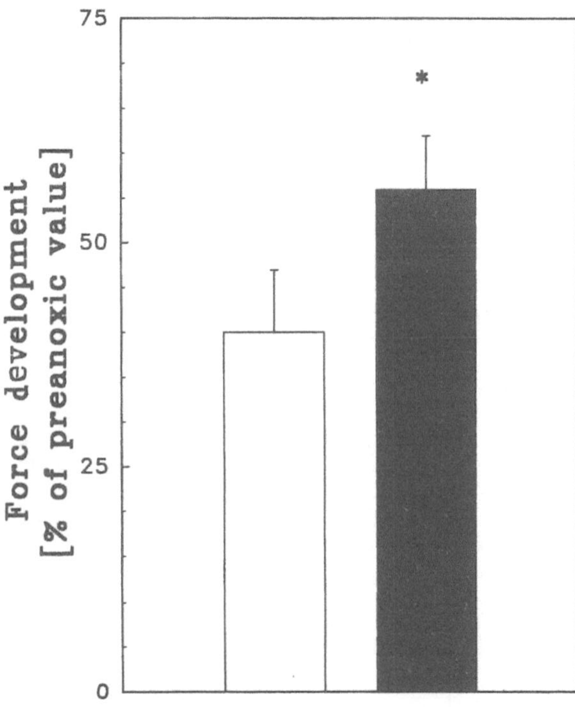

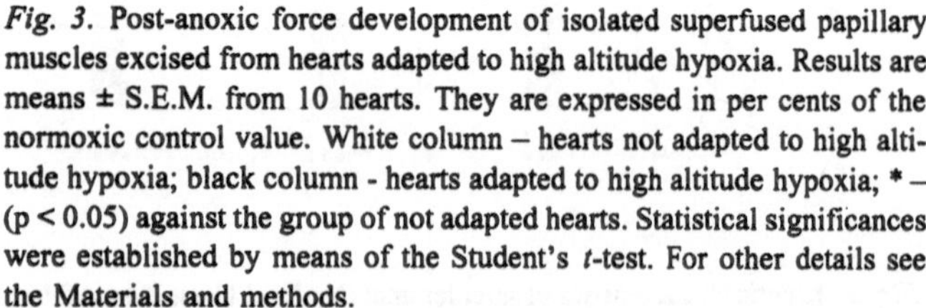
Fig. 3. Post-anoxic force development of isolated superfused papillary muscles excised from hearts adapted to high altitude hypoxia. Results are means ± S.E.M. from 10 hearts. They are expressed in per cents of the normoxic control value. White column – hearts not adapted to high altitude hypoxia; black column - hearts adapted to high altitude hypoxia; * – ($p < 0.05$) against the group of not adapted hearts. Statistical significances were established by means of the Student's *t*-test. For other details see the Materials and methods.

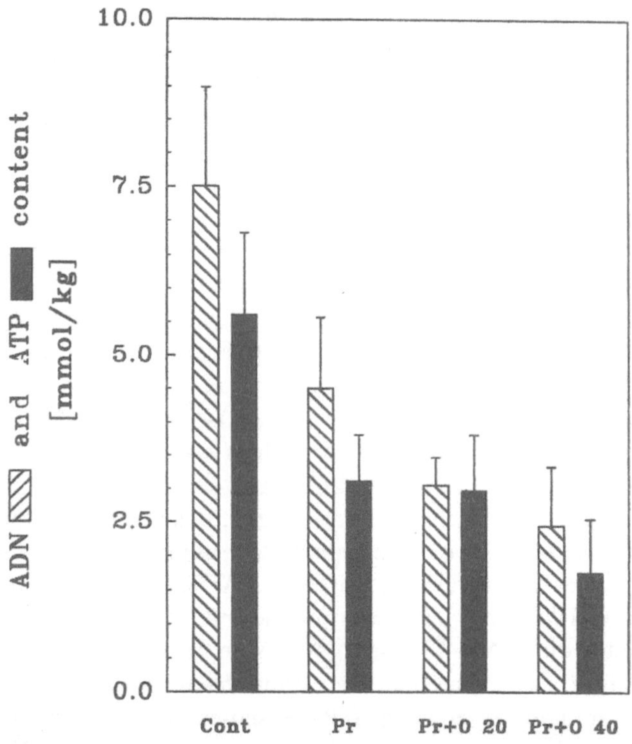

Fig. 4. Adenosine triphosphate (ATP) and total adenine nucleotides (ADN) content in non-ischemic areas of the left ventricle of dog hearts preconditioned with repeated 4 min occlusions of the left anterior descendent coronary artery (LAD) *in situ*. Cont – control values *in situ*, prior to onset of coronary occlusion; Pr – value after preconditioning with three successive four min occlusions of LAD followed by 5 min reperfusion of the ischemic area; Pr+O 20 – following 20 min testing ischemia started 60 min after preconditioning; Pr+O 40 – following 40 min testing ischemia started 60 min after preconditioning. Results are means + S.E.M. from 4–5 experiments in each group. Statistical significances for changes in both, ATP and ADN contents: Cont vs. Pr – $p < 0.05$; Pr vs. Pr+O 20 and/or Pr+O 40 $p > 0.05$ (Student's *t*-test).

indicated the creation of a new equilibrium in energy metabolism, on lower energy level. Essentially the same equilibrium (characterised by non-significant changes in ADN/ATP ratio, $p < 0.05$) was found preserved after both, the 20 and 40 min lasting testing-ischemic episodes. This indicated that a high level of ischemic preconditioning was achieved.

A gradual increase of protein content in HS fraction of coronary sinus blood serum from dog hearts was observed in the time course of ischemic preconditioning (Fig. 5). This increase involved particularly two high molecular and one low molecular (14 kD) glycoproteins as revealed by means of SDS-PAGE and Western blot analysis with Con A and anti-Con A antibody (Fig. 6). These glycoproteins seem not to belong to the shock proteins described up to now. Their level in coronary sinus blood was gradually increasing in the time course of preconditioning and it reached its maximum 120 min after the third 4 min lasting occlusion, i.e., coincidentally with the peak preconditioning. In addition to latter glycoproteins, also the accumulation of a Con A-insensitive protein (approximately 20 kD) was observed in the HS fraction after preconditioning. When the HS fraction isolated from the blood of exsanguinated animal was later concentrated, freeze-dried, than again dissolved in saline (10 ml) and finally administered intracoronarily to an other animal 60 min prior to a 40 min coronary occlusion, it induced a cardioprotective action resembling that of the ischemic preconditioning.

Discussion

In physiological conditions the energy metabolism of the myocardium is always in a state of dynamic equilibrium characterised by balanced energy production and energy utilisation. Any impulse leading to decrease in energy production or to increase in energy consumption induces a transient disbalance between them. Then in turn mechanisms are activated which are shifting the capacity of myocardial energy production to match the energy demands or vice versa and

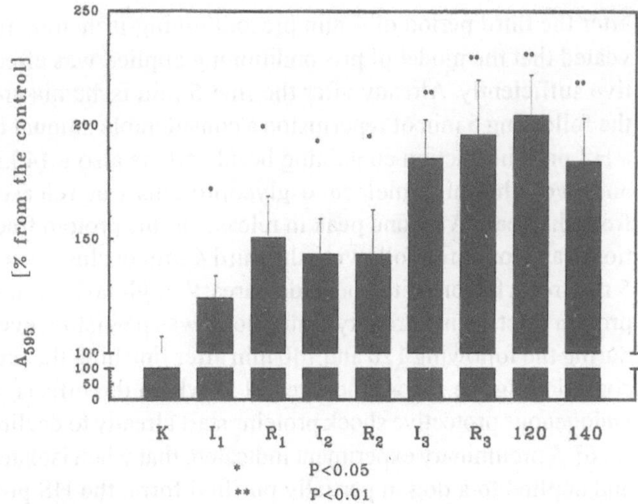

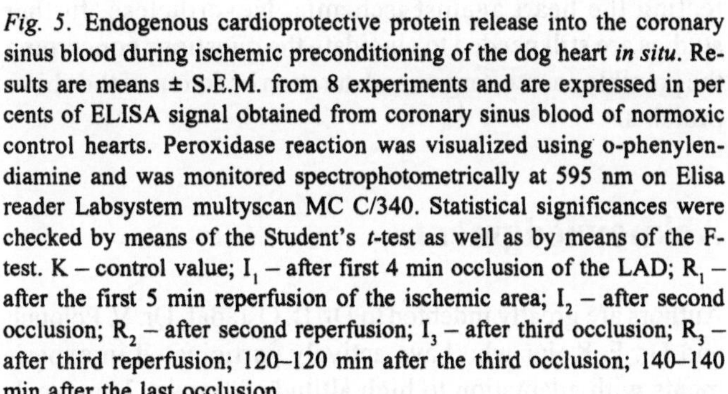

Fig. 5. **Endogenous cardioprotective protein release into the coronary sinus blood during ischemic preconditioning of the dog heart *in situ*. Results are means ± S.E.M. from 8 experiments and are expressed in per cents of ELISA signal obtained from coronary sinus blood of normoxic control hearts. Peroxidase reaction was visualized using o-phenylendiamine and was monitored spectrophotometrically at 595 nm on Elisa reader Labsystem multyscan MC C/340. Statistical significances were checked by means of the Student's *t*-test as well as by means of the F-test. K – control value; I_1 – after first 4 min occlusion of the LAD; R_1 – after the first 5 min reperfusion of the ischemic area; I_2 – after second occlusion; R_2 – after second reperfusion; I_3 – after third occlusion; R_3 – after third reperfusion; 120–120 min after the third occlusion; 140–140 min after the last occlusion.**

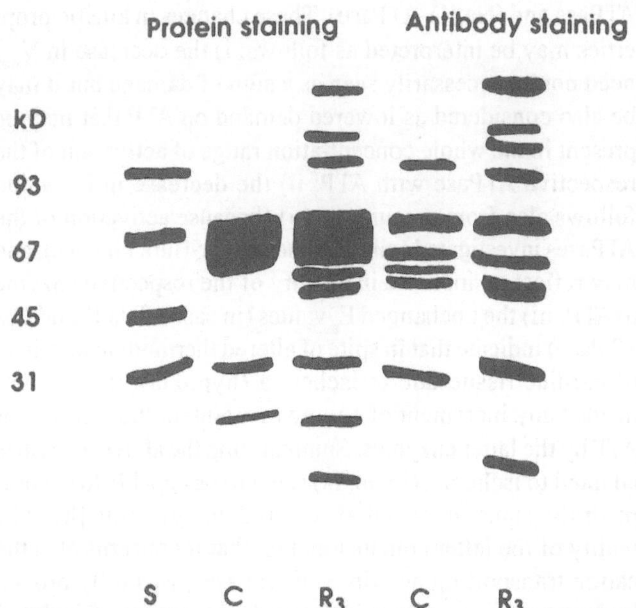

Fig. 6. **Protein and glycoprotein patern of HS fraction prior (C) and after (R_3) three times repeated short occlusion and reperfusion cycles (see Materials and methods). Protein bands were separated on SDS-PAGE using the gradient 5–25% of polyacrylamide gel, further transferred electrophoretically to nitrocellulose membrane and stained using Ponceau red (protein staining) and Con A – mouse anti-Con A antibody – pig anti-mouse antibody coupled with horseradish peroxidase (antibody staining).**

to establish a new equilibrium state. This is what is expected to occur in ischemic preconditioning of the heart in consequence of repeated short ischemic insults followed by reperfusion of the ischemic area [29–32]. However, some studies have shown that already one ischemic impulse may be sufficient to start trigger the mechanisms involved in adaptation of myocardial energetics to ischemia [4, 16]. Nevertheless, in previous studies with rabbit hearts we have demonstrated that a near-to-equilibrium state in mutual transformation of ATP to ADP and AMP in both, the direction of synthesis and/or breakdown, may be achieved with repeated ischemic impulses only [2, 33, 34]. Similarly to previous results, our present data obtained with rat hearts indicate that the above near-to-equilibrium state is characterized by: i) tissue contents of ATP, ADP and AMP that, when plotted semilogarithmically, are lying on a straight line; ii) by similar ratios of ATP/ADP and ADP/AMP. This follows from i) and testifies for constant rate of ATP, ADP and AMP interconversion (Fig. 1). In the same [33, 34] as well as in further studies [35–37] it was also investigated the molecular basis of adaptation of cardiac energy utilizing systems to decreased availability of ATP in ischemia. In this respect, the interest was focused on sarcolemmal ATPases owing to their regulatory role for heart function [38]. An other reason for investigation of ATPases in these circumstances was their particular suitability for involvement into the adaptation to decreased availability of ATP, because they have been shown to utilize predominantly ATP produced in glycolysis [39]. The latter is namely believed to be the predominating source of energy production in the subsarcolemmal space, where relatively little mitochondria are present. The dependence of sarcolemmal ATPases on glycolytic-ATP might contribute to increasing lack of ATP in the vicinity of latter enzymes already in the time interval between onset of ischemia and the time when glycolysis becomes really accelerated. The last is coupled with its conversion from aerobic to anaerobic one. Electrophysiological investigations in ischemia of short duration as well as in the early phase of reperfusion seem to support this notion (for review see Parratt and Vegh [4]). As another support may serve the findings on ischemic dog hearts [40] where considerable diminution in tissue content of ATP was found already during the first 5 min of ischemia. Nevertheless, this finding concerns only the bulk concentration of ATP. However, owing to the overall accepted internal compartmentalization of ATP in heart cells, the actual ATP contents in the vicinity of sarcolemma may be expected to change even more dramatically.

As it appears from Fig. 2 the adaptation of sarcolemmal ATPases to both, hypoxia as well as ischemia is manifested in decrease of their V_{max} and K_m values for ATP and unchanged, if not decreased activation energies E_a (the Ca-

ATPase and (Na,K)-ATPase). These changes in kinetic properties may be interpreted as follows: i) the decrease in V_{max} need not be necessarily seen as a sign of damage but it may be also considered as lowered demand on ATP that may be present in the whole concentration range of activation of the respective ATPase with ATP; ii) the decrease in K_m value follows also from presumption i) (because activation of the ATPases investigated follows Lineweaver-Burk kinetics), and may reflect an increase in affinity of the respective enzyme to ATP; iii) the unchanged E_a values (in case of Na,K and Ca-ATPase) indicate that in spite of altered thermodynamic state of cardiac tissue due to ischemia (hypoxia), there is not needed any increment of energy to maintain the splitting of ATP by the latter enzymes. Summarizing the above: enzymes adapted to ischemia (hypoxia) seem to be capable to accomplish the same work but at lowered energy costs [8]. The reality of the latter conclusion, i.e., that it concerns also the cation transporting activities of ATPases, was fully proved by testing the post-anoxic force development of isolated superfused papillary muscles excised from hearts adapted to high altitude (Fig. 3).

It is also important to be aware of the fact that cardiac sarcolemmal ATPases are *in situ* acting at bulk concentrations of ATP amounting approximately 10–30% of their K_m value. However, in consequence of compartmentalization in tissue, the real ATP concentrations in vicinity of the ATPases may be even lower. Therefore, even small ischemia- or hypoxia-induced changes in availability of ATP may be for sarcolemmal ATPases of keen importance. The relative resistance of Mg-ATPase towards hypoxia and ischemia is not very surprising because this enzyme is not involved essentially in any sarcolemmal transport process.

Validity of our conclusions driven from experiments on isolated perfused rat heart preconditioned with global ischemia and on hearts of rats preconditioned with intermittent high altitude hypoxia was checked also on dog hearts in situ preconditioned with repeated acute occlusions of the LAD. In previous studies we have reported that acute occlusion of the LAD and following reperfusion of the occluded area are changing the glutathione status and induce accumulation of malondialdehyde as well as activation of γ-glutamyl transpepetidase to almost similar extent in all parts of the heart [41]. This indicated that the effect of preconditioning with local ischemia can not be confined to ischemic and reperfused part of the heart only. For this reason, and also because of little information available about the metabolism in nonischemic areas of the heart preconditioned with occlusions of the LAD, in present experiments we focused our interest on the energy status of the left ventricular myocardium adjacent to the ischemized and reperfused area (Fig. 4). The unchanged or only not significantly altered ATP and total adenine nucleotide contents in non-ischemic areas of the left ventricle after the 20 min and 40 min test insults, in comparison to the state after the third period of 4 min precoditioning-ischemia, revealed that the model of preconditioning applied was effective sufficiently. Already after the first 5 min ischemia and the following 5 min of reperfusion a considerable amount of a HS protein fraction containing beside others also a 14 kD and two other high molecular glycoproteins was released from the heart. A second peak in release of this protein fraction was monitored following the third 4 min occlusion and 5 min reperfusion of the occluded area. A high level of this protein fraction in coronary sinus blood was persisting even during the following 120 and 140 min after finishing the preconditioning, i.e., in a time interval in which the effects of endogenous protective shock proteins start already to decline [2, 6]. A preliminary experiment indicated, that when isolated and applied to a dog in partially purified form, the HS protein fraction seems to act as an endogenous substance protecting the heart against ischemia. Nevertheless, further studies are still needed to elucidate the questions concerning the possible source, targets and duration of action of the above proteins.

Acknowledgment

Authors are greatly indebted to Dr. B. Ošťádal, Dr. V. Pelouch and Dr. F. Kolář who have actively participated in experiments with adaptation to high altitude hypoxia. The excellent technical assistance of Mrs. M. Hýbelová, E. Havránková, Z. Hradecká, L. Belešová and J. Hal'áková is also gratefully appreciated.

This work was supported, in part, by the Slovak Grant Agency for Science 9, grants No 23793, 240/93, 241/93 and 242/93.

References

1. Murry CE, Jennings RB, Reimer KA: Preconditioning with ischemia. A delay of lethal cell injury in ischemic myocardium. Circulation 74: 1124–1136, 1986
2. Reimer KA, Murry CE, Jennings RB: Cardiac adaptation to ischemic preconditioning increases myocardial tolerance to subsequent ischemic episodes. Circulation 82: 2266–2268, 1990
3. Downey JM: Ischemic preconditioning. Nature's own cardioprotective intervention. Trends Cardiovasc Med 2: 170–176, 1992
4. Parratt JR: Vegh A: Pronounced antiarrhythmic effects of ischemic preconditioning. Cardioscience 5: 9–18, 1994
5. Walker DM, Yellon DM: Ischemic preconditioning: from mechanism to exploatation. Cardiovasc Res 26: 734–739, 1992
6. Marber MS, Latchman DS, Walker DM, Yellon DM: Cardiac stress protein elevation 24 h after brief ischemia or heat stress is associated with resistance to myocardial infarction. Circulation 88: 1264–1272, 1993
7. Lawson CS: Preconditioning in man: progress and prospects. In: S. Haunso and K. Kjeldsen (eds). International Society for Heart

Research, European Section Meeting, Copenhagen (Denmark), June 8–11, 1994. Mondzzi Editore, Bologna, 1994, pp 81–85

8. Ziegelhöffer A, Grunermel J. Džurba A, Procházka F. Kolář F. Vrbjar N, Pelouch V, Ošt'ádal B. Szekeres L: Sarcolemmal cation transport systems in rat hearts acclimatized to high altitude hypoxia, influence of 7-oxo prostacyclin, In: B. Ošťádal and N.S. Dhalla (eds). Heart Function in Health and Disease. Kluwer Academic Publishers. Norwell, Massachusetts, 1993, pp 219–228
9. Krause EG, Szekeres L: On the mechanism and possible therapeulic application of delayed adaptation of the heart to stress situations. Mol Cell Biochem 1994 in press
10. Šiška K, Ziegelhöffer A, Fedelešová M, Holec V, Slezák J, Styk J, Pancza D, Gabauer I: Effect of Intra-aortic balloon counterpulsation in experimental myocardial injury following acute coronary occlusion. Biochemical, ultrastructural and physiological aspects. Cardiovasc Res 8: 404–414, 1974
11. Deutsch E, Berger M, Kussmaul WG, Hirschfield JW, Hermann HC, Laskey WK: Adaptation to ischemia during percutaneous transluminal coronary angioplasty: clinical metabolic and hemodynamic features. Circulation 82: 2044–2051, 1990
12. Ambrosio G, Tritto I, Chiariello M: Oxygen free radicals and preconditioning. In: S. Haunso and K. Kjeldsen (eds). International Society for Heart Research. European Section Meeting, Copenhagen, Denmark, June 8–11, 1994. Monduzzi Editore, Bologna, 1994, pp 87–91
13. Liu GS, Thornton J, Van Winkle DM, Stanley AWH, Olsson RA, Downey JM: Protection against infarction afforded by preconditioning is mediated by A_1 adenosine receptors in rabbit heart. Circulation 84: 350–356, 1991
14. Mullane K: Myocardial preconditioning. Part of the adenosine revival. Circulation 85: 845–847, 1992
15. Kitazake M, Hori M, Takashima S, Sato H, Inone M, Kamada T: Ischemic preconditioning increases adenosine release and 5'-nucleotidase activity during myocardial ischemia and reperfusion in dogs. Circulation 87: 208–215, 1993
16. Ravingerová T, Pyne NJ, Parratt JR: Ischaemic preconditioning in the rat heart: the role of G-proteins and adrenergic stimulation. Mol Cell Biochem 1994, in press
17. Végh A, Szekeres L, Parratt JR: Preconditioning of the ischemic myocardium: involvement of the L-arginine nitric oxide pathway. Br J Pharmacol 107: 648–652, 1992
18. Végh A, Szekeres L, Parratt JR: Protective effects of preconditioning of the ischemic myiocardium involve cyclo-oxygenase products. Cardiovasc Res 24: 1020–1023, 1990
19. Parratt JR: Endogenous myocardial protective substances. Cardiovasc Res 27: 698–702, 1993
20. Vegh A, Papp JGy, Szekeres L, Parratt JR: Evidence that bradykinin contributes to the pronounced effects of ischemic preconditioning. Br J Pharmacol 1993, in press
21. Brand T. Sharma HS, Fleischmann KE, Duncker DJ, McFalls EO, Verdow PD, Schaper W: Proto-oncogene expression in porcine myocardium subjected to ischemia and reperfusion. Cardiovasc Res 71: 1351–1360, 1992
22. Gross GJ, Auchampach JA: Blockade of ATP sensitive potassium channels prevents myocardial preconditioning in dogs. Circulation Res 70: 223–235, 1992
23. Mitchell MB, Parker CG, Meng X, Brew EG, Ao L, Brown J. Harken A, Banjeree A: Protein kinase C mediates preconditioning in isolated rat heart. Circulation 884: 1–633, 1993
24. Fulton RM, Hutchinson EC, Jones AN: Ventricular weight in cardiac hypertrophy. Brit Heart J 14: 413–420, 1952
25. Ziegelhöffer A, Procházka J, Pelouch V, Ošťádal B, Džurba A, Vrbjar N: Increased affinity to substrate in sarcolemmal ATPases from hearts acclimatized to high altitude hypoxia. Physiol bohemoslov 36: 404–415, 1987
26. Vrbjar N, Soos J, Ziegelhöffer A: Secondary structure of heart sarcolemmal proteins during interaction with metallic cofactors of (Na,K) ATPase. Gen Physiol Biophys 3: 317–325, 1984
27. Taussky HH, Shorr EE: A microcolorimetric method for determination of inorganic phosphorus. J Biol Chem 202: 575–585, 1953.
28. Lowry OH, Rosebrough NJ, Farr AL, Randall RJ: Protein measurement with the folin phenol reagent. J Biol Chem 193, 265–275, 1953
29. Reimer KA, Murry CE, Yamasawa I, Hill ML, Jennings RB: Four brief episodes of myocardial ischemia cause no cumulative ATP loss or necrosis. Am J Physiol 251: H1306–H1316,1986
30. Lange R, Ingwall JS, Hale SL, Alker KJ, Kloner RA: Effects of recurrent ischemia on myocardial high energy phosphate content in canine hearts. Basic Res Cardiol 79: 469–478, 1984
31. Swain JL, Sabina ML, Hines JJ, Greenfield JC Jr, Holmes EW: Repetitive episodes of brief ischemia (12 min) do not produce a cumulative depletion of high energy phosphate compounds. Cardiovasc Res 18: 264–269, 1984
32. Hoffemeister HM, Mauser M, Schaper W: Repeated episodes of regional myocardial ischemia: Effect on local function and high energy phosphate levels. Basic Res Cardiol 81: 361–372, 1986
33. Ziegelhoffer A, deJong JW, Ferrari R. Turi Nagy L: Ischemic preconditioning of the myocardium as a result of adaptation of enzymes catalyzing energy consuming processes to decreased accessibility of metabolic energy. A theoretical study based upon real measurements. I. J Mol Cell Cardiol 24 (Supplement I): S.150, 1992
34. Ziegelhöffer A, deJong JW, Ferrari R, Turi Nagy L: Ischemic preconditioning of the myocardium as a result of adaptation of enzymes catalyzing energy consuming processes to decreased accessibility of metabolic energy. A theoretical study based upon real measurements. II. J Mol Cell Cardiol 24 (Supplement I): S.151, 1992
35. Vrbjar N, Slezák J, Ziegelhöffer A, Tribulová N: Features of the (Na,K)-ATPase of cardiac sarcolemma with particular reference to myocardial ischemia. Europ Heart J 12 (Supplement F): 149–152, 1991
36. Vrbjar N, Džurba A, Ziegelhöffer A: Kinetic and thermodynamic properties of membrane bound Ca-ATPase with low affinity to calcium in cardiac sarcolemma; response to global ischemia of the heart. Life Sci 53: 1875–1973, 1993
37. Vrbjar N, Džurba A, Ziegelhöffer A: Enzyme kinetics and activation energy of (Na,K)-ATPase in ischemic hearts: Influence of the duration of ischemia. Gen Physiol Biophys 13: 405–411,1994
38. Dhalla NS, Ziegelhöffer A, Harrow JAC: Regulatory role of membrane sysytems in heart function. Review Canad J Physiol Pharmacol 55: 1211–1234, 1977
39. Fedelešová M, Dhalla NS, Balasubramanian V, Ziegelhöffer A: Energy dependent stimulation of membrane bound Mg^{2+}- and K^+-Na^+-ATPase by glucose. In: P. Hatt (ed). Les Surcharges Cardiaques (Heart Overloading). Colloque INSERM, Paris, 1972, pp 217–221
40. Reimer AK, Jennings RB: Myocardial ischemia, hypoxia and infarction. In: H.A. Fozard *et al.* (eds). The Heart and Cardiovascular System, Second Edition. Raven Press Ltd., New York, 1992, pp 1875–1973
41. Ondrejičková O, Ziegelhöffer A, Gabauer I, Sotníková R, Styk J, Gibala P, Sedlák J, Horáková L: Evaluation of ischemia-reperfusion injury by malondialdehyde, glutathione, and gamma-glutamyl transpeptidase: lack of specific local effects in diverse parts of the heart following acute coronary occlusion. Cardioscience 4: 225–230, 1993

Molecular and Cellular Biochemistry **147**: 139–144, 1995.

Inhibition of cardiac sarcolemma Na^+-K^+ ATPase by oxyradical generating systems

Qiming Shao, Taku Matsubara, Sunil K. Bhatt and Naranjan S. Dhalla
Division of Cardiovascular Sciences, St Boniface General Hospital Research Centre, Faculty of Medicine, University of Manitoba, Winnipeg, Canada, R2H 2A6

Abstract

The Na^+-K^+ ATPase activity and SH group content were decreased whereas malondialdehyde (MDA) content was increased upon treating the porcine cardiac sarcolemma with xanthine plus xanthine oxidase, which is known to generate superoxide and other oxyradicals. Superoxide dismutase either alone or in combination with catalase and mannitol fully prevented changes in SH group content but the xanthine plus xanthine oxidase-induced depression in Na^+-K^+ ATPase activity as well as increase in MDA content were prevented partially. The Lineweaver-Burk plot analysis of the data for Na^+-K^+ ATPase activity in the presence of different concentrations of MgATP or Na^+ revealed that the xanthine plus xanthine oxidase-induced depression in the enzyme activity was associated with a decrease in V_{max} and an increase in K_m for MgATP; however, K_a value for Na^+ was decreased. Treatment of sarcolemma with H_2O_2 plus Fe^{2+}, an hydroxyl and other radical generating system, increased MDA content but decreased both Na^+-K^+ ATPase activity and SH group content; mannitol alone or in combination with catalase prevented changes in SH group content fully but the depression in Na^+-K^+ ATPase activity and increase in MDA content were prevented partially. The depression in the enzyme activity by H_2O_2 plus Fe^{2+} was associated with a decrease in V_{max} and an increase in K_m for MgATP. These results indicate that the depressant effect of xanthine plus xanthine oxidase on sarcolemmal Na^+-K^+ ATPase may be due to the formation of superoxide, hydroxyl and other radicals. Furthermore, the oxyradical-induced depression in Na^+-K^+ ATPase activity may be due to a decrease in the affinity of substrate in the sarcolemmal membrane. (Mol Cell Biochem **147**: 139–144, 1995)

Key words: sarcolemmal Na^+-K^+ ATPase, lipid peroxidation, oxyradicals, cardiac membrane, oxidative stress

Introduction

It is now well known that Na^+-K^+ ATPase, which is localized in the sarcolemmal membrane, serves as a Na^+-pump and maintains the electrolyte homeostasis as well as electrical characteristics of cardiomyocytes [1–3]. On the basis of the inhibitory action of digitalis glycosides on this enzyme, a depression in the activity of Na^+-K^+ ATPase is considered to result in an increase in the cardiac contractile force development [1]. However, depression in the Na^+-K^+ ATPase activity has also been observed in several pathological conditions such as ischemia-reperfusion or hypoxia-reoxygenation in which the contractile force development is markedly decreased [4–8] and thus the significance of alterations in Na^+-K^+ ATPase with respect to changes in contractile force development is poorly understood. Since a wide variety of free radicals are formed during the ischemia-reperfusion injury in the heart [9–12], some investigators have attempted to examine the role of oxygen-free radicals in causing depression of the Na^+-K^+ ATPase activity in the ischemic-reperfused myocardium [13]. Accordingly, different free radical scavengers were shown to prevent the depression in Na^+-K^+ ATPase activity in ischemia-reperfused myocardium [13]. In fact various species of oxyradicals and oxidants were found to decrease the cardiac sarcolemma Na^+-K^+ ATPase activity [14–18]. Nonetheless, results with respect to the effects of xanthine plus xanthine oxidase system, which is considered to generate superoxide radicals and other oxygen-free radicals, on the cardiac sarcolemma Na^+-K^+ ATPase activity are conflicting [16, 17]. Thus in view of this controversy as well as lack of information concerning mechanisms of the oxyradical-induced changes in cardiac sarcolemma Na^+-K^+ ATPase activity, the

Address for offprints: N.S. Dhalla, Division of Cardiovascular Sciences, St Boniface General Hospital Research Centre, 351 Taché Avenue, Winnipeg, Manitoba, R2H 2A6 Canada present study was undertaken to examine the effect of xanthine plus xanthine oxidase system on sarcolemmal Na^+-K^+ ATPase by using different oxygen-free radicals scavengers. In addition, low concentrations of H_2O_2 plus $FeSO_4$ were employed to examine the action of hydroxyl radicals on the Na^+-K^+ ATPase activity. It should be noted that the Na^+-K^+ ATPase activity in these experiments was measured in the presence of various concentrations of substrate (MgATP) and Na^+ in order to gain some insight into the mechanism of oxyradical-induced depression in the enzyme activity.

Materials and methods

Isolation of sarcolemmal membranes

The isolation method for sarcolemma according to Pitts [19] was somewhat modified. Briefly the frozen porcine heart ventricle was cut into small pieces and homogenized in 0.6 M sucrose, 10 mM imidazole-HCl, pH 7.0 (3–3.5 ml/g tissue) with a Polytron PT-20 (5 × 20 sec, setting 5). The resulting homogenate was centrifuged at 12,000 g for 30 min and the pellet was discarded. After diluting (3-fold) with 140 mM KCl, 20 mM 3-(N-morpholino) propanesulfonic acid (MOPS), pH 7.4, the supernatant was centrifuged at 95,000 g for 60 min. This pellet was suspended in 140 mM KCl, 20 mM MOPS buffer (pH 7.4) and layered over 0.6 M sucrose solution containing 0.3 M KCl, with 50 mM $Na_4P_2O_7$ and 0.1 mM Tris (hydroxymethyl)aminomethane (Tris-HCl, pH 8.3). After centrifugation at 95,000 g for 90 min (utilizing a Beckman swinging bucket rotor), the band at the sucrose-buffer interface was taken and diluted with 3 vol of KCl-MOPS solution. A final centrifugation at 95,000 g for 30 min resulted in a pellet rich in sarcolemma. The pellet was suspended in 140 mM KCl-10 mM MOPS (pH 7.4) except for the estimation of sulfhydryl groups, 10 mM Tris-HCl (pH 7.4) solution was used in the final centrifugation as well as for the final suspension instead of KCl-MOPS solution. All these steps for the isolation of sarcolemma were carried out at 0–4°C. All preparations were used immediately after isolation. Protein concentration was estimated by the method of Lowry *et al.* [20].

Free radical generating systems

Two different sources of oxyradicals were employed: a) Superoxide anion radicals were generated by the xanthine oxidase (23 µg or 0.03 U/ml) (Calbiochem-Behring, Ltd.) reaction using xanthine (2 mM) as a substrate [21]. It should be mentioned that xanthine oxidase was pretreated with 0.4 mM phenylmethylsulforyl fluoride (PMSF) in order to inhibit a trypsin-like activity which is present in the commercial product as a contaminant. PMSF was found to exert no effect on the ATPase activity being measured in this study; b) For hydroxyl radical generating system, 0.1 mM H_2O_2 plus 0.5 mM $FeSO_4$ was used [21]. Superoxide dismutase (Sigma Diagnostics), catalase (Sigma Diagnostics), and D-mannitol (Sigma Diagnostics) were used to scavenge superoxide anions, H_2O_2 and hydroxyl radicals, respectively [21]. It is understood that neither the free radical generating systems nor the scavengers are very specific for any given species of free radicals.

Biochemical assays

Sarcolemma 400 µg protein/ml was incubated separately with desired concentrations of xanthine plus xanthine oxidase or H_2O_2 plus $FeSO_4$ in absence or presence of scavengers (in a total volume of 0.5 ml) for 10 min. The reaction was initiated at the end of incubation by addition of a portion (100 µl) to the cuvette equilibrated with the components for the determination of Na^+-K^+ ATPase (in the total volume of 2 ml) as described earlier [22]. Methods for the determination of malondialdehyde (MDA) as well as for sulfhydryl content in the membrane were same as used before [21, 23].

Statistical analysis

Results are presented as means ± SE and were analyzed statistically by using the Students *t* test. P level < 0.05 was taken to reflect a significant difference between the control and treated preparations.

Results

Treatment of the sarcolemmal membrane with xanthine plus xanthine oxidase was found to depress Na^+-K^+ ATPase activity and SH group content markedly whereas MDA content was increased significantly (Table 1). The presence of superoxide dismutase, a well known scavenger of superoxide radicals, either alone or in combination with catalase, a scavenger of H_2O_2, in the incubation medium with xanthine plus xanthine oxidase prevented the depression in SH group content completely whereas the decrease in Na^+-K^+ ATPase activity and the increase in MDA content were prevented partially (Table 1). The addition of D-mannitol, a scavenger of hydroxyl radicals, into the above oxyradical scavenger mixture also showed partial prevention of the xanthine plus xanthine oxidase-induced depression in Na^+-K^+ ATPase activity (Table 1). It was interesting to observe that the

Table 1. Effect of xanthine plus xanthine oxidase on porcine cardiac sarcolemma Na^+-K^+ ATPase, malondialdehyde (MDA) content and sulfhydryl (SH group) content

	Na^+-K^+ ATPase activity (μmol Pi/mg protein/h)	MDA content (nmol/mg protein)	SH group content (nmol/mg protein)
Control	14.27 ± 1.07	61.67 ± 3.67	72.61 ± 2.37
X + XO	6.81 ± 1.03*	81.85 ± 2.54*	37.59 ± 4.14*
X + XO + SOD	9.61 ± 1.02*	72.54 ± 2.69*	66.01 ± 5.75
X + XO + SOD + CAT	9.87 ± 1.02*	71.99 ± 2.39*	65.11 ± 7.64
X + XO + SOD + CAT + MAN	10.89 ± 1.08*	—	—
X + XO + DTT	9.24 ± 1.24*	—	—

Each value is a mean ± SE of five experiments. Sarcolemmal vesicles were incubated with or without 2 mM xanthine (X) plus 0.03 U/ml xanthine oxidase (XO). The concentrations of Superoxide Dismutase (SOD), Catalase (CAT), D-Mannitol (MAN) and Dithiothreitol (DTT) were 80 μg/ml, 10 μg/ml, 20 mM and 1 mM, respectively. *$p < 0.05$ compared with control.

presence of dithiothreitol, which is known to protect changes in SH groups, exerted partial protection against the xanthine plus xanthine oxidase-induced decrease in Na^+-K^+ ATPase activity (Table 1).

In order to examine if the xanthine plus xanthine oxidase-induced depression in Na^+-K^+ ATPase activity is due to any alterations in the affinities of the enzyme for its ligands, the Na^+-K^+ ATPase activities in the control and xanthine plus xanthine oxidase treated preparations were measured in the presence of different concentrations of MgATP or Na^+. The results in Fig. 1 indicate that the xanthine plus xanthine oxidase induced decrease in Na^+-K^+ ATPase activity was evident at both low and high concentrations of the substrate (MgATP) whereas significant depressions in the enzyme activity were seen only at high concentrations of Na^+ (Fig. 2). The Lineweaver-Burk analysis of the data (Figs 1 and 2; and Table 2) reveal that the xanthine plus xanthine oxidase-induced depression in Na^+-K^+ ATPase activity was associated with a decrease in V_{max} for the enzyme activity. Furthermore, the K_m value for MgATP was increased (Fig. 1 and Table 2) whereas the K_a value for Na^+ was decreased (Fig. 2 and Table 2) in the xanthine plus xanthine oxidase treated preparations.

Since hydroxyl radicals are considered to be the most reactive species of the oxyradicals, membranes were treated with a mixture containing low concentrations of H_2O_2 and low concentrations of $FeSO_4$ to examine the effects of hydroxyl

Table 2. Kinetic characteristics of porcine cardiac sarcolemma Na^+-K^+ ATPase treated with or without xanthine (X) plus xanthine oxidase (XO)

	Control	X + OX treated
MgATP:		
V_{max} (μmol Pi/mg/h)	20.00 ± 1.75	11.75 ± 1.11*
K_m (mM)	0.31 ± 0.04	0.62 ± 0.05*
Na^+:		
V_{max} (μmol Pi/mg/h)	28.57 ± 2.05	10.00 ± 0.90*
K_a (mM)	10.00 ± 1.21	16.67 ± 1.35*

Each value is a mean ± SE of five experiments. *$p < 0.05$ compared with control.

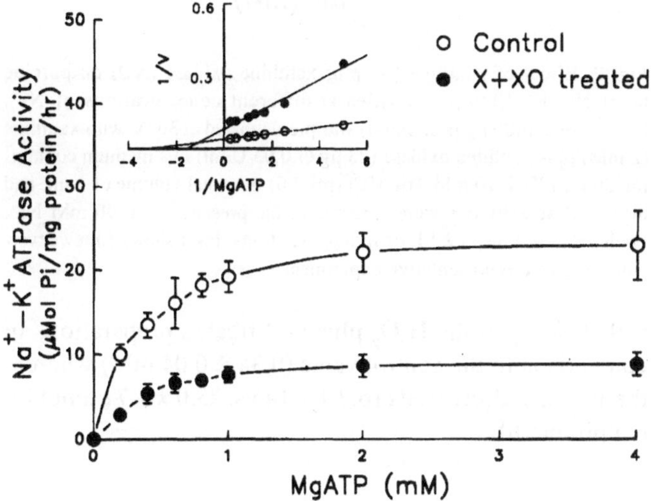

Fig. 1. Effect of xanthine (X) plus xanthine oxidase (XO) on porcine heart Na^+-K^+ ATPase activities at different concentrations of MgATP. Sarcolemma (400 μg protein/ml) was preincubated at 30°C with xanthine (2 mM) plus xanthine oxidase (XO; 23 μg or 0.03 U/ml) in a medium containing 50 mM KCl, 10 mM Tris-HCl (pH 7.0) in a total volume of 1 ml, and then ATPase activities were assessed in the presence of 0.2–4 mM MgATP. Each value is a mean ± SE of five preparations. Inset shows Lineweaver-Burk plot of a representative experiment.

radicals on the Na^+-K^+ ATPase activity. The data in Table 3 reveal that low concentrations of H_2O_2 plus Fe^{2+} produced marked depressions in both Na^+-K^+ ATPase activity and SH group content whereas MDA content was increased. The effects on Na^+-K^+ ATPase and MDA content were partially prevented whereas that on SH group content was fully prevented by the presence of D-mannitol alone or in combination with catalase (Table 3). The Na^+-K^+ ATPase activity in the control and H_2O_2 plus Fe^{2+} treated preparations was also studied in the presence of different concentrations of MgATP. A significant depression in the treated membrane Na^+-K^+ ATPase activity was observed at all concentrations of the substrate (Fig. 3). The Lineweaver-Burk plot analysis (Fig. 3) of the data revealed a marked increase in the K_m value (1.17

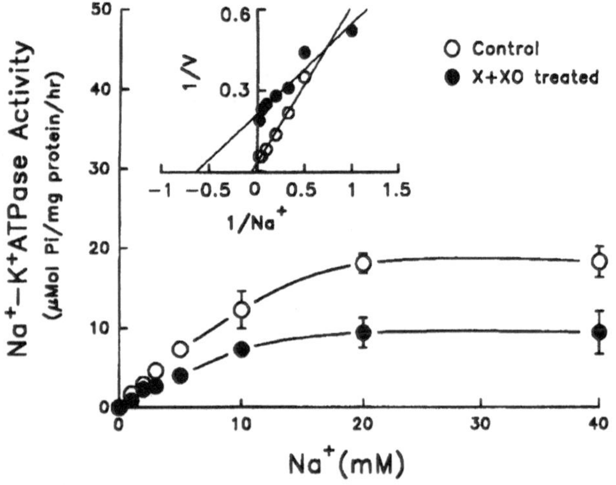

Fig. 2. Effect of xanthine (X) plus xanthine oxidase (XO) on porcine heart Na$^+$-K$^+$ ATPase activities at different concentrations of Na$^+$. Sarcolemma (400 μg protein/ml) was preincubated at 30 °C with xanthine (2 mM) plus xanthine oxidase (23 μg or 0.03 U/ml) in a medium containing 50 mM KCl, 10 mM Tris-HCl (pH 7.0) in a total volume of 1 ml, and then ATPase activities were assessed in the presence of 1–40 mM Na$^+$. Each value is a mean ± SE of five preparations. Inset shows Lineweaver-Burk plot of a representative experiment.

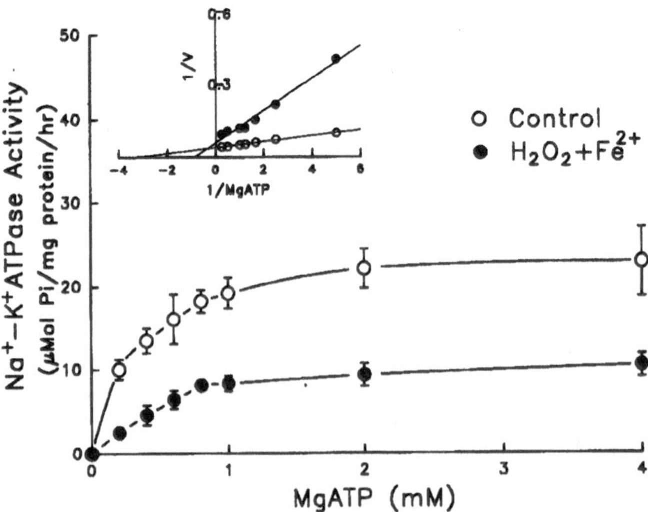

Fig. 3. Effect of H_2O_2 plus $FeSO_4$ on porcine heart Na$^+$-K$^+$ ATPase activities at different concentrations of MgATP. Sarcolemma (400 μg protein/ml) was preincubated at 37°C with 0.1 mM H_2O_2 plus 0.05 mM $FeSO_4$ in a medium containing 50 mM KCl, 10 mM Tris-HCl (pH 7.0) in a total volume of 1 ml, and then ATPase activities were assessed in the presence of 0.2–4 mM MgATP. Each value is a mean ± SE of five preparations. Inset shows Lineweaver-Burk plot of a representative experiment.

± 0.19 mM) in the H_2O_2 plus Fe^{2+} treated preparations in comparison to the control value (0.34 ± 0.04 mM) whereas the V_{max} was decreased (16.7 ± 1.14 vs. 25.0 ± 1.72 μmol Pi/mg protein/h).

Discussion

In this study we have shown that treatment of porcine heart sarcolemma with xanthine plus xanthine oxidase resulted in a depression of the Na$^+$-K$^+$ ATPase activity. This finding confirms the data reported by Xie *et al.* [16] with bovine heart sarcolemma but is in contrast to the results reported by Kukreja *et al.* [17] with dog heart. The inability of Kukreja *et al.* [17] to observe the action of xanthine plus xanthine oxidase on cardiac sarcolemmal Na$^+$-K$^+$ ATPase may be either due to species difference or the preconditioning effect which may have occurred when these investigators subjected the sarcolemmal vesicles to 15 freeze-thaw cycles before exposing to the oxyradical generating system. Nonetheless, exposure of rat cardiomyocytes to xanthine plus xanthine oxidase was found to depress the Na$^+$-K$^+$ ATPase as well as Na$^+$-pump (as monitored by $^{86}Rb^+$ uptake) activities [16]. Likewise, reductions in Na$^+$-pump and p-nitrophenyl phosphatase (which represents the catalytic site of Na$^+$-K$^+$ ATPase) activities as well as the formation of acylphosphates were observed in pig coronary artery upon exposure to xanthine plus xanthine oxidase [24]. It should be pointed out that the results in this study demonstrate that the inhibition of heart sarcolemmal Na$^+$-K$^+$ ATPase is not limited to the oxyradical generation by xanthine plus xanthine oxidase because another oxyradical generating system, low H_2O_2 plus low Fe^{2+}, also

Table 3. Effect of H_2O_2 plus Fe^{2+} porcine cardiac sarcolemma Na$^+$-K$^+$ ATPase, malondialdehyde (MDA) content and sulfhydryl (SH group) content

	Na$^+$-K$^+$ ATPase activity (μmol Pi/mg protein/h)	MDA content (nmol/mg protein)	SH group content (nmol/mg protein)
Control	13.85 ± 0.95	55.67 ± 5.63	66.52 ± 3.91
H_2O_2	12.51 ± 1.12	59.58 ± 2.50	61.74 ± 3.94
MAN	12.53 ± 1.28	59.79 ± 6.33	71.55 ± 5.01
H_2O_2 + Fe^{2+}	6.93 ± 0.13*	88.26 ± 3.07*	40.86 ± 3.54*
H_2O_2 + Fe^{2+} + MAN	9.90 ± 0.98*	76.42 ± 3.47*	67.27 ± 1.47
H_2O_2 + Fe^{2+} + MAN + CAT	10.06 ± 0.84*	—	63.21 ± 5.40

Each value is a mean ± SE of five experiments. Sarcolemmal vesicles were incubated with 0.1 mM H_2O_2 plus 0.05 mM Fe^{2+} with or without D-mannitol (MAN) and catalase (CAT). The concentrations of MAN and CAT were 20 mM and 10 μg/ml, respectively. *$p < 0.05$ compared with control.

produced similar effects. Other oxyradical generating systems and oxidants have also been reported to inhibit Na^+-K^+ ATPase activities in cardiac [14–18], coronary [24] and kidney [25–27] preparations. Nonetheless, the present study extends the previous information that the inhibitory effect of oxyradical generating systems on Na^+-K^+ ATPase is associated with a depression in the V_{max} as well as the affinity of substrate for this enzyme. The observed reduction in the substrate affinity (increased K_m value) for Na^+-K^+ ATPase appears to be specific since the affinity of this enzyme to Na^+ was increased (decreased K_a value) upon treating the sarcolemmal membrane with xanthine plus xanthine oxidase.

We have observed that the xanthine plus xanthine oxidase-induced decrease in Na^+-K^+ ATPase activity was partially prevented by superoxide dismutase either alone or in combination with catalase whereas the depression in SH group content was fully prevented. Furthermore, addition of D-mannitol to these scavenging systems did not exert any further protection. Since superoxide dismutase, catalase and D-mannitol are known to scavenge superoxide radicals, H_2O_2 and hydroxyl radicals, respectively [10, 13, 17, 21, 23], it appears that a part of the inhibitory effect of xanthine plus xanthine oxidase on Na^+-K^+ ATPase may be due to the generation of superoxide radicals whereas the formation of H_2O_2 and hydroxyl radicals may not be contributing significantly in the system employed in this study. Furthermore, hydroxyl radicals generated in the H_2O_2 plus Fe^{2+} system may also explain a part of the inhibitory effect of H_2O_2 plus Fe^{2+} because the presence of D-mannitol, which prevented SH group change completely in this system, only partially prevented the depression in Na^+-K^+ ATPase activity. Because both xanthine plus xanthine oxidase and H_2O_2 plus Fe^{2+} are not specific for the generation of superoxide radicals and hydroxyl radicals, respectively, the contribution of other oxyradicals and non-radicals in causing the inhibitory effect on Na^+-K^+ ATPase under the experimental conditions employed in this study cannot be ruled out. In this regard, it should be noted that singlet oxygen, which is known to be produced in these systems, has been shown to depress cardiac sarcolemmal Na^+-K^+ ATPase activity [18]. It was interesting to observe that addition of superoxide dismutase either alone or in combination with catalase in the xanthine plus xanthine oxidase system prevented the increase in lipid peroxidation (as monitored by changes in MDA content) only partially. Likewise, mannitol alone or in combination with catalase partially prevented the H_2O_2 plus Fe^{2+} induced increase in lipid peroxidation. These observations raise the possibility that a part of the inhibitory effect of these oxyradical generating systems on Na^+-K^+ ATPase activity may be due to the formation of lipid peroxide products. However, further studies on this aspect are needed before making any meaningful conclusion.

Acknowledgements

The research reported in this study was supported by a grant from the Medical Research Council of Canada (MRC Group in Experimental Cardiology).

References

1. Schwartz A, Lindenmayer GE, Allen JC: The sodium-potassium adenosine triphosphate: pharmacological, physiological and biochemical aspects. Pharmacol Rev 27: 3–134, 1975
2. Jorgensen PL: Mechanism of the Na^+,K^+ pump. Protein structure and conformations of the pure (Na^+,K^+) ATPase. Biochim Biophys Acta 694: 27–68, 1982
3. Apell H-J: Electrogenic properties of the Na,K pump. J Membr Biol 110: 1103–1114, 1989
4. Balasubramanian V, McNamara DB, Singh JN, Dhalla NS: Biochemical basis of heart formation. X. Reduction in the Na^+-K^+ stimulated ATPase activity in failing rat heart due to hypoxia. Can J Physiol Pharmacol 51: 504–510, 1973
5. Beller GA, Conroy J, Smith TW: Ischemia-induced alterations in myocardial (Na^+ + K^+)-ATPase and cardiac glycoside binding. J Clin Invest 57: 341–350, 1976
6. Schwartz A, Wood JM, Allen J, Bornet EP, Entman ML, Goldstein MA, Sordahl LA, Suzuki M, Lewis RM: Biochemical and morphological correlates of cardiac ischemia. Am J Cardiol 32: 46–61, 1973
7. Samouilidon EC, Lewis GM, Darsinos JT, Pistevos AC, Karli JN, Tsiganos CP: Effect of low calcium on high-energy phosphates and sarcolemmal Na^+-K^+ ATPase in the infarcted-reperfused heart. Biochim Biophys Acta 1070: 343–348, 1991
8. Dhalla NS, Panagia V, Singal PK, Makino N, Dixon IMC, Eyolfson DA: Alterations in heart membrane calcium transport during the development of ischemia-reperfusion injury. J Mol Cell Cardiol 20 (Suppl II): 3–13, 1988
9. Hammond B, Hess ML: The oxygen free radical system: potential mediator of myocardial injury. J Am Coll Cardiol 6: 215–220, 1985
10. Hess ML, Manson NH, Okabe E: Involvement of free radicals in the pathophysiology of ischemic heart disease. Can J Physiol Pharmacol 60: 1382–1389, 1982
11. Ambrosio G, Flaherty JT, Dullio C, Tritto I, Santoro G, Elia PP, Condorelli M, Chiariello M: Oxygen radicals generated at reflow induce peroxidation of membrane lipids in reperfused hearts. J Clin Invest 87: 2056–2066, 1991
12. Bolli R, Pater BS, Jerondi MO, Lai EK, McCoy PB: Demonstration of free radical generation in "stunned" myocardium of intact dogs with the use of the spin trap alpha-phenyl N-tert-butryl nitrone. J Clin Invest 82: 476–485, 1988
13. Kim MS, Akara T: O_2 free radicals: cause of ischemia-reperfusion injury to cardiac Na^+-K^+ ATPase. Am J Physiol 252: H252–H257, 1987
14. Kramer JH, Mak IT, Weglicki WB: Differential sensitivity of canine cardiac sarcolemmal and microsomal enzymes to inhibition by free radical-induced lipid peroxidation. Circ Res 55: 120–124, 1984
15. Mak IT, Kramer JH, Weglicki WB: Potentiation of free radical-induced lipid peroxidative injury to sarcolemmal membranes by lipid amphiphiles. J Biol Chem 261: 1153–1157, 1986
16. Xie Z, Wang Y, Askari A, Huang W-H, Klaumig JE, Askari A: Studies on the specificity of the effects of oxygen metabolites on cardiac sodium pump. J Mol Cell Cardiol 22: 911–920, 1990
17. Kukreja RC, Weaver AB, Hess ML: Sarcolemmal Na^+-K^+-ATPase:

inactivation by neutrophil-derived free radicals and oxidants. Am J Physiol 259: H1330–H1336, 1990

18. Vinnikova AK, Kukreja RC, Hess ML: Singlet oxygen-induced inhibition of cardiac sarcolemmal Na^+-K^+ ATPase. J Mol Cell Cardiol 24: 465–470, 1992
19. Pitts BJR: Stoichiometry of sodium–calcium exchange in cardiac sarcolemmal vesicles. J Biol Chem 254: 6232–6235, 1979
20. Lowry OH, Rosenbrough NJ, Farr AL, Randall RJ: Protein measurement with the Folin phenol reagent. J Biol Chem 193: 265–275, 1951
21. Kaneko M, Beamish RE, Dhalla NS: Depression of heart sarcolemmal Ca^{2+}-pump activity by oxygen free radicals. Am J Physiol 256: H368–H374, 1989
22. Dixon IMC, Hata T, Dhalla NS: Sarcolemmal Na^+-K^+ ATPase activity in congestive heart failure due to myocardial infarction. Am J Physiol 262: C664–C671, 1992
23. Kaneko M, Elimban V, Dhalla NS: Mechanism of depression of heart sarcolemmal Ca^{2+} pump by oxygen free radicals. Am J Physiol 257: H804–H811, 1989
24. Elmoselli AB, Butcher A, Samson SE, Grover AK: Free radicals uncouple the sodium pump in pig coronary artery. Am J Physiol 266: C720–C728, 1994
25. Kako K, Kato M, Matsuoka T, Mustaphe A: Depression of membrane-bound Na^+-K^+ ATPase activity by free radicals and by ischemia of kidney. Am J Physiol 254: C330–C337, 1988
26. Matsuoka T, Kato M, Kako KJ: Effect of oxidants on Na,K ATPase and its reversal. Basic Res Cardiol 85: 330–341, 1990
27. Thomas CE, Reed DJ: Radical-induced inactivation of kidney Na^+,K^+-ATPase: sensitivity of membrane lipid peroxidation and the protective effect of vitamin E. Arch Biochem Biophys 281: 96–105, 1990

PART III

SIGNAL TRANSDUCTION

Molecular and Cellular Biochemistry **147**: 147–160, 1995.

Alterations of β-adrenoceptor-G-protein-regulated adenylyl cyclase in heart failure

Michael Böhm
Klinik III für Innere Medizin der Universität zu Köln, Joseph-Stelzmann Straße 9, 50924 Köln, Germany

Abstract

Alterations of receptor-G-protein-regulated adenylyl cyclase activity have been suggested to represent an important alteration leading to contractile dysfunction in the failing human heart. Recent experiments suggest that the β_1-adrenoceptor (β_1AR) density and mRNA levels are reduced, while β_2-adrenoceptors and stimulatory G-proteins are unchanged (mRNA and protein level). Functional assays demonstrated that the catalyst of the adenylyl cyclase is not different between failing and nonfailing myocardium. Inhibitory G-proteins are increased (pertussis toxin substrates, protein and mRNA) and correlate to the reduced inotropic effects of β-adrenoceptor agonists and of cAMP-PDE inhibitors. Giα-coupled m-cholinoceptors and A_1-adrenergic receptors are unchanged in density and affinity. Stimulation of these receptors resulted in an unchanged antiadrenergic effect on force of contraction. In conclusion, a downregulation of β_1AR and an increase of Giα have been observed as signal transduction alteration in failing human myocardium. These alterations are due to alterations of gene expression in the failing heart and are related to a defective regulation of force of contraction in heart failure. (Mol Cell Biochem **147**: 147–160, 1995)

Key words: adenylyl cyclase, β-adrenoceptors, G-proteins, heart failure, cardiomyopathy

Introduction

Since several years, it is well established that sympathetic activation occurs in early stages of heart failure to maintain oxygen and substrate supply to the periphery [1]. Consistently, the circulating norepinephrine levels were found to be increased [2] and to be related to the prognosis of the syndrome [2, 3]. In addition to the increased circulating norepinephrine levels, Swedberg *et al.* [4] reported an increased release of norepinephrine from the heart which leads to reduced cardiac norepinephrine levels [5]. These findings have been suggested to be due to an activation of cardiac sympathetic nerves and demonstrates that the heart itself contributes to the elevated catecholamine levels in heart failure. The released norepinephrine produces a stimulation of the postsynaptic β-adrenoceptor-adenylyl cyclase system (cf. Fig. 1). This excessive sympathetic activation in heart failure produces a desensitization of the β-adrenoceptor-adenylyl cyclase system which represents a clinically important pathophysiological alteration resulting in catecholamine refractoriness of the failing myocardium. The subcellular mechanisms will be reviewed herein.

β-Adrenoceptor system

Mechanisms of β-adrenergic desensitization

β-Adrenoceptor desensitization occurs in two steps. The short term desensitization is triggered by two kinases, namely β-adrenoceptor kinase, (β-ARK) and the cAMP-dependent protein kinase A (PKA) [6]. Following long-term agonist exposure of the β-adrenoceptors to agonists, the receptors become downregulated. This means that the receptor molecules cannot be detected in any cell compartment which is most likely due to a breakdown of the protein [7, 8]. At present, it is not known whether β-adrenoceptor desensitization is a necessary prerequisite for the β-adrenoceptor downregulation. β-Adrenoceptor densitization is a rapid process occuring within seconds to minutes. β-ARK phosphorylates the agonist occupied receptor (homologous densitization). This rapid process ($t_{1/2} \approx 20$s) permits the binding of the cytosolic protein β-arrestin to the receptor and results in uncoupling of the β-adrenoceptor from the stimulatory guanine-nucleotide binding protein (Gsα) preventing β-adrenoceptor mediated adenylyl cyclase stimulation [9]. PKA produces a slower ($t_{1/2} \approx 3.5$ min) phosphorylation of

Address for offprints: M. Böhm, Klinik III für Innere Medizin der Universität zu Köln, Joseph-Stelzmann Straße 9, 50924 Köln, Germany

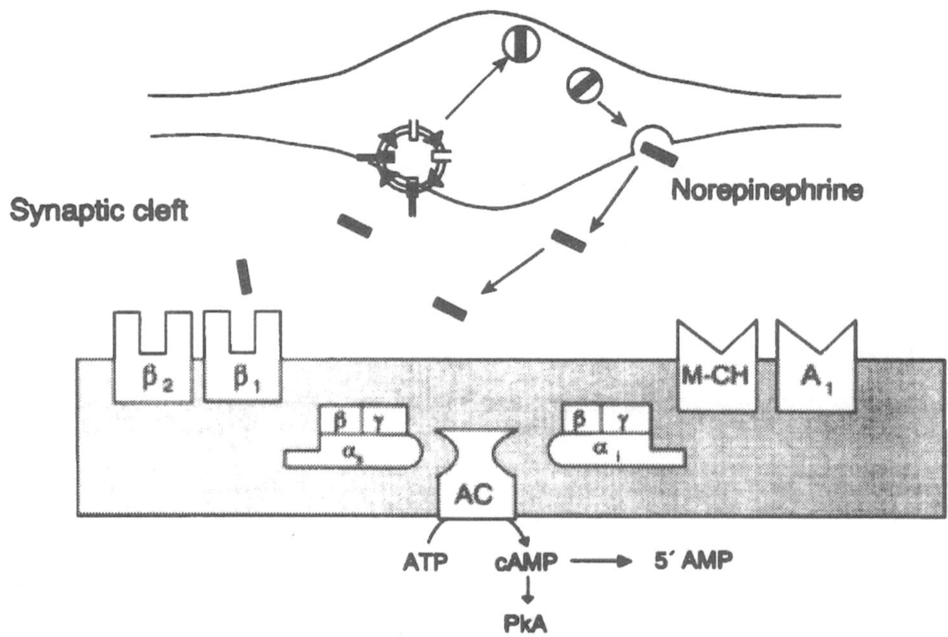

Fig. 1. Scheme of the receptor-G-protein regulated adenylyl cyclase of the myocardial cell. Norepinephrine is released from sympathetic nerve terminals into the synaptic cleft. The catecholamine stimulates cardiac β-adrenoceptors (selectively $\beta_1:\beta_2 \approx 30:1$). The action of norepinephrine is (predominantly) terminated by reuptake into presynaptic stores. Stimulatory β-adrenoceptors (β-AR, β_1- and β_2-subtypes) couple through stimulatory G-proteins to the adenylyl cyclase, while inhibitory m-cholinoceptors (m-Ch, also A_1-adenosine receptors) mediate adenylyl cyclase inhibition and antiadrenergic effects. G-proteins are heterotrimeric complexes (αsβγ, αiβγ) the α-subunits of which are subject for cholera toxin- (αs ≈ 45, 52 kDa) or pertussis toxin- (αi ≈ 40–42 kDa) catalysed ADP-ribosylation. The catalyst forms cAMP from ATP which in turn activates cAMP-dependent protein kinase.

the β-adrenoceptor. This occurs at rather low β-adrenoceptor agonist concentrations and also following stimulation of other receptors coupled to adenylyl cyclase (heterologous desensitization) [6, 7, 10, 11]. The mechanisms of β-adrenergic desensitization by phosphorylation processes are reviewed in Fig. 2. Finally, desensitization of the adenylyl cyclase system can also occur at sites distinct from the β-adrenoceptors.

β-Adrenoceptors and β-adrenoceptor-mediated effects in heart failure

Bristow *et al.* [12] were the first to describe a reduction of the density of β-adrenoceptors in the failing myocardium. This finding has been confirmed by several other groups [13–15]. In later studies, the alterations of β_1- and β_2-adrenoceptor subtypes were analysed separately. It has been shown that the density of total β-adrenoceptors of β_1-adrenoceptors is reduced in myocardial membranes from patients with terminal heart failure due to dilated and ischemic cardiomyopathy [14]. In cardiomyopathic tissue from patients with ischemic cardiomyopathy, the density of β_2-adrenoceptors has been reported to be unchanged or slightly reduced, while the number of β_2-adrenoceptors has constantly been described to be unchanged in dilated cardiomyopathy (for review 14). The reduction of β-adrenoceptors is apparently dependent on the severity of heart failure, but not on the underlying cause of contractile dysfunction. Figure 3 shows a series of experiments on myocardial samples of a group of patients with very heterologous cardiac diseases. The decline of the number of myocardial β-adrenoceptors was only related to the clinical judgement of heart failure state rather than on the underlying disease. This becomes most evident when samples from patients suffering from predominant stenosis and mitral incompetence are compared. In these groups, the numbers of left ventricular β-adrenoceptors are similar whereas the hemodynamic load imposed on the left ventricle is entirely different. In agreement with the reduced β-adrenoceptor density, the stimulatory effects on adenylyl cyclase [12] and force of contraction [12–14] have been observed to be reduced. However, in myocardial samples with unchanged numbers of β-adrenoceptors, the effects of β_2-adrenoceptor agonists like dopexamine [16] and fenoterol [17] were reduced. These findings and those observed with the β_2-adrenoceptor agonist zinterol on adenylyl cyclase activity in membranes [18] strongly suggest an uncoupling of β_2-adrenoceptors from postreceptor events.

In this respect, it appears noteworthy that beside β-adrenoceptor agonists cAMP-dependent positive inotropic agents like the cAMP-phosphodiesterase inhibitors milrinone [13, 19, 20] or pimobendan [21] also exert reduced positive inotropic effects in failing myocardium (Fig. 4). As shown in Fig. 4, the reduction of the effects of isoprenaline and milrinone can amount to 60–70% compared to nonfailing

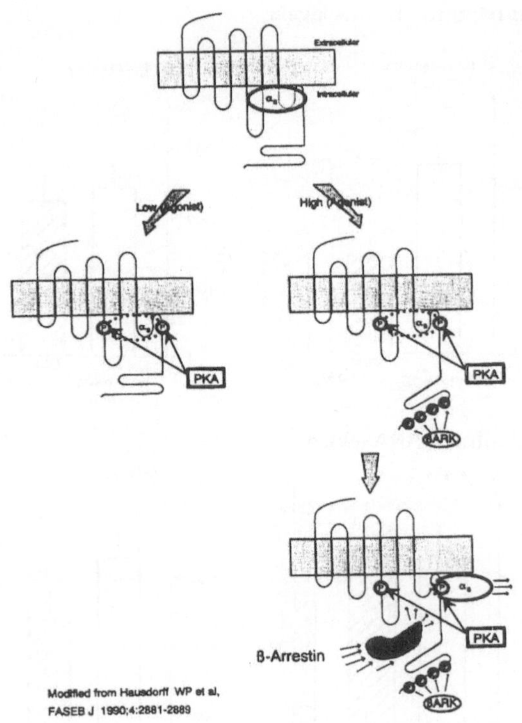

Fig. 2. Mechanisms of β-adrenergic desensitization involving phosphorylation of receptors. PKA: cAMP-dependent protein kinase. β-ARK: β-adrenoceptor kinase. For a detailed description see text. Modified from Hausdorf *et al.* [7].

myocardium. This impairment of inotropic responsiveness has been related to postreceptor events as described below.

β-Adrenoceptor mRNA levels in heart failure

The decrease in the number of β-adrenoceptors can be influenced by two mechanisms. In response to agonist stimulation, β-adrenoceptors could be more rapidly degradated because the PKA-phosphorylated form could be a better substrate receptor for degrading enzymes [22]. However, the downregulation can also occur in the absence of cAMP-elevation [23] or PKA-activation [24], although a preserved coupling of β-adrenoceptors to Gsα appears to be necessary for this process [25, 26]. Second, stimulation of β-adrenoceptors by agonists has been suggested to reduce mRNA levels of β-adrenoceptors [27], which could be due to a decreased half life of mRNA [24, 27]. In order to address the question of whether the second mechanism plays a role in heart failure, the steady state mRNA levels of β_1- and β_2-adrenoceptors were determined. The expression levels of β_1- and β_2-adrenoceptors mRNA were too low to allow a direct determination of receptor mRNA. Therefore, a quantitative PCR technique was used (experimental details described elsewhere, 28). As demonstrated in previous experiments with different amounts of template mRNA, the technique was able to quantify changes of mRNA levels smaller than 30% [28]. As shown in Fig. 5, the steady state levels of quantitative PCR of β_1-adrenoceptor mRNA were similarly reduced by 45–

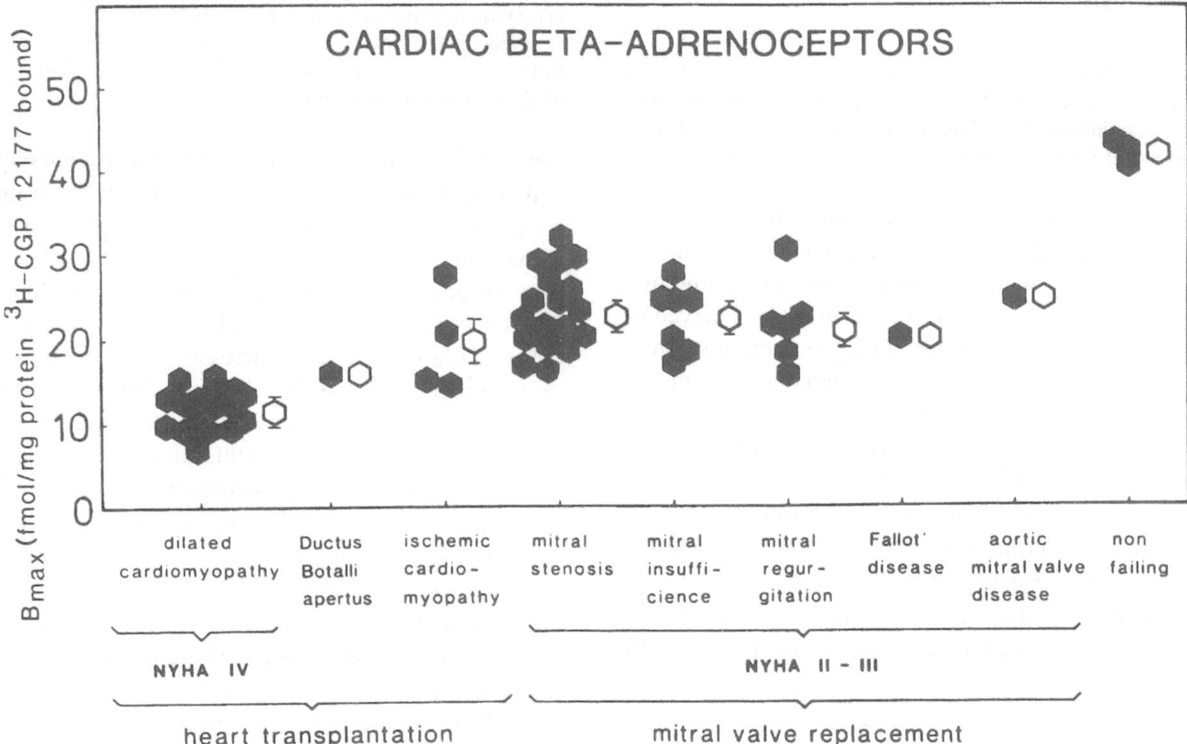

Fig. 3. Number of β-adrenoceptors in myocardial homogenates from left ventricular samples of patients with different cardiac diseases and stages of heart failure.

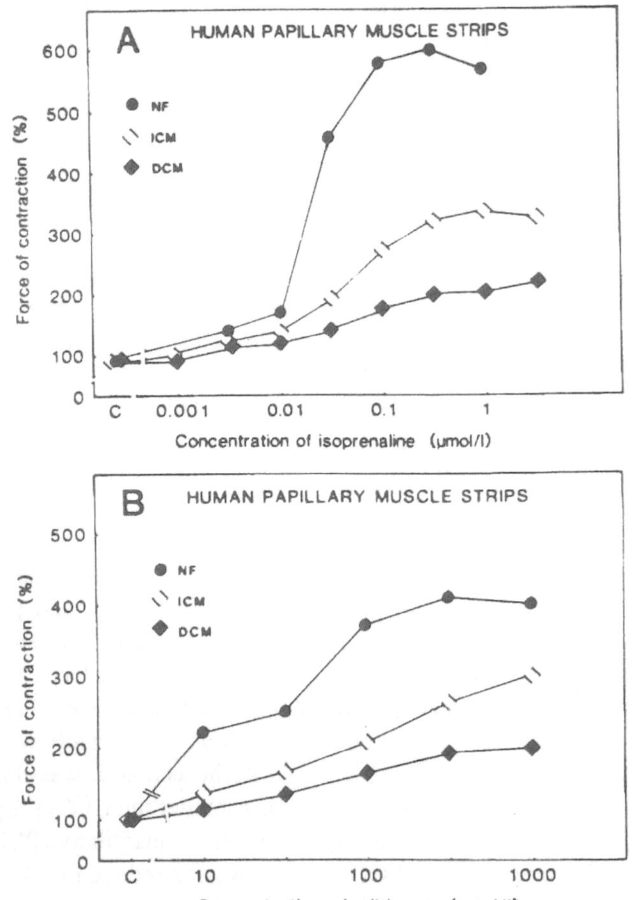

Fig. 4. Concentration-response-curves for the effects of isoprenaline (A) and milrinone in severely (NYHA IV) ICMs: ischemic cardiomyopathy; DCM: dilated cardiomyopathy failing myocardium as well as in nonfailing controls. Modified from previous work [32].

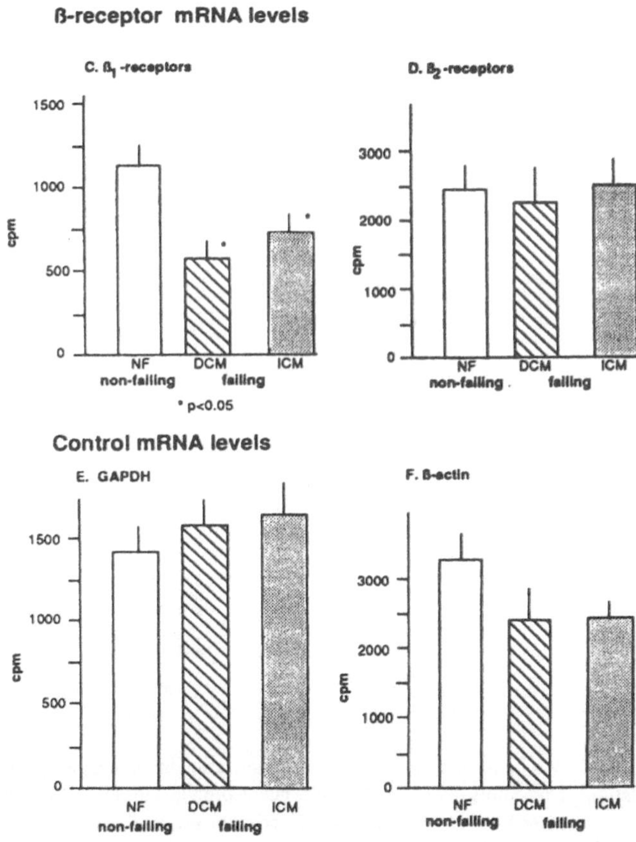

Fig. 5. β1- (left panel) and β2- (right panel) adrenoceptor mRNA levels in samples from patients with ischemic (ICM, n = 10) or dilated (DCM, n = 8) cardiomyopathy and nonfailing controls (NF, n = 6). Data were similar when numbers were related to GAPDH mRNA as standard (not shown). Modified from previous work [28].

55% in ischemic and dilated cardiomyopathy. In contrast, the mRNA levels of β_2-adrenoceptors were not significantly altered in ischemic or dilated cardiomyopathy. These findings on the mRNA levels closely correspond to those observed in radioligand binding experiments. Thus, one might suggest that the reduced β_1- mRNA levels contribute to the reduced number of β_1-adrenoceptors, whereas β_2-adrenoceptors are unchanged on the protein [14–16] and mRNA level. The latter finding was observed with quantitative PCR-techniques [28] and later confirmed with a nuclease protection assays [29]. Since norepinephrine has about 30 times higher potency at β_1-adrenoceptors than at β_2-adrenoceptors [30] it is likely that the alterations of steady-state mRNA are also agonist-dependent and due to the sympathetic activation occuring in heart failure.

β-Adrenoceptor kinase activity and expression

As discussed above, one potential mechanism of β_2-adrenoceptor uncoupling as demonstrated in biochemical [18] and functional [16, 17] studies could be an increased expression and activity of β-ARK. This would result in an increased receptor phosphorylation and uncoupling from Gsα. As shown in Fig. 6, β-ARK activity (measured by phosphorylation of rhodopsin) was increased in failing myocardium (from 28). This effect was not dependent on the underlying cause of heart failure, because the increases were similar in ischemic or dilated cardiomyopathy. Since in CHO-cells transfected with β-ARK, β-adrenoceptor desensitization is enhanced [31], one might speculate that this alteration could play a role in the uncoupling of β_2-adrenoceptors and the remaining β_1-adrenoceptor population. Figure 7 shows mRNA levels of β-ARK. Using different primer pairs, two PCR products of 334 bp (Fig. 6A, 5'end of the mRNA) and 1050 bp (Fig. 6B, middle part of the mRNA) were obtained. Both products were enhanced in ischemic or dilated cardiomyopathy. Thus, it is likely that the increased β-ARK activity could be due to changes on the level of transcription or mRNA processing.

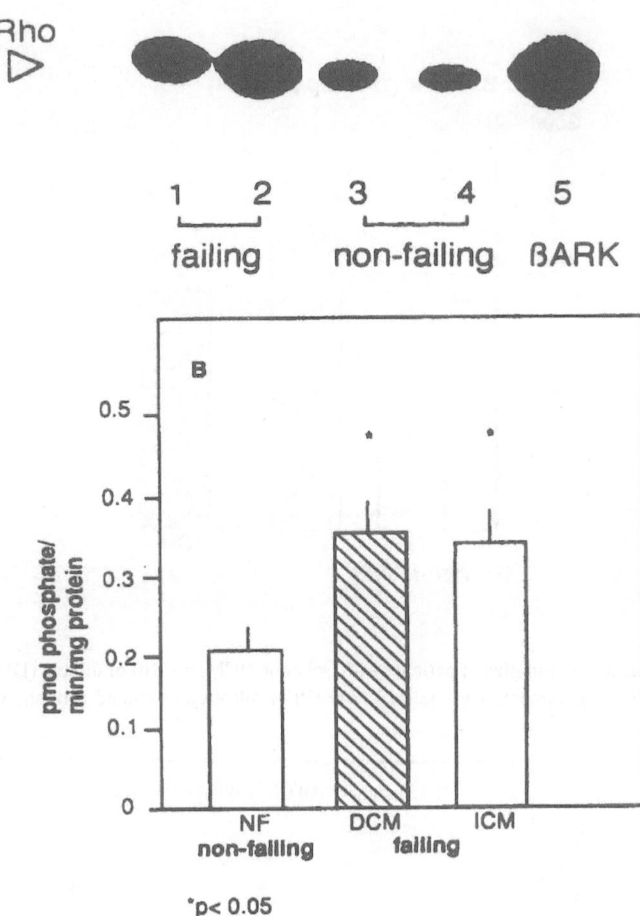

Fig. 6. A: Autoradiogram showing phosphorylation of rhodopsin by β-ARK from failing (lane 1, 2), nonfailing (lane 3, 4) myocardium and purified β-ARK from bovine heart (lane 5). B: β-ARK activity in left ventricular myocardium from nonfailing hearts (NF, n = 6), dilated (DCM, n = 8) and ischemic (ICM, n = 10) cardiomyopathy. Asterisks denote significant differences vs. NF. Modified from previous work [28].

Postreceptor-mechanisms

General considerations

Besides the reduced effects of β-adrenoceptor agonists, the positive inotropic responses to cAMP-phosphodiesterase inhibitors like milrinone [19, 20, 32] or pimobendan [21] have been reported to be reduced. Since these agents increase cAMP-levels and force of contraction with mechanisms occuring beyond receptor occupation, this observation has been explained by a reduced basal rate of cAMP-formation in the failing myocardium [34]. Potential mechanisms could involve a reduced activity of the catalyst, a depressed level or function of stimulatory guanine-nucleotide binding proteins (Gsα) or increased expression or function of inhibitory guanine-nucleotide binding proteins (Giα).

Catalyst activity of the adenylyl cyclase

The catalyst activity of the adenylyl cyclase activity has been studied in enzyme assays in membrane preparations using the diterpen derivative forskolin which directly activates the catalyst of the adenylyl cyclase [35]. The stimulation of adenylyl cyclase by forskolin in membranes was not different from failing and nonfailing hearts, when guanine-nucleotides were omitted from the assay medium [32]. However, in the presence of GTP, the effect of forskolin to activate adenylyl cyclase was reduced in failing human hearts [36]. This observation could be due to the fact that the effect of forskolin can be dependent on G-proteins [35]. When the experiments were performed in the presence of manganese chloride, which is capable of uncoupling the catalyst from the effects of G-proteins [37, 38], the difference was abolished [5, 39]. Taken together, there is no evidence for a change of the catalyst activity in the failing human heart.

Stimulatory guanine-nucleotide binding protein (Gsα)

Stimulatory guanine nucleotide binding proteins (Gsα) mediate the effects of β-adrenoceptors to the catalyst of the adenylyl cyclase [40, 41]. Gsα is a single-gene product with four mRNA isoforms due to alternative splicing [42]. Two protein isoforms have been characterized with the electrophoretic mobility of approximately 45-kDa and 52-kDa proteins [43]. Cholera toxin is able to ADP-ribosylate an arginine residue located in the GTPase site of Gsα [43]. By using ^{32}P-NAD as substrate, one can identify Gsα by autoradiography following electrophoretic separation of membrane proteins. Figure 8 shows ^{32}P-ADP ribosylation in membranes from failing and nonfailing myocardium. HL60 cells and S49 mouse lymphoma cells are also shown. The 45-kDa substrate is lacking in the S49 cyc-variant, which is deficient in Gsα. It can be seen that no difference occurs in the amount of cholera toxin substrates in failing myocardium [36, 44]. Insel and Ransnäs [46] reported that only 5% of immunodetectable Gsα is substrate for cholera toxin labeling. Thus, immunochemical detection is necessary to quantify the amount of Gsα. Figure 9 shows that immunochemical quantification of Gsα is not different in failing and nonfailing human myocardium. This was measured with an antiserum raised against the synthetic C-terminus of Gsα RMHLRQYELL. Finally, Feldman *et al.* [36] reconstituted Gsα from nonfailing and failing myocardium into Gsα-deficient S49 cyc-mouse lymphoma cells and observed similar effects on adenylyl cyclase activity in both groups. The latter findings argue against a functional impairment of Gsα in heart failure in the presence of unchanged amount of protein. Taken together, hitherto, no change in Gsα has been observed using cholera toxin labeling, immunochemical quantification, or functional reconstitution experiments.

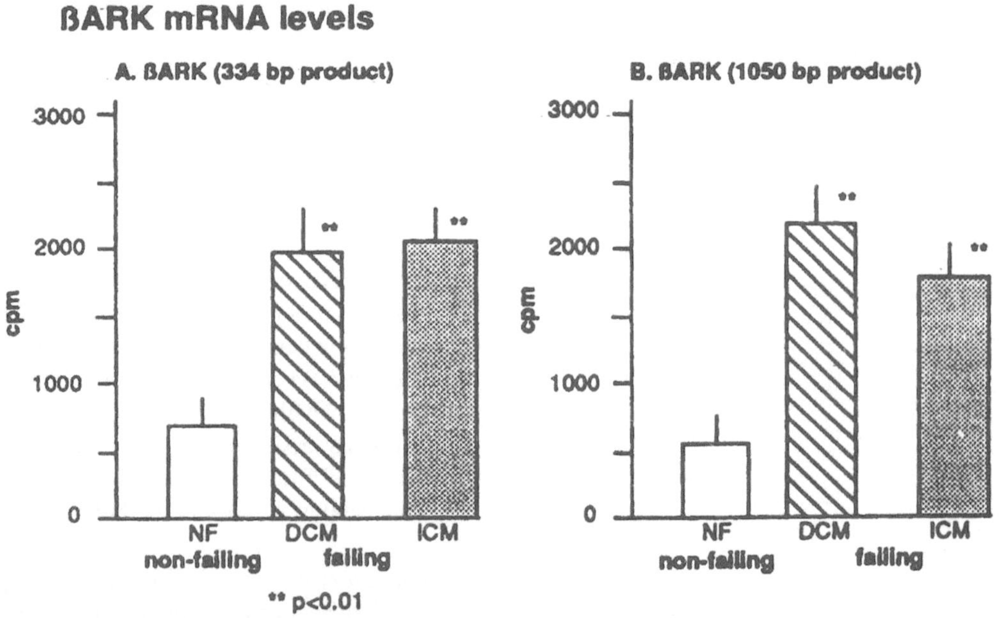

ßARK mRNA levels

A. ßARK (334 bp product)

B. ßARK (1050 bp product)

cpm

NF non-failing DCM ICM failing

** p<0.01

Fig. 7. β-ARK mRNA-levels (A: 5'region of message, B: middle region of message) in samples of patients with ischemic (ICM, n = 10) or dilated (DCM, n = 8) cardiomyopathy and nonfailing controls (NF, n = 6). Data were similar when numbers were related to GAPDH mRNA as standard (not shown). Modified from previous work [28].

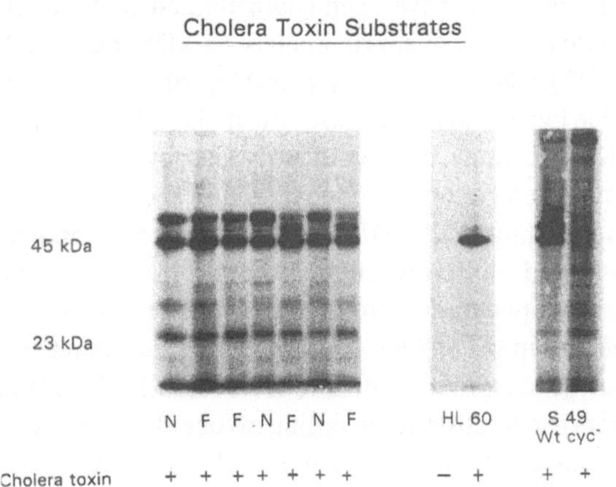

Fig. 8. Cholera toxin labeling of Gsα in membranes from the left ventricles of patients with heart failure (F; dilated cardiomyopathy) or from nonfailing hearts (NF). Cholera toxin substrates in HL60 cells and S49 wild type (wt) mouse lymphoma cells in comparison to the cyc-variant of S49 cells genetically lacking Gsα. Modified from previous work [44].

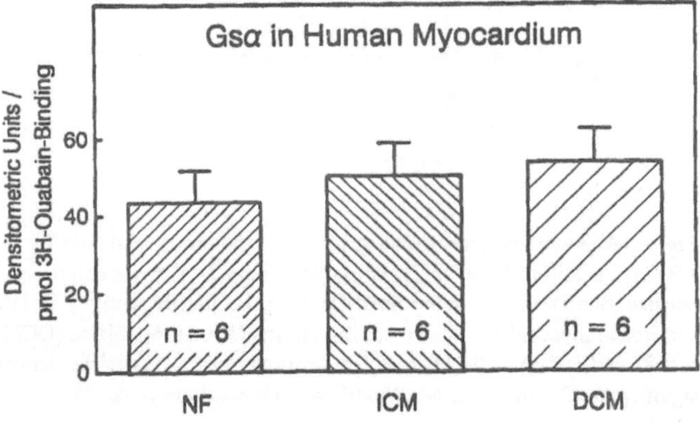

Fig. 9. Gsα content in membranes from the left ventricles of nonfailing hearts (NF) and from the left ventricles of patients with dilated (DCM) and ischemic (ICM) cardiomyopathy. Ordinate: Gsα content in densitometric units of Gsα (45 kDa) per pmol/^{3}H-ouabain binding as myocardial membrane marker. Abscissa: Studied conditions.

Inhibitory guanine-nucleotide binding proteins (Giα)

Pertussis toxin catalysed ^{32}P-ADP-ribosylation of Giα. Pertussis toxin-sensitive Giα-proteins occur as at least three different subtypes (αi_1, αi_2, αi_3) in various cells and tissues, each of which is a single gene product [42]. The α-subunits of one family G-proteins (Giα_1, Giα_2, Gia$_3$, Goα and retinal transducin) can be covalently modified by the mono ADP-ribosyltransferase activity of pertussis toxin [40, 47]. When the radioactively labeled substrate ^{32}P-NAD is used these Giα-proteins can be identified and quantified after separation of membrane proteins by SDS-PAGE and autoradiography. With this technique an increase by 35–40% of pertussis toxin substrates was observed in myocardial membranes from patients with dilated cardiomyopathy [32, 36, 39, 48] compared to nonfailing myocardium.

The ADP-ribosyltransferase activity of pertussis toxin covalently links an ADP-ribosyl moiety to the cysteine residue at the fourth amino acid position from the C-terminus [47, 49]. There are a number of factors which influence the pertussis toxin-catalysed (ADP-ribosylation of Giα-proteins and

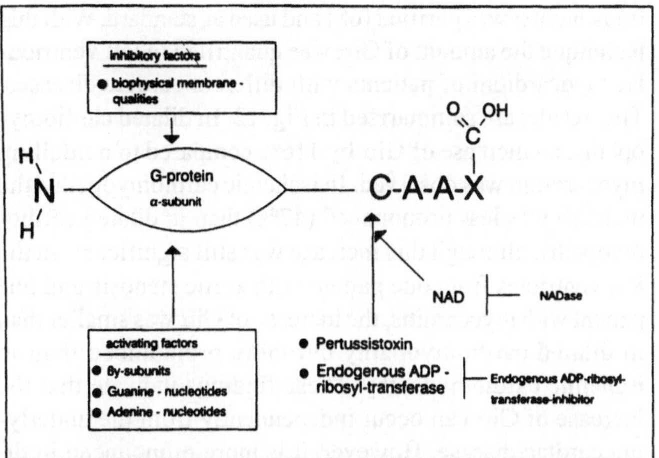

Fig. 10. Summary of factors influencing the pertussis toxin-catalysed ADP-ribosylation of G-protein α-subunits. Pertussis toxin covalently modifies the α-subunits by transferring ADP-ribose of NAD to a cysteine residue present at the fourth position from the C-terminus. Pertussis toxin-substrates possess a common CAAX-motif at their C-terminus where C is cysteine, A is an aliphatic amino acid and X is any amino acid. Exogenous pertussis toxin-catalysed ^{32}P-ADP-ribosylation may be impaired by biophysical membrane properties limiting the accessability of the G-protein α-subunit C-terminus for pertussis toxin, phosphorylation and NADase activity breaking down the substrate for the ADP-ribosylation reaction, and endogenous ADP-ribosyltransferase. On the other hand, ^{32}P-ADP-ribosylation is facilitated by βγ-subunits, guanine nucleotides and inhibitors of endogenous ADP-ribosyltransferases.

limit their usefulness to quantify Giα in membranes or intact cells. These factors are summarized in Fig. 10. Pertussis toxin-induced ADP-ribosylation is facilitated by GTP, GDP and GDPβS and least by nonhydrolysable GTP derivatives [50–52] as well as by β-subunits [53, 54]. This indicates that the GDP-liganded αβγ-complex is the most susceptable substrate for pertussis toxin-induced ADP-ribosylation. In addition, ATP enhances pertussis toxin labeling by activating the toxin itself [55] by binding to the B-oligomer [56] and facilitating dissociation of A-subunit and B-oligomer of the toxin [57]. Endogenous NADase activity metabolizing ^{32}P-NAD necessary for ADP-ribosylation can reduce pertussis toxin labeling of membrane proteins [58]. Moreover, the nonionic detergent Lubrol PX has been shown to concentration-dependently increase labeling in thyroid slices [50] indicating that the biophysical membrane properties play an important role. In addition, phosphorylation [59] or endogenous ADP-ribosylation might limit the ability of pertussis toxin to incorporate ^{32}P-ADP-ribose into Giα in membranes. In fact, an endogenous ADP-ribosyltransferase activity has been identified in human erythrocytes [60], which is capable to ADP-ribosylate G-proteins *in vitro*. Moreover, an endogenous inhibitor of ADP-ribosylation [61] and an endogenous ADP-ribosylhydrolase C activity cleaving mono ADP-ribosyl linkages from Giα [62] have been identified. Finally, lipid modifications, such as myristoylation of G-protein α-subunits can influence the affinity of the α-subunits to β-subunits and could indirectly influence pertussis-toxin labeling [63]. Taken together, all mentioned factors influence the capability of pertussis toxin to determine Giα in membranes. The possible differences between pertussis-toxin substrates and immunodetectable Giα content is demonstrated in Fig. 11. Pertussis toxin substrates comigrating with the α-subunit of Gi/Go from bovine brain in membranes from human heart membranes are more strongly labeled compared to the 40 kDa membrane protein in human lung. On the con-

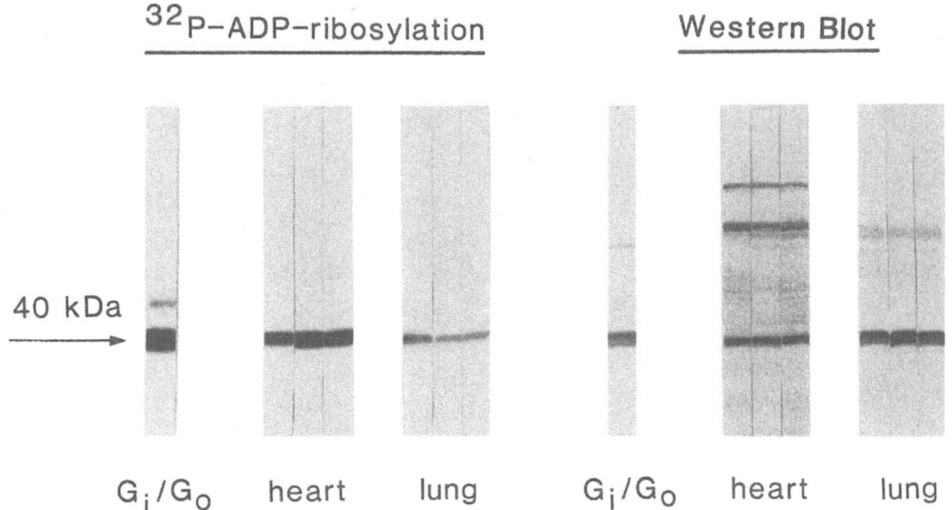

Fig. 11. Pertussis toxin-catalysed ^{32}P-ADP-ribosylation (left, autoradiogram) and immunochemical detection (right, Western blots) of Giα subunits (Mr ≈ 40 kDa) in human myocardial membranes and human lung membranes. Gi/Go (1.5 μg) purified from bovine brain is shown for comparison. Each lane contained 50 μg of membrane protein in ^{32}P-ADP-ribosylation experiments and 150 μg of membrane protein in Western blots. Samples from the same specimens were used. Note that ^{32}P-ADP-ribose incorporation by pertussis toxin was more pronounced in myocardial membranes compared to lung membranes, whereas the intensity of immunostaining of Giα was stronger in lung membranes than in myocardial membranes.

trary, immunodetectable Giα using an antiserum raised against the C-terminal decapeptide of retinal transducin α, clearly showed more immunodetectable Giα in the same samples in human lung than in human heart. These examples emphasize that the intensity of pertussis toxin labeling does not necessarily correspond to the amount of expressed Giα proteins. These findings show that Giα determination by pertussis toxin labeling is hampered by numerous technical and biological uncertainties and that at least one alternative technique should be used to reliably quantify Giα-proteins.

Immunochemical quantification of Giα

In order to study whether the increase by 35–40% of pertussis toxin-substrates in dilated cardiomyopathy is due an increased amount of Giα, immunoblotting techniques were employed in two previous studies to directly quantify Giα-proteins on immunoblots. Some studies reported an increase of immunoreactive material [32, 64], whereas one other report showed only a slight insignificant increase of Giα despite a significant increase of pertussis toxin labeling [65].

In order to improve the immunochemical quantification of Giα-protein, a radioimmunoassay was developed [66]. An antiserum (MB1) was raised against the C-terminal decapeptide of retinal transducin α, which strongly recognizes $G_{i\alpha_1}$ and $G_{i\alpha_2}$ but not $G_{i\alpha_3}$ and Goα [64, 67]. The transducin α C-terminus KENLKDCGLF was iodinated and the assay was performed as described earlier [66]. Isolated retinal transducin α was purified [68] and used as standard. With this technique the amount of Giα was quantified in left ventricular myocardium of patients with different cardiac diseases. The results are summarized in Fig. 12. In dilated cardiomyopathy, an increase of Giα by 116% compared to nonfailing myocardium was observed. In ischemic cardiomyopathy, the increase was less pronounced (47%) than in dilated cardiomyopathy, although this increase was still significant. In the left ventricles from one patient with aortic stenosis and one patient with myocarditis, the increase of Giα was smaller than in dilated cardiomyopathy but more pronounced than in ischemic cardiomyopathy. These findings indicate that the increase of Giα can occur independently from the underlying cardiac disease. However, it is more pronounced in dilated cardiomyopathy, myocarditis and aortic stenosis than in ischemic heart disease.

Relationship of Giα to positive inotropic responses

As discussed above, positive inotropic responses to cAMP-dependent positive inotropic agents have been reported to be reduced in isolated cardiac preparations from failing human hearts. Since both β-adrenoceptors are reduced and Giα-proteins are increased, it is not clear to which extent each biochemical alteration contributes to the impaired functional responses in human heart failure. Figure 13 shows the relation of the positive inotropic responses to the β-adrenoceptor agonists isoprenaline (Fig. 13A) and the cAMP-phosphodi-

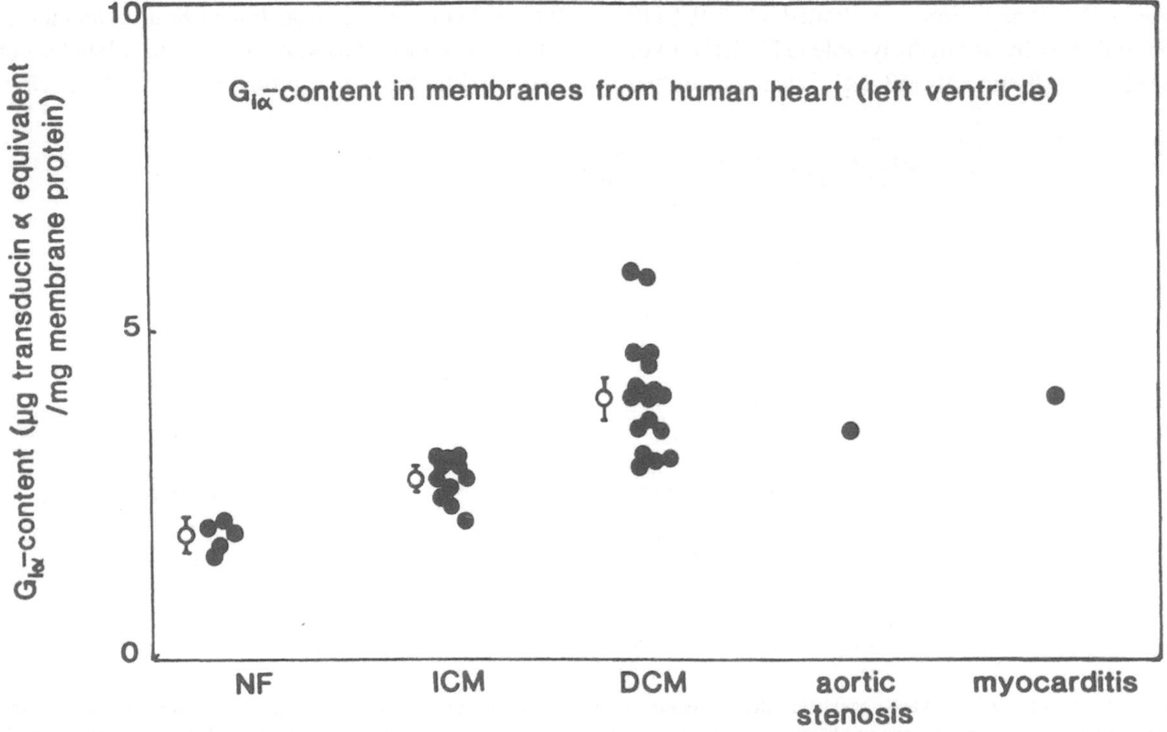

Fig. 12. Giα-content in detergent extracts of membranes from the left ventricles of human hearts with various cardiac diseases and nonfailing hearts (NF). Ordinate: Giα-content in µg transducin a equivalents/mg membrane protein. Abscissa: Studied conditions.

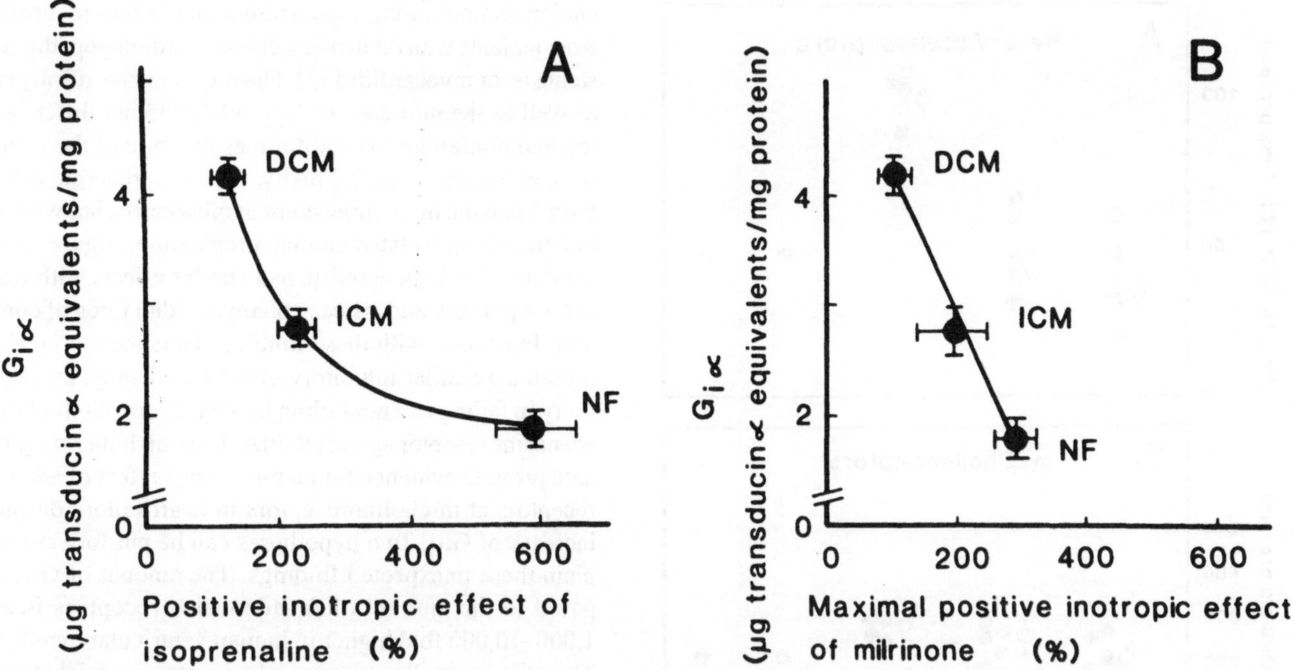

Fig. 13. Relation of immunochemically quantified Giα-content in myocardial membranes (ordinates) to the maximal positive inotropic effects (abscissae) of isoprenaline (A) or milrinone (B) an isolated papillary muscle strips from hearts of patients with dilated (DCM) or ischemic (ICM) cardiomyopathy as well as in nonfailing (NF) hearts.

esterase inhibitor milrinone (Fig. 13B) to the amount of immunodetectable Giα. The inotropic responses to milrinone were reduced in parallel with an increase of immunodetectable Giα levels. These observations might indicate that the reduced positive inotropic effects of cAMP-phosphodiesterase inhibitors are – at least in past – related to an increase of Giα-proteins. This conclusion is strengthened by observations of Brown and Harding [69], in isolated human cardiomyocytes. Pertussis toxin treatment markedly increased the contractile response of the myocytes from failing hearts to isoprenaline. In myocytes from nonfailing heart, pertussis toxin treatment had only small effects. Unlike in native cardiomyocytes, no difference of the inotropic effects of isoprenaline was observed between pertussin toxin-treated cells from failing or nonfailing hearts.

Mechanisms of Giα-regulation

The underlying mechanism of increased Giα-protein expression is not entirely resolved. Treatment of rats with isoprenaline [70] or neonatal rat cardiocytes with norepinephrine [71] produces an increase of Giα. Since the increase of Giα was sensitive to β-adrenoceptor antagonists but not to prazosin [72], this observation suggests that excessive β-adrenoceptor stimulation of the failing heart could also be relevant in this receptor-independent adenylate cyclase regulation. The cAMP-dependent increase of Giα is apparently due to an activation of transcription, as shown by Müller *et al.* [73] in nuclear run on assays. In this respect, it is noteworthy that the promoter region of the $Gi\alpha_2$ gene – which is the predominant Giα-subtype in the human heart possesses a binding domain for AP-2 [74]. This effector is involved in the effects of cAMP on transcription [75]. Interestingly, heterologous adenylyl cyclase desensitization could not be induced by epinephrine in $S49cyc^-$ mouse lymphoma cells, genetically lacking Gsα, or in $S49kin^-$ mouse lymphoma cells, genetically lacking the cAMP-dependent protein kinase [75]. Thus, a cAMP-dependent phosphorylation of cAMP-responsive element binding protein might be involved in the transcriptional effects of cAMP by activating the promoter region of the $Gi\alpha_2$-gene. Since Giα mRNA levels but no Gsα mRNA content was increased in failing hearts, it has been suggested that an increase of transcription could also occur in human heart failure [77], although alterations of the Giα mRNA processing cannot be excluded. Hence, not only myocardial β-adrenoceptors but also Giα-proteins could be altered by the activity of the sympathetic nervous system. Hence, it is tempting to speculate that the increase of Giα-proteins could represent a general mechanism of long term regulation of adenyl cyclase activity. Future studies will have to investigate, whether this mechanism can be pharmacologically reversed in the condition of heart failure.

Giα-coupled receptors

In the human ventricular myocardium. A_1-adenosine-receptors and m-cholinoceptors mediate antiadrenergic effects on force of contraction [32]. In laboratory animals the effects are pertussis toxin-sensitive indicating that A_1-adenosine receptors and m-cholinoceptors are coupled to the family of Giα

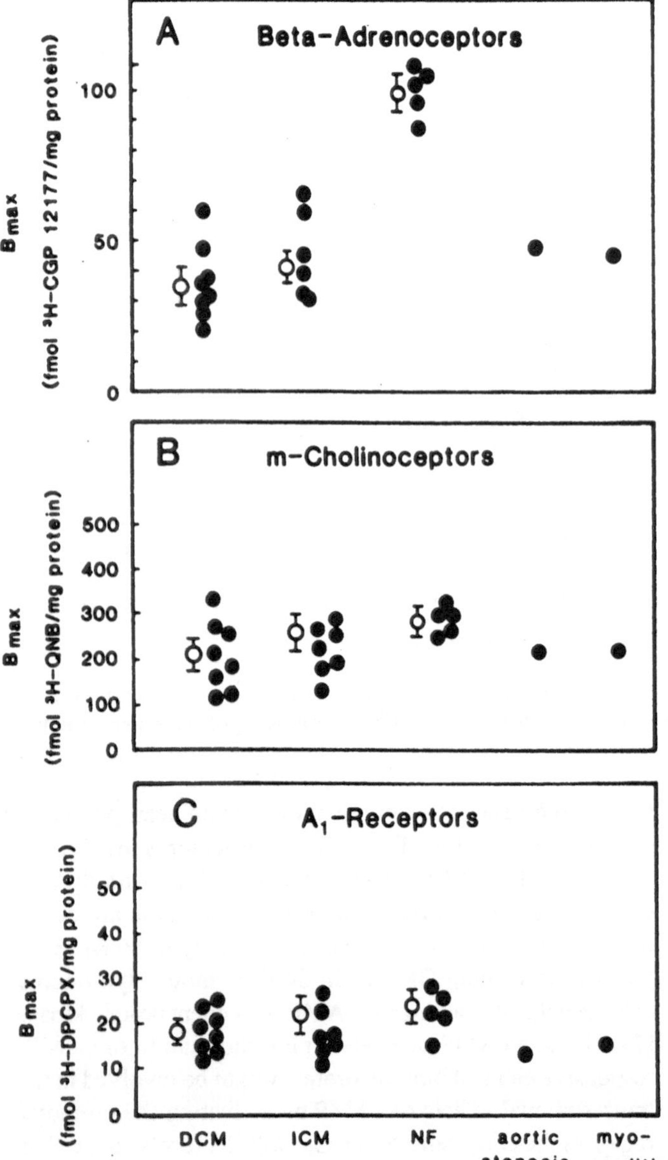

Fig. 14. Density of β-adrenoceptors (A), m-cholinoceptors (B), and A_1-adenosine receptors (C) in left ventricular membranes from nonfailing heart (NF) and from the heart of patients with dilated (DCM), ischemic (ICM) cardiomyopathy, aortic stenosis, and myocarditis. Error bars indicate SEM. Modified from previous work [32].

or Goα proteins [78]. Thus, one would expect that in the failing heart with increased amounts of Giα, coupling of m-cholinoceptor and A_1-adenosine receptors could be facilitated resulting in an augmented 'indirect' antiadrenergic effect thereby contributing to the impaired response to cAMP-dependent positive inotropic agents in human heart failure. Therefore, the density, affinity states, and coupling of A_1-adenosine receptors were investigated in nonfailing and failing human myocardium. Figure14 shows the number of β-adrenoceptors, A_1-adenosine receptors and m-cholinoceptors. Despite a marked reduction of β-adrenoceptors, the number of m-cholinoceptors and A_1-adenosine receptors was similar in non-failing myocardium and failing myocardium from patients with dilated or ischemic cardiomyopathy, aortic stenosis, or myocarditis [32]. The high and low affinity states as well as the influence of Gpp(NH)p did not differ in failing and nonfailing myocardium as measure of the coupling of these receptors [32]. Finally, the antiadrenergic effect of R-PIA and the m-cholinoceptor agonist carbachol were studied directly in isolated cardiac preparation. Figure 15 demonstrates that both agonists had similar effects with respect to their potency and efficacy on myocardial force of contraction. In concert with these findings, Hershberger *et al.* [39] reported a similar inhibitory effect on adenylyl cyclase activity in failing and nonfailing human myocardium of the A_1-adenosine receptor agonist R-PIA. Thus, all hitherto available data provide evidence for an unchanged effect of adenosine receptors at m-cholinoreceptors in heart failure despite an increase of Giα. Two hypotheses can be put forward to explain these unexpected findings. The amount of Giα compared to the amount of A_1-adenosine receptors is about 1,000–10,000 fold higher in human ventricular membranes. Therefore, a small proportion of Giα-proteins might be sufficient to produce an efficient receptor coupling. Thus, a further increase of Giα-proteins would not result in a further facilitation of adenosine receptor coupling in heart failure. In addition, A_1-adenosine receptors could couple to a Giα-subtype which is not increased. In human heart, $Gi\alpha_2$ is the most prominent G-protein subtype as judged from pertussis toxin labeling [32] and mRNA-studies [77]. Freissmuth *et al.* [79] reported that A_1-adenosine receptors purified from bovine brain selectively interact with recombinant $Gi\alpha_3$ proteins. Thus, an increase of a G-protein α-subunit in heart failure not involved in the coupling of adenosine receptors or m-cholinoceptors but with inhibitory effects on adenylyl cyclase could explain the unchanged effect of adenosine receptor and m-cholinoceptor agonists in heart failure.

Summary

In Fig. 16 the alterations of the β-adrenoceptor-adenylyl cyclase system are summarized. As observed in the most studies, the number of β_1-adrenoceptors is reduced whereas the β_2-adrenoceptors are unchanged in number but uncoupled from the adenylyl cyclase. The decrease of β_1 mRNA levels is likely to contribute to the reduced receptor number. Increased β-ARK activity and expression as suggested by elevated mRNA levels could be involved in receptor uncoupling. Receptor-independent adenylyl cyclase desensitization in the failing heart is due to an increased expression of Giα proteins. The catalyst, Gsα and M_2-muscarinic and A_1-adenosine receptors coupled to Giα are unchanged. The experimental observations on reduced cAMP-formation and diminished cAMP-dependent positive inotropic responses are in agree-

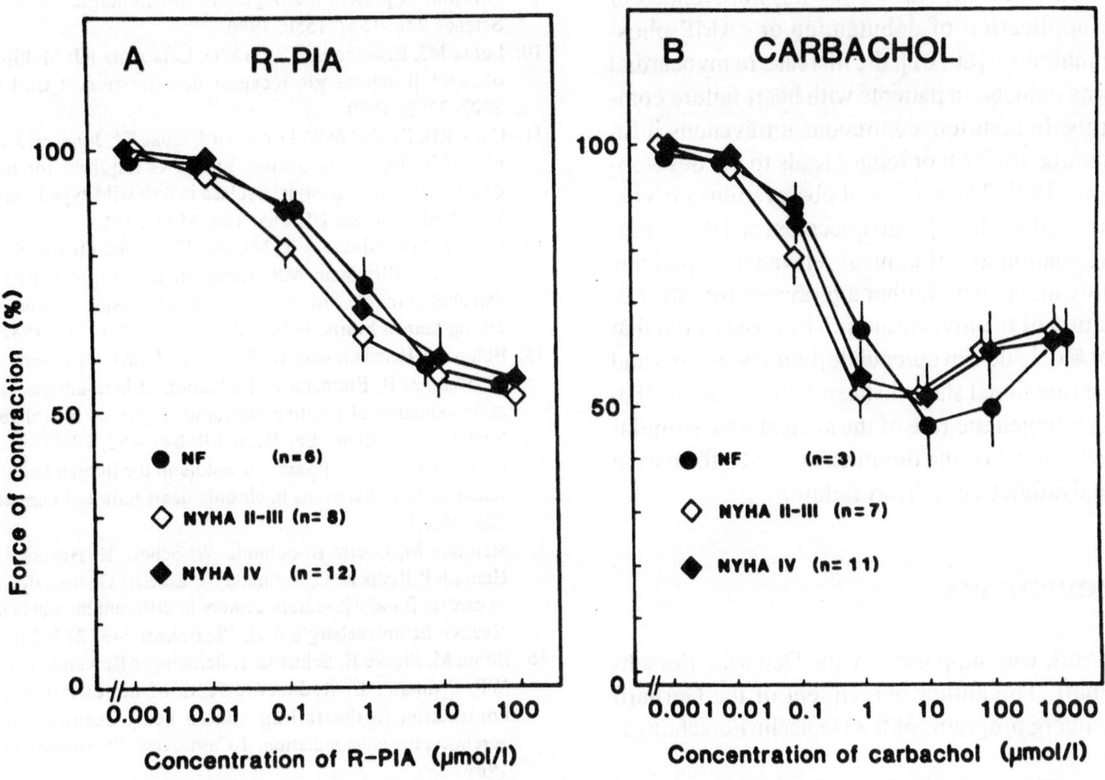

Fig. 15. Concentration response curve for the 'indirect' effect of R-PIA (A) and carbachol (B) in isolated, electrically driven papillary muscle strips from nonfailing hearts (NF) and hearts from patients with moderate (New York Association (NYHA) classes II-III, mitral valve replacement) or severe (NYHA class IV, ischemic and dilated cardiomyopathy) heart failure. Modified from previous work [32].

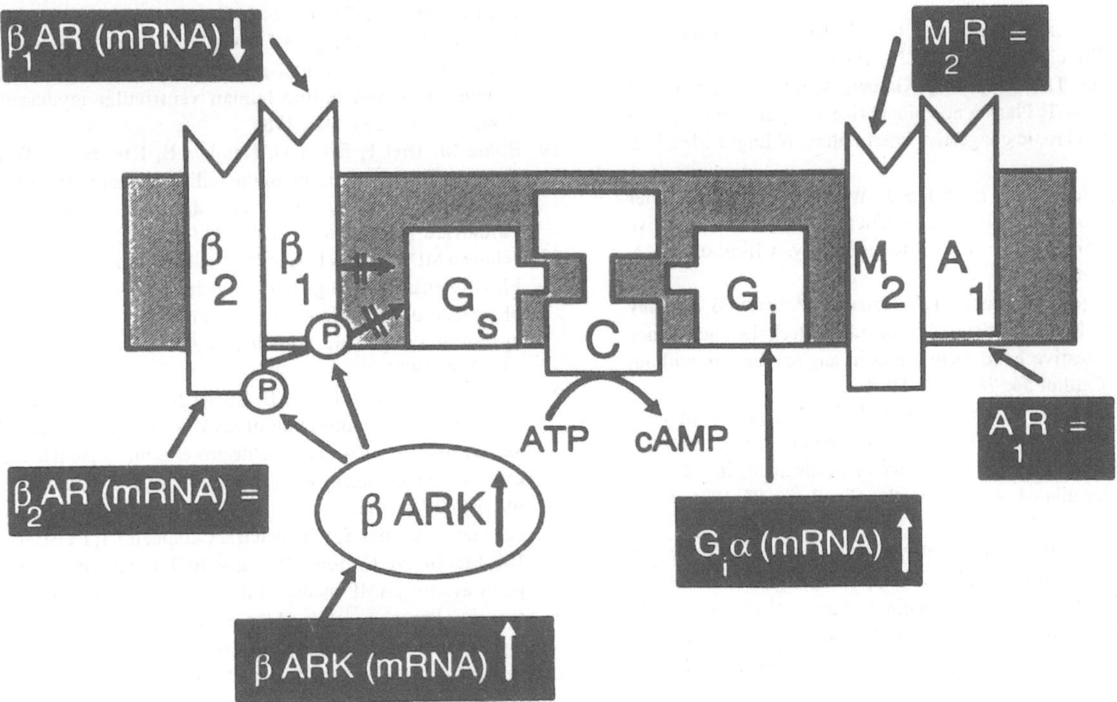

Fig. 16. Scheme summarizing the hitherto characterized defects of β-adrenoceptor adenylyl cyclase in the failing human heart. For the detailed description see text.

ment with clinical observations. Following intravenous or intracoronary application of dobutamine or cAMP phosphodiesterase inhibitors [80, 81], the increase in myocardial contractility was reduced in patients with heart failure compared to controls. In addition, continuous intravenous infusion of dobutamine for 72 h or longer leads to the development of tolerance [82]. These clinical observations provide evidence that the reduction in β-adrenoceptors and the diminished cAMP formation are of clinical relevance in patients with heart failure and can be further aggravated by catecholamine stimulation of the myocardium. The observation that β-adrenoceptor blockade can upregulate β-adrenoceptors and can improve the functional status in heart failure patients [83, 84] suggests a pathogenetic role of the sympathetic stimulation in the development of the diminished cAMP formation and contractile dysfunction in heart failure.

Acknowledgements

Experimental work was supported by the Deutsche Forschungsgemeinschaft. The author is recipient of the Gerhard Hass and Heisenberg programs of the Deutsche Forschungsgemeinschaft.

References

1. Packer M: Neurohormonal interactions and adaptations in congestive heart failure. Circulation 77: 721–730, 1988
2. Cohn JN, Levine TB, Olivari MT, Garberg V, Lura D, Francis GS, Simon AB, Rector T: Plasma norepinephrine as a guide to prognosis in patients with chronic congestive heart failure. N Engl J Med 311: 819–823
3. Swedberg K, Eneroth P, Kjekshus J, Wilhelmsen L: Hormones regulating cardiovascular function in patients with severe congestive heart failure and their relation to mortality. Circulation 82: 1730–1736, 1990
4. Swedberg K, Viquerat C, Rouleau JL, Roizen M, Atherton B, Parmley WW, Chatterjee K: Comparison of myocardial catecholamine balance in chronic congestive heart failure and in angina pectoris without failure. Am J Cardiol 54: 783–789, 1984
5. Bristow MR, Anderson FL, Port DP, Skerl L, Hershberger RS, Larabee P, O'Conell JB, Renlund DG, Volkman K, Murray J, Feldman AM: Differences in β-adrenergic neuroeffector mechanisms in ischemic versus idiopathic dilated cardiomyopathy. Circulation 84: 1024–1039, 1991
6. Lohse MJ: Mechanisms of β-adrenergic receptor desensitization. In: P.A. Hargrave, K.P. Hofmann, U.B. Kaupp (eds). Signal Transmission in Photoreceptor Systems. Berlin, Springer-Verlag, 1992, pp 160–171
7. Hausdorff WP, Caron MG, Lefkowitz RJ: Turning off the signal: Desensitization of β-adrenergic receptor function. FASEB J 4: 2881–2889, 1990
8. Hadcock JR, Malbon CC: Regulation of receptor expression by agonists: Transcriptional and posttranscriptional controls. Trends Neurosci 14: 242–247, 1991
9. Lohse MJ, Benovic JL, Codina J, Caron MG, Lefkowitz RJ: β-Arrestin: A protein that regulates β-adrenergic receptor function. Science 248: 1547–1550, 1990
10. Lohse MJ, Benovic JL, Caron MG, Lefkowitz RJ: Multiple pathways of rapid β_2-adrenergic receptor densitization. J Biol Chem 265: 3202–3209, 1990
11. Clark RB, Kunkel MW, Friedman J, Goka TJ, Johnson JA: Activation of cAMP-dependent protein kinase is required for heterologous desensitization of adenylyl cyclase in S49 wild-type lymphoma cells. Proc Natl Acad Sci USA 85: 1442–1446, 1988
12. Bristow MR, Ginsburg R, Minobe W, Cubiciotti RS, Sageman WS, Lurie K, Billingham ME, Harrison DE, Stinson EB: Decreased catecholamine sensitivity and beta-adrenergic receptor density in failing human hearts. N Engl J Med 307: 205–211, 1982
13. Böhm M, Beuckelmann D, Brown L, Feiler G, Lorenz B, Näbauer M, Kemkes B, Erdmann E: Reduction of beta-adrenoceptor density and evaluation of positive inotropic responses in isolated, diseased human myocardium. Eur Heart J 9: 844–852, 1988
14. Brodde OE: β_1- and β_2-adrenoceptors in the human heart: Properties, function, and alterations in chronic heart failure. Pharmacol Rev 43: 203–242, 1991
15. Steinfath M, Geertz B, Schmitz W, Scholz H, Haverich A, Breil I, Hanrath P, Reupcke C, Sigmund M, Lo HB: Distinct downregulation of cardiac β_1- and β_2-adrenoceptors in different human heart diseases. Naunyn Schmiedeberg's Arch Pharmacol 343: 217–220, 1991
16. Böhm M, Pieske B, Schnabel P, Schwinger R, Kemkes B, Klövekorn WP, Erdmann E: Reduced effects of dopexamine on force of contraction in the failing human heart despite preserved β_2-adrenoceptor subpopulation. J Cardiovasc Pharmacol 14: 549–559, 1989
17. Steinfath M, Danielsen W, von der Leyen H, Mende U, Meyer W, Neumann J, Nose M, Reich T, Schmitz W, Scholz H, Starbatty J, Stein B, Döring V, Kalmar P, Haverich A: Reduced α_1- and β_2-adrenoceptor-mediated positive inotropic effects in human end-stage heart failure. Br J Pharmacol 105: 463–469, 1992
18. Bristow MR, Hershberger RE, Port JD, Minobe W, Rasmussen R: β_1- and β_2-Adrenergic receptor-mediated adenylate cyclase stimulation in nonfailing and failing human ventricular myocardium. Mol Pharmacol 35: 295–303, 1989
19. Böhm M, Diet F, Feiler G, Kemkes B, Kreuzer E, Weinhold C, Erdmann E: Subsensitivity of the failing human heart to isoprenaline and milrinone is related to beta-adrenoceptor downregulation. J Cardiovasc Pharmacol 12: 726–732, 1988
20. Feldman MD, Copelas L, Gwathmey JK, Philips P, Warren SE, Schoen FJ, Grossman W, Morgan JP: Deficient production of cyclic AMP: pharmacologic evidence of an important cause of contractile dysfunction in patients with end-stage heart failure. Circulation 75: 331–339, 1987
21. Böhm M, Morano I, Pieske B, Rüegg JC, Wankerl M, Zimmermann R, Erdmann E: Contribution of cAMP-phosphodiesterase inhibition and sensitization of the contractile proteins for calcium to the inotropic effect of pimobendan in the failing human myocardium. Circ Res 68: 689–701, 1991
22. Bouvier M, Collins S, O'Dowd BF, Campbell PT, DeBlasi A, Kobilka BK, MacGregor C, Irons GP, Caron MG, Lefkowitz RJ: Two distinct pathways for cAMP-mediated down-regulation of the β_2-adrenergic receptor: Phosphorylation of the receptor and regulation of its mRNA level. J Biol Chem 264: 16786–16792, 1989
23. Mahan LC, Koachman AM, Insel PA: Genetic analysis of β-adrenergic receptor internalization and down-regulation. Proc Natl Acad Sci USA 82: 129–133, 1985
24. Allen JM, Abrass IB, Palmiter RD: β_2-Adrenergic receptor regulation after transfection into a cell line deficient in the cAMP-dependent protein kinase. Mol Pharm 36: 248–255, 1989

25. Hadcock JR, Ros M, Malbon CC: Agonist regulation of β-adrenergic receptor mRNA: Analysis in S49 mouse lymphoma mutants. J Biol Chem 264: 13956–13961, 1989
26. Campbell PT, Hantowich M, O'Dowd BF, Caron MG, Lefkowitz RJ, Hausdorff WP: Mutations of the human β_2-adrenergic receptor that impair coupling to Gs interfere with receptor down-regulation but not sequestration. Mol Pharm 39: 192–198, 1991
27. Hadcock JR, Wang H, Malbon CC: Agonist-induced destabilization of β-adrenergic receptor mRNA: Attenuation of glucocorticoid-induced up-regulation of β-adrenergic receptors. J Biol Chem 264: 19928–19933, 1989
28. Ungerer M, Böhm M, Elce JS, Erdmann E, Lohse MJ: Altered expression of β-adrenergic receptor kinase and β_1-adrenergic receptors in the failing human heart. Circulation 87: 454–463, 1993
29. Bristow MR, Minobe WA, Raynolds MV, Port JD, Rasmussen R, Ray PE, Feldman AM: Reduced β_1-receptor messenger RNA abundance in the failing human heart. J Clin Invest 92: 2737–2745, 1993
30. Gille E, Lemoine H, Ehle B, Kaumann AJ: The affinity of (-)-propranolol for β_1- and β_2-adrenoceptors of human hearts. Differential antagonism of the positive inotropic effects and adenylate cyclase stimulation by (-)-noradrenaline and (-)-adrenaline. Naunyn-Schmiedeberg's Arch Pharmacol 331: 60–70, 1985
31. Pippig S, Andexinger S, Daniel K, Puzicha M, Caron MG, Lefkowitz RJ, Lohse MJ: Overexpression of β-arrestin and β-adrenergic receptor kinase augment desensitization of β_2-adrenergic receptors. J Biol Chem 268: 3201–3208, 1993
32. Böhm M, Gierschik P, Jakobs KH, Pieske B, Schnabel P, Ungerer M, Erdmann E: Increase of Giα in human hearts with dilated but not ischemic cardiomyopathy. Circulation 82: 1249–1265, 1990
34. Schmitz W, von der Leyen H, Meyer W, Neumann J, Scholz H: Phosphodiesterase inhibition and positive inotropic effects. J Cardiovasc Pharmacol 14 (Suppl 3): S11–S14, 1989
35. Seamon KB, Daly JW: Forskolin: its biological and chemical properties. Adv Cyclic Nucleotide Protein Phosphorylation Res 20: 1–150, 1989
36. Feldman AM, Cates AE, Veazey WB, Hershberger RE, Bristow MR, Baughman KL, Baumgartner WA, Van Dop C: Increase in the 40,000-mol wt pertussis toxin substrate (G-protein) in the failing human heart. J Clin Invest 82: 189–197, 1988
37. Limbird LE, Hickey AR, Lefkowitz RJ: Unique uncoupling of the frog erythrocyte adenylate cyclase system by manganese: loss of hormone and guanine nucleotide-sensitive enzyme activities without loss of nucleotide-sensitive, high affinity agonist binding. J Biol Chem 254: 2677–2683, 1979
38. Cech SY, Broaddus WC, Maguire ME: Adenylate cyclase: the role of magnesium and other divalent cations. Mol Cell Biochem 33: 67–92, 1980
39. Hershberger RE, Feldman AM, Bristow MR: A_1-Adenosine receptor inhibition of adenylate cyclase in failing and nonfailing human ventricular myocardium. Circulation 83: 1343–1351, 1991
40. Gilman AG: G proteins: Transducers of receptor-generated signals. Ann Rev Biochem 56: 615–649, 1987
41. Cassel D, Selinger Z: Mechanism of adenylate cyclase activation by cholera toxin: inhibition of GTP hydrolysis at the regulatory site. Proc Natl Acad Sci USA 74: 3307–3311, 1977
42. Bray P, Carter A, Guo V, Puckett C, Kamholz J, Spiegel A, Nirenberg M: Human cDNA clones for an α subunit of Gi signal-transduction protein. Proc Natl Acad Sci USA 84: 5115–5119, 1987
43. Moss J, Vaughan M: ADP-ribosylation of guanyl nucleotide-binding regulatory proteins by bacterial toxins. Adv Enzymol 61: 13303–13379, 1988
44. Schnabel P, Böhm M, Gierschik P, Jakobs KH, Erdmann E: Improvement of cholera toxin-catalyzed ADP-ribosylation by endogenous ADP-ribosylation factor from bovine brain provides evidence for an unchanged amount of Gsα in failing human myocardium. J Mol Cell Cardiol 22: 73–82, 1990
45. Insel PA, Ransnäs LA: G proteins and cardiovascular disease. Circulation 78: 1511–1513, 1988
46. Ransnäs LA, Insel PA: Quantification of the stimulatory guanine nucleotide binding protein Gs in S49 cell membranes, using anti-peptide antibodies to αs. J Biol Chem 263: 9482–9485, 1988
47. Milligan G: Techniques used in the identification and analysis of function of pertussis toxin-sensitive guanine nucleotide binding proteins. Biochem J 255: 1–13, 1988
48. Neumann J, Scholz H, Döring V, Schmitz W, v. Meyerinck L, Kalmar P: Increase in myocardial Gi-proteins in heart failure. Lancet II: 936–937, 1988
49. West RE Jr, Moss J, Vaughan M, Liu T, Liu TY: Pertussis toxin-catalyzed ADP-ribosylation of transducin. Cysteine 347 is the ADP-ribose acceptor site. J Biol Chem 260: 14428–14430, 1985
50. Ribeiro-Neto F, Matera R, Grenet D, Sekura RD, Birnbaumer L, Field JB: Adenosine disphosphate ribosylation of G proteins by pertussis and cholera toxin in isolated membranes. Different requirement for and effects of guanine nucleotides and Mg^{2+}. Mol Endocrinol 1: 472–481, 1987
51. Ribeiro-Neto F, Mattera R, Hildebrandt JD, Codina J, Field JB, Birnbaumer L, Sekura RD: ADP-ribosylation of membrane components by pertussis and cholera toxin. Meth Enzymol 109: 566–582, 1985
52. Tsai SC, Adamik R, Kanaho Y, Hewlett EL, Moss J: Effects of guanyl nucleotides and rhodopsin on ADP-ribosylation of the inhibitory GTP-binding component of adenylate cyclase by pertussis toxin. J Biol Chem 259: 15320–15323, 1984
53. Neer EJ, Lok JM, Wolf LG: Purification and properties of the inhibitory guanine nucleotide regulatory unit of brain adenylate cyclase. J Biol Chem 259: 14222–14229, 1984
54. Ohguro H, Fukada Y, Yoshizawa T, Saito T, Akino T: A specific β-subunit of transducin stimulates ADP-ribosylation of the α-subunit by pertussis toxin. Biochem Biophys Res Commun 167: 1235–1241, 1990
55. Lim LK, Sekura RD, Kaslow HR: Adenine nucleotides directly stimulate pertussis toxin. J Biol Chem 260: 2585–2588, 1985
56. Hausman SZ, Manclark CR, Burns DL: Binding of ATP by pertussis toxin and isolated toxin subunits. Biochemistry 29: 6128–6131, 1990
57. Burns DL, Manclark CR: Adenine nucleotides promote dissociation of pertussis toxin subunits. J Biol Chem 261: 4324–4327, 1985
58. Longabaugh JP, Vatner DE, Graham RM, Homcy CJ: NADP improves the efficiency of cholera toxin catalyzed ADP-ribosylation in liver and heart membranes. Biochem Biophys Res Commun 137: 328–333, 1986
59. Watanabe Y, Imaizumi T, Misaki N, Iwankura K, Yoshiba H: Effects of phosphorylation of inhibitory GTP-binding protein by cyclic AMP-dependent protein kinase on its ADP-ribosylation by pertussis toxin, islet activating protein. FEBS Lett 236(2): 372–374, 1988
60. Tanuma S, Kawashima K, Endo H: Eukaryotic mono(ADP-ribosyl)transferase that ADP-ribosylates GTP-binding regulatory Giα protein. J Biol Chem 263: 5485–5489, 1988
61. Hara-Yokoyama M, Furuyama S: Endogenous inhibitor of the ADP-ribosylation of (a) G-protein(s) as catalyzed by pertussis toxin is present in rat liver. FEBS Lett 234: 27–30, 1988
62. Tanuma S, Endo H: Identification in human erythrocytes of mono(ADP-ribosyl)protein hydrolase that cleaves a mono (ADP-ribosyl) Gi linkage. FEBS Lett 261: 381–384, 1990
63. Linder ME, Pang IH, Duronio RJ, Gordon JI, Sternweiss PC, Gilman AG: Lipid modifications of G protein subunits. Myristolation of Goα increases its affinity for βγ. J Biol Chem 266: 4654–4659, 1991

64. Böhm M, Eschenhagen T, Gierschik P, Larisch K, Lensche H, Mende U, Schmitz W, Schnabel P, Scholz H, Steinfath M, Erdmann E: Radioimmunochemical quantification of Giα in right and left ventricles from patients with ischaemic and dilated cardiomyopathy and predominant left ventricular failure. J Mol Cell Cardiol 26: 133–149, 1994
65. Feldman AM, Jackson DG, Bristow MR, Van Dop C: Immunologic quantification of G proteins in failing and nonfailing human heart. Circulation 80(Suppl): II–293, 1989
66. Böhm M, Larisch K, Erdmann E, Camps M, Jakobs KH, Gierschik P: Failure of (^{32}P)-ADP-ribosylation by pertussis toxin to determine Giα content in membranes from various human tissues. Improved radioimmunological quantification using the ^{125}I-labeled C-terminal decapeptide of retinal transducin. Biochem J 277: 223–229, 1991
67. Goldsmith P, Gierschik P, Milligan G, Unson CG, Vinitsky R, Malech HL, Spiegel AM: Antibodies directed against synthetic peptides distinguish between GTP-binding proteins in neutrophils and brain. J Biol Chem 262: 14683–14688, 1987
68. Pines M, Gierschik P, Milligan G, Klee W, Spiegel AM: Antibodies against the carboxyl-terminal 5-kDa peptide of the α-subunit of transducin crossreact with the 40 kDa but not the 39-kDa guanine nucleotide binding protein from brain. Proc Natl Acad Sci USA 82: 4095–4099, 1985
69. Brown LA, Harding SE: The effect of pertussis toxin on β-adrenoceptor responses in isolated cardiac myocytes from noreadrenaline-treated guinea-pigs and patients with cardiac failure. Br J Pharmacol 106: 115–122, 1992
70. Eschenhagen T, Mende U, Diederich M, Nose M, Schmitz W, Scholz H, Schulte am Esch J, Warnholtz A, Schäfer H: Long-term β-adrenoceptor-mediated up-regulation of Giα and Goα mRNA levels and pertussis toxin-sensitive guanine nucleotide binding proteins in rat heart. Mol Pharm 42: 773–783, 1992
71. Reithmann C, Gierschik P, Müller U, Werdan K, Jakobs KH: Pseudomonas exotoxin A prevents β-adrenoceptor-induced up-regulation of Gi protein α-subunits and adenylyl cyclase desensitization in rat heart muscle cells. Mol Pharm 37: 631–638
72. Reithmann C, Gierschik P, Jalobs KH, Werdan K: Regulation of adenylyl cyclase by noradrenaline and tumor necrosis factor α in rat cardiomyocytes. Eur Heart J 12: (Suppl F): 139–142, 1991
73. Müller FU, Boheler KR, Eschenhagen T, Schmitz W, Scholz H: Isoprenaline stimulates gene transcription of the inhibitory G protein alpha-subunit Gi alpha-2 in rat heart. Circ Res 72: 696–700, 1993
74. Weinstein LA, Spiegel AM, Carter AD: Cloning and characterization of the human gene for the α-subunit of Gi_2, a GTP-binding signal transducin protein. FEBS Lett 232: 333–340, 1988
75. Imagawa M, Chiu R, Karin M: Transcription factor AP-2 mediates induction of two different signal-transduction pathways: protein kinase C and cAMP. Cell 51: 251–260, 1987
76. Clark RB, Kunkel MW, Friedman J, Goka TJ, Johnson JA: Activation of cAMP-dependent protein kinase is required for heterologous desensitization of adenylyl cyclase in S49 wild-type lymphoma cells. Proc Natl Acad Sci USA 85: 1442–1446, 1988
77. Eschenhagen T, Mende U, Nose M, Schmitz W, Scholz H, Haverich A, Hirt S, Döring V, Kalmar P, Höppner W, Seitz HJ: Increased messenger RNA level of the inhibitory G-protein α-subunits $Gi\alpha_2$ in human end-stage heart failure. Circ Res 70: 688–696, 1992
78. Böhm M, Ungerer M, Erdmann E: Adenosine receptors in the human heart: pharmacological characterization in nondiseased and cardiomyopathic tissue. Drug Development Research 28: 268–276, 1993
79. Freissmuth M, Schütz W, Linder ME: Interactions of the bovine brain A_1-adenosine receptor with recombinant G protein α-subunits. J Biol Chem 266: 17778–17783, 1991
80. Colucci WS, Leatherman GF, Ludmer PL, Gauthier DF: β-Adrenergic inotropic responsiveness of patients with heart failure: studies with intracoronary dobutamine infusion. Circ Res 61 (Suppl 1): 82–86, 1987
81. Gage J, Rutman H, Lucido D, Le Jemtel TH: Additive effects of dobutamine and amrinone on myocardial contractility and ventricular performance in patients with severe heart failure. Circulation 74: 367–373, 1986
82. Unverferth DV, Blanford M, Kates RE, Leier CV: Tolerance to dobutamine after a 72 h continuous infusion. Am J Med 69: 262–268, 1980
83. Heilbrunn SM, Shah P, Bristow MR, Valantine HA, Ginsburg R, Fowler MG: Increased β-receptor density and improved hemodynamic response to catecholamine stimulation during long-term meoprolol therapy in heart failure from dilated cardiomyopathy. Circulation 79: 483–490, 1989
84. Waagstein F, Caidahl K, Wallentin I, Bergh CH, Hjalmarson A: Long-term β-blockade in dilated cardiomyopathy: effects of short- and long-term metoprolol treatment followed by withdrawal and readministration of metoprolol. Circulation 80: 551–563, 1989

Molecular and Cellular Biochemistry **147**: 161–168, 1995.

Immunocytochemical studies of the G_i Protein mediated muscarinic receptor-adenylyl cyclase system

Wolfgang Schulze, Wolf-Peter Wolf, Michael L. X. Fu[1], Rosemarie Morwinski, Igor B. Buchwalow and Liane Will-Shahab

Max-Delbrück-Centre for Molecular Medicine, Division of Molecular Cardiology, 13125 Berlin, Germany;
[1]Wallenberg Laboratories for Cardiovascular Research, Sahlgren's Hospital, 41345 Göteborg, Sweden

Abstract

The localization of three key signal transduction components was indicated in rat heart tissue by immunocytochemical and histochemical experiment. It was shown that:

1. The M2 muscarinic receptors are localized along outer cell membranes and T-tubule membranes of cardiomyocytes but additionally at membranes of endothelial cells and fibroblasts.
2. $G_{i\alpha}$ was found along outer cell membranes of cardiomyocytes and other cells of the heart and also inside the cells of the perinuclear space in close contact to the nuclei envelope and the endoplasmic reticulum membranes. $G_{o\alpha}$ were found to be associated mainly in atrial tissue, especially at the nerval (neuronal) endings located among the cardiac muscle cells. This was shown in parallel incubation with specific neuronal antibody as a marker for these structures.
3. Adenylyl cyclase was localized along the sarcolemma and the T-tubule membranes in normal cardiomyocytes of rat and guinea pig hearts. Under ischemic conditions, the adenylyl cyclase was also seen in junctional sarcoplasmic reticulum membranes. The reasons for this changed localization need further elucidation. Binding of the adenylyl cyclase within the molecular structure of the membrane or variation of the marker penetration remain to be clarified. (Mol Cell Biochem **147**: 161–168, 1995)

Key words: adenylyl cyclase, muscarinic receptors, G-proteins, immunocytochemistry

Introduction

Signals that control heart functions are initiated at the cell surface by the interaction of hormones and neurotransmitters with their specific receptors. The signals were amplified and sorted by the guanine nucleotide regulatory binding proteins (G proteins) and were sent forward to a variety of cellular effectors such as adenylyl cyclase, phospholipase C, phospholipase A_2 and several ion channels. Numerous G-protein-linked receptors are present in the heart and are involved in regulation of the heart and blood vessel contractility in response to neurotransmitters or hormones.

This report deals with the localization of the main components of the G_i-protein mediated muscarinic-receptor adenylyl cyclase signal pathway. Several reviewers discussed the function of this adenylyl cyclase inhibitory way [1–3]. The localization of the signal components described by these authors, as far as it was mentioned, is known from pharmacological binding studies of corresponding antagonists to the receptor. For G-proteins the pertussis toxin induced ribosylation are measured in several cellular fractions. And finally the adenylyl cyclase activities were localized by determining the cAMP or P_i generation in enriched cell fragments. The heart is a multicellular organ with different cell types. The

Address for offprints: W. Schulze, Max-Delbrück-Centre for Molecular Medicine, Division of Molecular Cardiology, Robert-Rössle-Straße 10, 13125 Berlin, Germany

Fig. 1

Fig. 2

Fig. 3

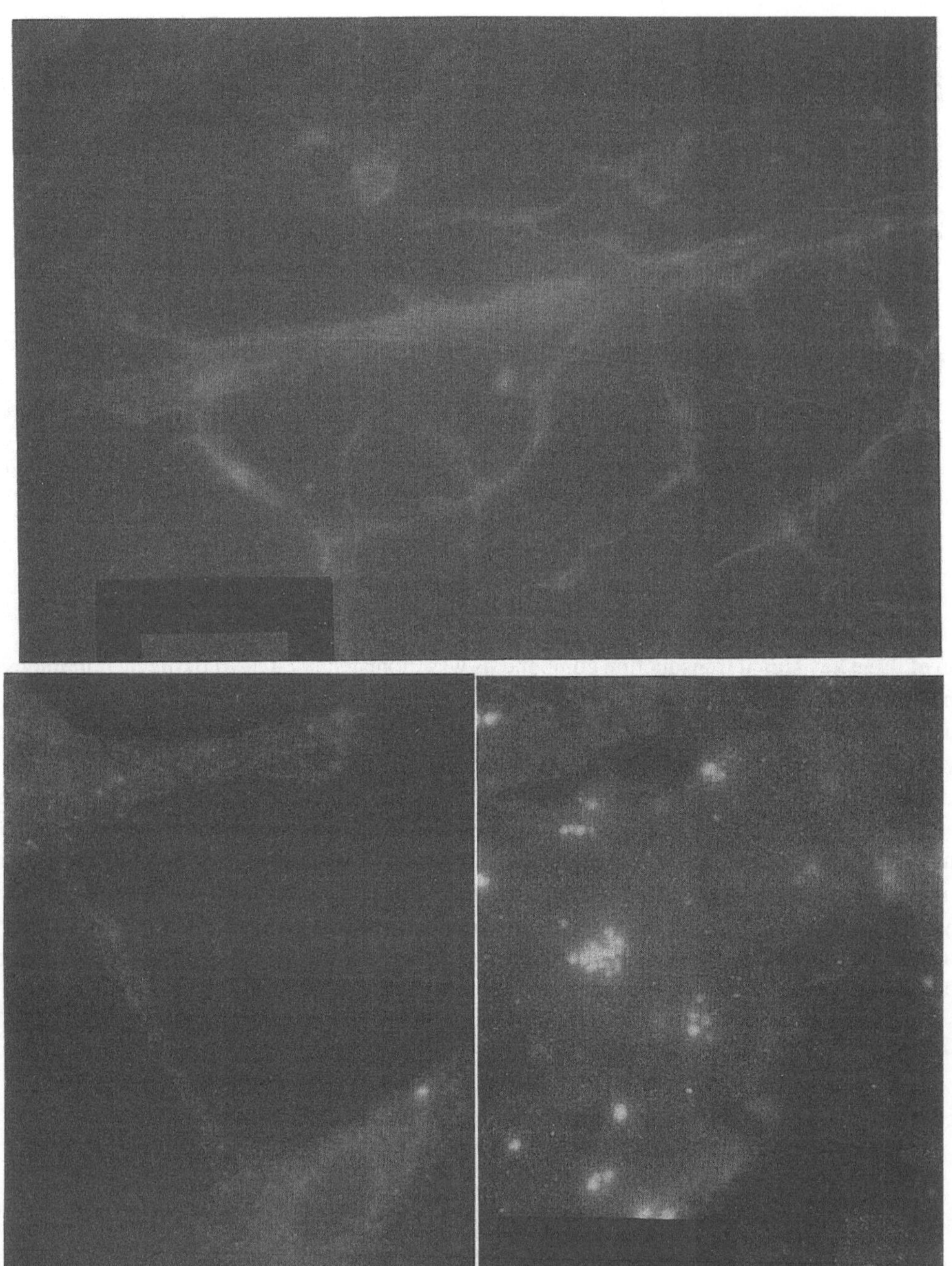

Fig. 1. M2 receptors localized by indirect immunofluorescence using affinity-purified anti-peptide antibodies (10 nM) raised against a peptide to the sequence (residue 168–192) of the second extracellular loop of the human M2 muscarinic receptor. Cryostat section of rat left ventricle fixed in acetone incubated overnight (12 hours, 4°C) with the first antibody. After washing, the slides were incubated with DTAF (dichlortriazinyl aminofluorescin)-conjugated goat anti-rabbit antibodies (1:100) for 2 h at room temperature. Fluorescence was seen at sarcolemma of cardiomyocytes and at outer cell membranes of endothelial cells. Bar = 20 µm.

Fig. 2. Semithin sections (1 µm) of bioptate from human right septial region embedded in LR White resin mounted on glass slides incubated with the affinity purified M2 antibodies (10 nM) for 2 h at room temperature then as described for Fig. 1. Dot-like fluorescence signals, the marker for the M2 receptor were seen along the surface of the cardiomyocytes. Bar = 20 µm.

Fig. 3. M2-receptor localization within isolated cultivated non-cardiomyocytes presumable fibroblasts from adult rat heart. Dot-like fluorescence signals over the surface of the cells. The highest density of receptor antibodies are localized in extensions of the non-cardiomyocytes. Bar = 20 µm.

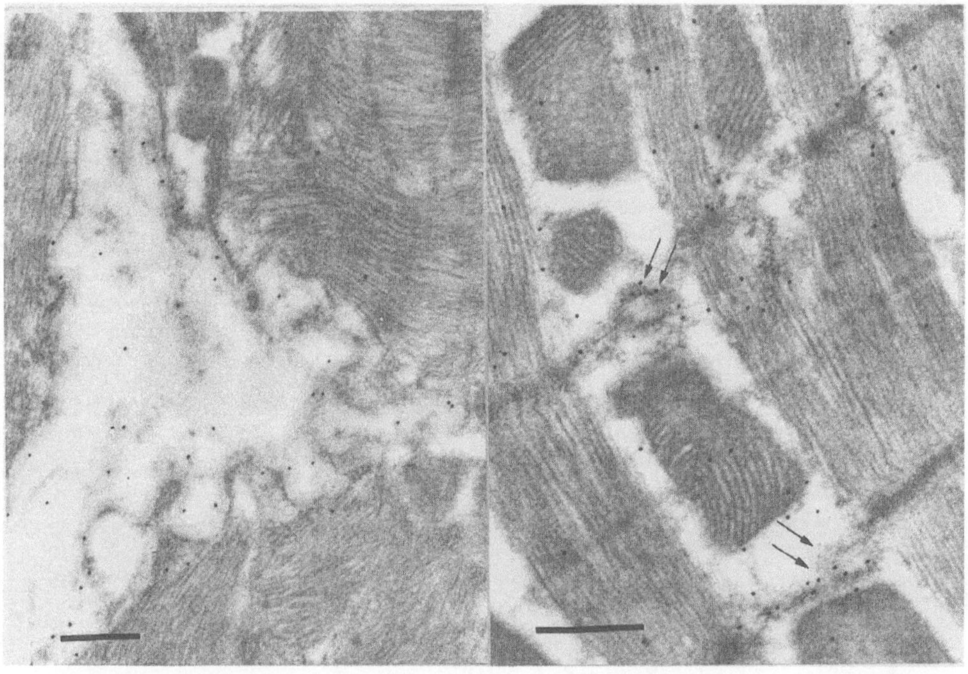

Fig. 4 Fig. 5

Fig. 4. Electronmicroscopic immunogold labeling of ultra-thin section from adult rat left ventricle embedded in LR White. Incubation with anti M2 muscarinic receptor peptide antibodies and protein A-gold (15 nm). Immersion fixed tissue with 4% formaldehyde and 0.025% glutaraldehyde in PBS for 1 h at room temperature. Bar = 0.5 μm.

Fig. 5. Muscarinic M2 receptor antibodies stained by EM immunogold. Adult rat heart left ventricle. Gold particles are localized near T-tubule membranes (arrows). Fixation and incubation protocol as described for Figs 1 and 4. Bar = 0.5 μm.

single cardiomyocyte consists of a large network of intracellular membranes (SR) closely connected with the sarcolemma and the T-tubule membranes. In most cases it has not been possible to isolate a membrane structure from heart muscle that is not contaminated by other cell components.

This report describes the localization of the muscarinic M2 receptor, G proteins α_i and α_0 subunits and the adenylyl cyclase activity by cytochemical and immunocytochemical methods without destroying the cellular organization of the heart. The microscope allows the exact identification of the special cell component in this study.

The localization of M2-muscarinic acetylcholine receptors

Muscarinic cholinergic receptors are integrate membrane proteins and belong to the G-protein coupled seven membrane spanning receptor family [4]. By molecular cloning, Peralta *et al.* [5] has been able to identify five subtypes of muscarinic receptors. In the heart the M2-subtype is the most dominant form. Among others it couples to G_i-protein and decreases the cAMP level by inhibition of the adenylyl cyclase. The localization of M2 muscarinic receptors has been attempted using conventional pharmacological approaches by receptor subtype-specific molecules and radioligands of high specific activity. The successful cloning and sequencing of the subtypes of muscarinic receptors have provided molecular biological tools to induce antibodies specific for each subtype [5, 6]. To produce antibodies with well-characterized antigenic determinants we preferred to raise them against a synthetic peptide of the amino acid sequence from the second extracellular loop of human M2 receptor [7].

The localization of muscarinic receptors by these M2 antibodies against the second extracellular loop was studied by immunofluorescence and electron microscopic techniques on rat and human heart tissue.

The antibody used was produced in rabbits and was shown to recognize muscarinic receptor specifically on immunoblot [7, 8]. It was shown that there exists a 100% homology between the second extracellular loop of rat and human M2 receptor subtype.

For fluorescence microscopy, rat heart cryostat sections (10 μm) fixed in acetone or semithin sections (1 μm) from formaldehyde fixed human bioptates embedded in LR White (London Resin Company, SCI Science Services GmbH) were used. Sections were incubated overnight at 4°C with affinity-

Fig. 6

Fig. 7

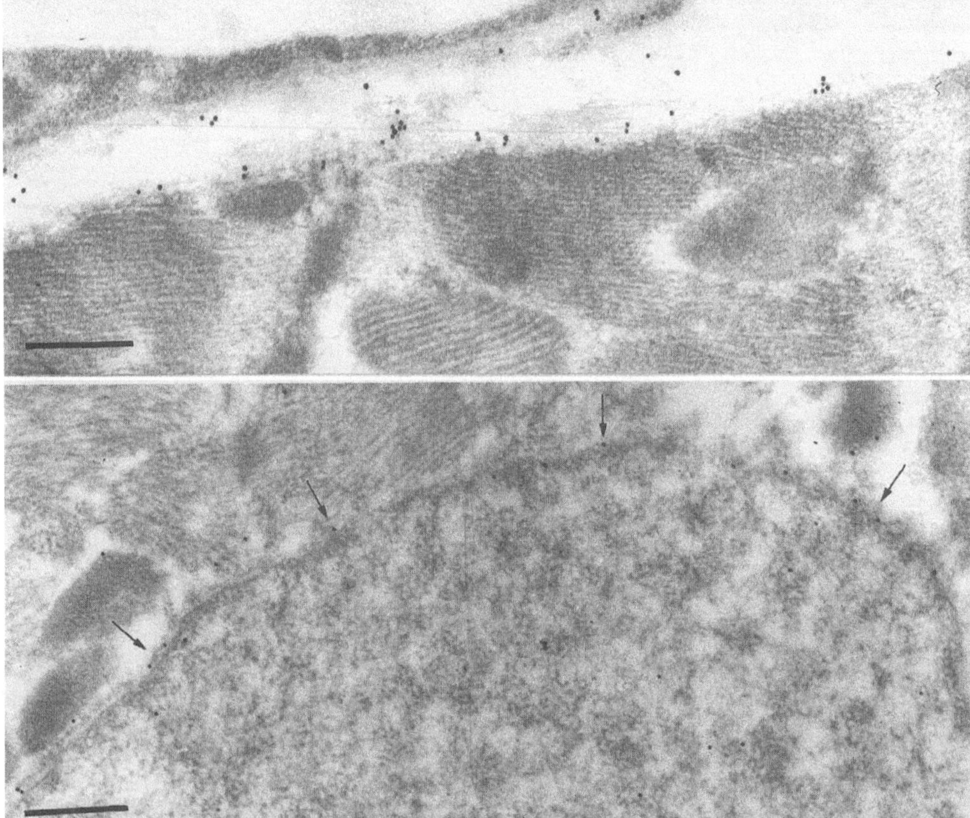

Fig. 6. Electron microscopic localization of $G_{i\alpha}$ subunits by immunogold technique with protein A-gold conjugated marker. Pieces from left ventricle rat heart tissue fixed with 3% (para) formaldehyde for 1 h at room temperature were embedded in Lowicryl K4M according to instructions from Chemische Werke Lowi GmbH, Waldkraiburg, 1982. Anti $\alpha_{i\ common}$ peptide antibodies diluted 1:50 were incubated for 2 h at room temperature and after washing treated with protein A-gold (purchased from G. Posthuma and J.W. Slot, Utrecht University) for 1 h. The slices were counterstained with uranyl acetate and lead citrate. Gold particles are associated with sarcolemma and surface of the endothelial cells. Bar = 0.5 µm.

Fig. 7. Immunogold labeling of $G_{i\alpha 1,2}$ antipeptide antibodies at nuclear envelope of a cardiomyocyte (arrows). Same incubation as in Fig. 6. Bar = 0.5 µm.

purified antibodies (10 nM). After washing they were incubated with DTAF (dichlortriazinyl aminofluorescin)-conjugated goat anti-rabbit antibodies (Dianova, Hamburg, Germany). The semithin sections of human heart bioptates are incubated in the same procedure without removing the resin. For electron microscopic detection ultra-thin sections were used according to the above incubation method. In this study protein-A-gold conjugates (10 or 14 nm gold particles) were used as the marker for the anti M2 antibodies.

Our results demonstrated that the M2 muscarinic receptor was located mainly on the surface of the cardiomyocytes (Fig. 1). The immunofluorescence in both rat left ventricle cryostat sections and semithin sections of human bioptates from right septial region was seen along the sarcolemma of the cardiomyocytes. The human semithin (1 µm) heart sections show dot-like fluorescence signals (Fig. 2) in contrast to the continuous fluorescence of the 10 µm thick rat heart cryosection. A specific fluorescence of the membranes were identified also in non-cardiomyocytes. Isolated and cultivated rat non-cardiomyocytes (presumable fibroblasts) show immunofluorescence (Fig. 3) over the cell surface. The highest amount of the dot-like fluorescence was found to be associated with extension of the non-cardiomyocytes. The specific immunostaining was abolished with non-immune serum as the first antibody and could be blocked by pre-incubation with the corresponding antigenic peptide.

Beside the labeling of the sarcolemma, the EM gold technique shows gold particles as the marker for the antibody binding near the T-tubule membranes (Figs 4 and 5). Endothelial cell surface are likewise marked with gold.

It should be taken into consideration that M2-receptors, as one of the main receptors of the heart, are not only localized at the sarcolemma of the cardiomyocytes but also found at the T-tubule membranes and that they are additionally associated with non-muscle cells e.g. endothelial cell membranes of capillaries and presumable outer membranes of the fibroblasts.

Fig. 8

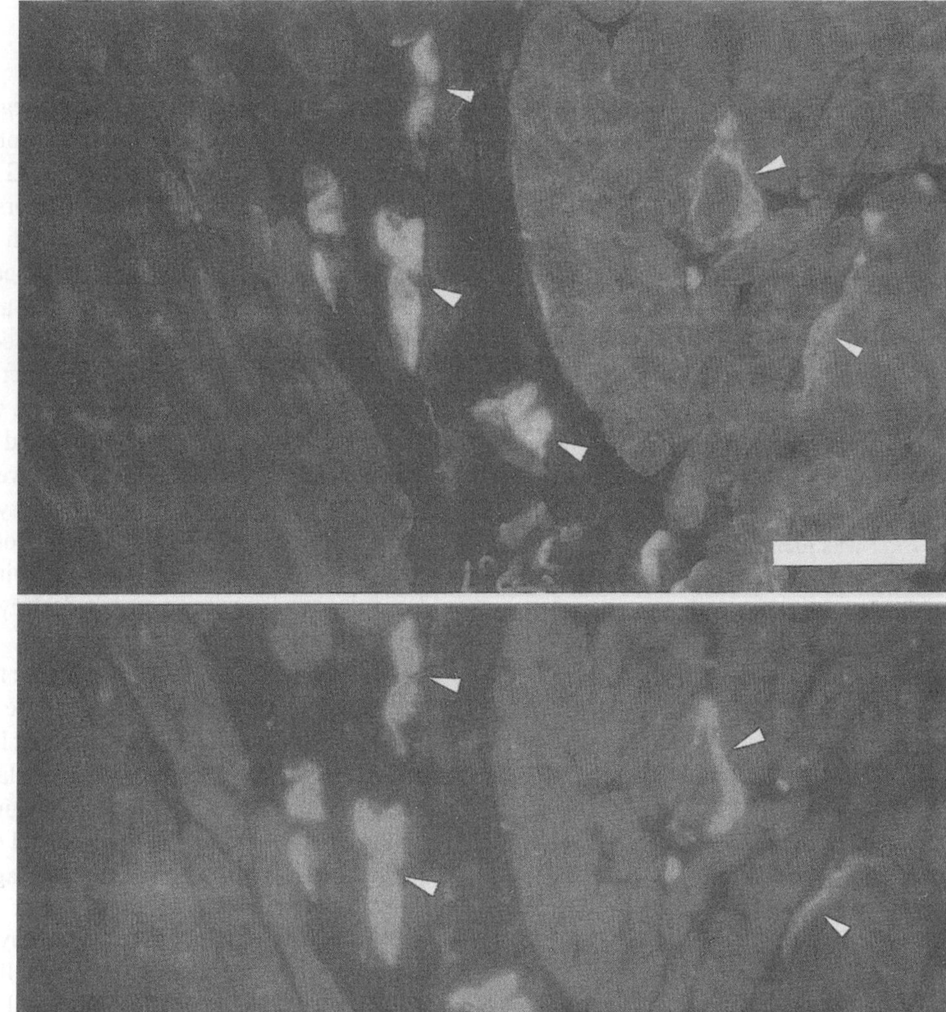

Fig. 9

Fig. 8. Immunofluorescence staining of $G_{o\alpha}$. Anti $G_{o\alpha}$ subunit was purchased by UBI. According to the certificate it is a polyclonal antibody and does not react with $G_{i\alpha\ 1\text{-}3}$ subunits. The immunogen was prepared from bovine brain. Cryosections of rat heart right atrium and ventricle were incubated overnight with 2 µg/ml anti α antibodies and stained with the fluorescence conjugated anti rabbit IgG. Fluorescence was seen in the sarcolemma of the right atrium and especially in neuronal endings located between right atrium and right ventricle. Neuronal tissue with intensive fluorescence between cardiomyocytes of the right atrium (arrows). Bar = 20 µm.

Fig. 9. Same section as in Fig. 8. Incubation with a monoclonal anti-neurofilament 200 (Sigma). This antibody stains neural specific antigens (neurofilaments of 200 kD using an immunoblotting technique). It does not crossreact with other intermediate filament proteins. The fluorescence shown in a red colour is associated with the neuronal endings indicate the identical cells as the $G_{o\alpha}$ antibodies. Bar = 20 µm.

Localization of $G_{i\alpha}$

The role of G_i-protein-linked pathways in mediating response to signals for M2-receptors is well established in the heart. It has been clearly documented that adenylyl cyclase, phospholipase C and cardiac potassium channels are primary effectors for the M2-signals and the G_i-proteins are the essential coupling elements [2, 9]. In previous publications it was shown that the highest amount of G-proteins in the heart belongs to the G_i-family [10, 11]. Since the general function of G-proteins is associated with the plasma membrane where receptors and some effectors are located, it was accordingly assumed that the G-proteins are bound with these membranes. But it is now apparent that some G-proteins may have variable subcellular localizations [12–14]. To prove whether this is true for the heart, we used immunocytochemical techniques for the localization of G_i-proteins with anti-peptide antibodies against α-subunits.

Two antisera were used. α_{common} antiserum recognizes the α-subunits of all $G_{i\alpha}$ subtypes (α_{1-3}) and the $\alpha_{i1,2}$ antiserum recognizes only $G_{i\alpha1,2}$. The specificity of these antibodies was tested by immunoblot and by pertussis toxin catalysed ADP ribosylation [15, 16]. In crude membranes of rat heart one broad band of 40–41 kD was identified.

In ventricles and atria from rat heart after using anti $\alpha_{icommon}$ and $\alpha_{i1,2}$-subunits antibodies, immunostaining was detected on the sarcolemma and in perinuclear space of cardiomyocytes (Figs 6 and 7). Labeling was seen near cell membranes of endothelial cells and fibroblasts and intracellularly in the endoplasmic reticulum and nuclei envelopes of these cells.

Immunostaining was found in all regions of the heart including atria and ventricles. Staining differences among these regions are not found. It seems to us that cardiomyocytes from right atrium containing ANF granules are more intensively stained.

Localization of $G_{o\alpha}$

In cardiac tissue, most of the known G-proteins have been detected [17]. The expression of $G_{o\alpha}$ subtypes, however, is controversial. G_o is a neuron specific G-protein [18]. It was identified by both pertussis toxin induced ADP ribosylation and immunoassay in the myocardium of rat [19] and dog [20, 21]. Eschenhagen *et al.* [22, 23] described a low level of $G_{o\alpha}$-mRNA in human atrium and rat heart. Böhm *et al.* [24] however, could not recognize $G_{o\alpha}$ in human myocardial membranes.

Using specific polyclonal antibody raised against purified $G_{o\alpha}$ subunit from bovine brain (Upstate Biotechnology Incorporated), we tried to localize α_o in the rat heart. We found a signal in the right atrium and a very low or no fluorescence in the ventricles (Fig. 8). Parallel incubations with a monoclonal neurofilament specific antibody (Monoclonal Anti-Neurofilament 200, Sigma) reveal a localization in the same structure in which α_o was seen (Fig. 9). Therefore, we conclude that $G_{o\alpha}$ is localized in neuronal endings which are situated between atrial and ventricular walls and on membranes of atrial cells. It might be possible that most of the $G_{o\alpha}$ measured in fractions of cardiac membranes come originally from nerval endings.

Localization of adenylyl cyclase

The activity of the adenylyl cyclase regulated by hormones acts via G_s and G_i proteins. From physiological aspect, it was assumed that the hormone receptors, G proteins and adenylyl cyclase, as one of the main effectors, form a membrane associated unit [3, 10, 25, 26]. Although the adenylyl cyclase has been cloned [27, 28] and two specific isoforms have been expressed in the heart, antibodies are not commercially available up to now. Therefore, we used the metal precipitation technique to mark the site where cAMP and Pi are generated. The detailed cytochemical techniques are published by Slezak and Geller [29, 30] and by Schulze [31, 32].

By quantitative measurements, we were able to show that in cardiac tissue the adenylyl cyclase system and its capacity to react with hormones and some other specific activators (e.g. NaF, forskolin) and inhibitors (e.g. alloxane, 3′5-dideoxyadenosine) was not destroyed, although its activity was reduced.

With the incubation method used we found a reaction pattern along the plasma membrane of the sarcolemma and on the T-tubuli membranes (Fig. 10) in rat heart sections. Some parts of the intercalated discs were similarly covered with the activity marker. In the normal rat and guinea pig heart, we did not see precipitates on the longitudinal system of the sarcoplasmic reticulum. Not even higher magnification (Fig. 11) shows reaction at the junctional SR or the subsarcolemmal cisterns. In these parts of the SR, adenylyl cyclase activity, however, was seen under experimentally induced ischemia (Fig. 12). It was discussed that the adenylyl cyclase may partly loose its tight coupling to the molecular structure of the membrane and that the membrane penetration for the reaction products decreased during ischemia [33].

Conclusion

A rapidly growing number of publications describing changes of the signal transducing system in cardiovascular disorders have been documented. Alterations of the hormonal receptors, the G proteins and the adenylyl cyclase have been shown in heart failure, cardiomyopathies and ischemia induced changes in both humans and animals. Up to now, the mechanism and the pathophysiological meaning of these alterations have not been fully understood. Numerous aspects have been taken

Fig. 10

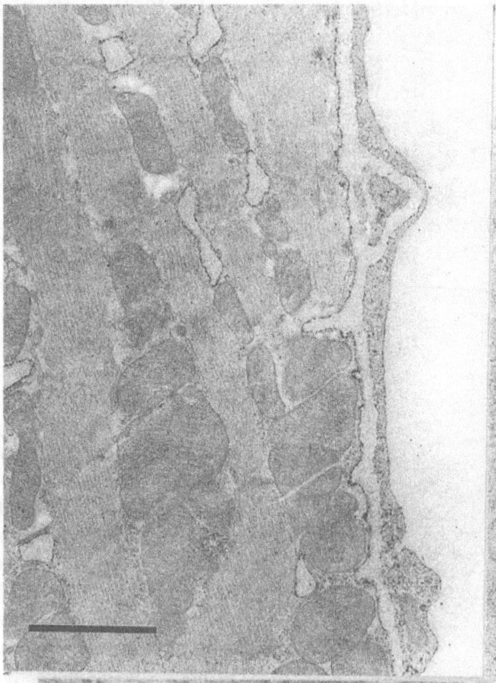

Fig. 11

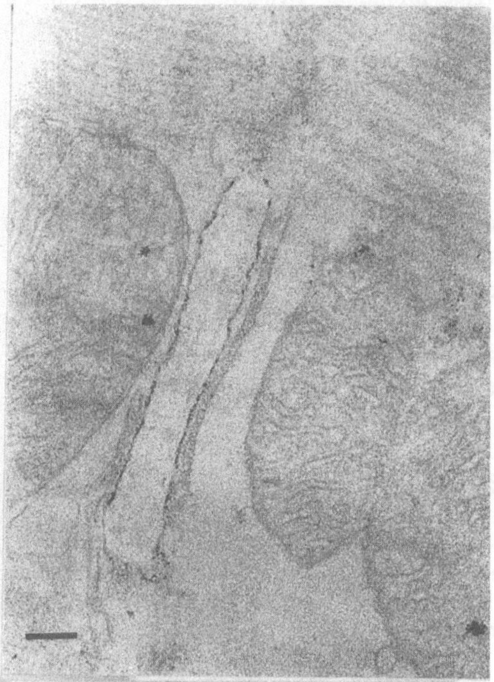

Fig. 12

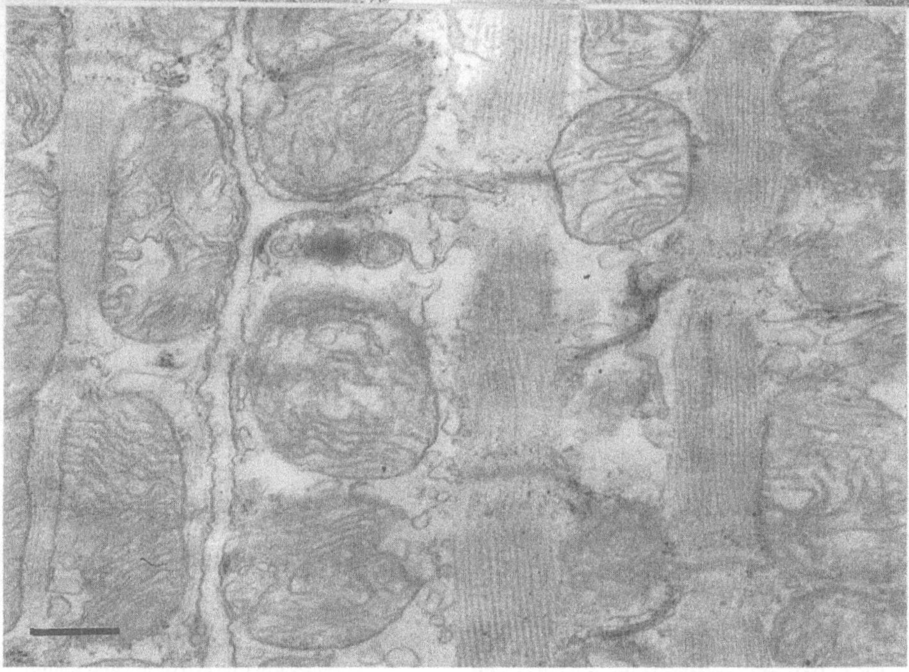

Fig. 10. Guinea pig myocardium incubated for 30 min in the adenyl cyclase assay medium containing tris-maleate buffer adenylylimidodiphosphate (AMP-PNP) and lead nitrate (see Ref. 32). Reaction products are located on the plasma membrane of cardiomyocytes and endothelial cells of capillaries. A strong reaction is located on the T-tubule membranes. Bar = 1 μm.

Fig. 11. Higher magnification shows that adenylyl cyclase activity is exactly localized on T-tubule membranes and not on membranes of the sarcoplasmic reticulum. Same incubation as in Fig. 10. Bar = 0.1 μm.

Fig. 12. Rat myocardium after 20 min global ischemia. Incubation protocol as for Fig. 10. The cardiomyocytes show mitochondria alterations and serated sarcolemma as signs for moderate ischemia injuries. Adenylyl cyclase activity was decreased on the sarcolemma. Reaction products are located on junctional sarcoplasmic reticulum. Bar = 0.5 μm.

into consideration. The applications of a great variety of methods are necessary for clarifying these disorders. As shown here, morphological immunohistochemical and cytochemical techniques provide an additional suitable tool for helping to define clinically important cardiovascular disorders.

Acknowledgements

We thank Marianne Vannauer, Ingeborg Ley, Klas-Göran Sjögren for skillful technical assistance, and Drs Peter Gierschik and Karsten Spicher for the generous gift of the anti-Giα subunits antibodies. Parts of this work were supported by grants of the Deutsche Forschungsgemeinsschaft and the Swedish Heart and Lung Foundation (Fu, 21021).

References

1. Jacobs KH, Aktories K, Schultz G: Mechanisms and components involved in adenylate cyclase inhibition by hormones. Adv Cyclic Nucl Res and Prot Phosphoryl 17: 135–143; P. Greengard *et al.* (eds). Raven Press, New York, 1985
2. Martin MW, Boyer JL, May JM, Birnbaumer L, Harden TK: Muscarinic receptors and their interactions with G proteins. In: R. Iyengar and L. Birnbaumer (eds). G proteins. Academic Press Inc, 1990, pp 317–354
3. Neer EJ, Clapham DE: Signal transduction through G proteins in the cardiac myocyte. Trends Cardiovasc Med 2: 6–11, 1992
4. Brown AM: Regulation of heartbeat by G protein-coupled ion channels. Am J Physiol 259: 1621–1628, 1993
5. Peralta EG, Ashkenazi A, Winslow JW, Smith D, Ramachandran J, Capon DJ: Distinct primary structures, ligand-binding properties and tissue specific expression of four human muscarinic acetylcholine receptors. EMBO J 13: 3923, 1987
6. Peralta EG, Winslow JW, Peterson GL, Smith DH, Ashkenazi A, Ramachandran J, Schimerlik MI, Capon DJ: Primary structure and biochemical properties of a muscarinic receptor. Science 230: 600, 1987
7. Fu M, Schulze W, Wolf WP, Hjalmarson A, Hoebeke J: Immunocytochemical localization of M2 muscarinic receptors in rat ventricles with anti-peptide antibodies. J Histochem Cytochem 42: 337–343, 1994
8. Fu M, Wallukat G, Hjalmarson A, Hoebeke J: Agonist-like activity of anti-peptide antibodies to autoimmune epitope on the second extracellular loop of human muscarinic receptors. Receptors and Channels 2: 121–130, 1994
9. Doods HN, Dämmgen J, Mayer N, Rinner I: Muscarinic receptors in the heart and vascular system. Progress in Pharmacol and Clinic Pharmacol 7/1: 47–72, 1989
10. Robishaw JD, Foster KA: Role of G proteins in the regulation of the cardiovascular system. Annu Rev Physiol 51: 229–244, 1989
11. Holmer SR, Homcy CJ: G proteins in the heart. A redundant and diverse transmembrane signaling network. Circulation 84: 1891–1902, 1989
12. Barr FA, Leyte A, Mollner S, Pfeuffer T, Tooze SA, Huttner WB: Trimeric G-proteins of the trans-Golgi network are involved in the formation of constitutive secretory vesicles and immature secretory granules. FEBS Lett 294: 239–243, 1991
13. Leyte A, Barr FA, Kehlenbach RH, Huttner WB: Multiple trimeric G-proteins on the trans-Golgi network exert stimulatory and inhibitory effects on secretory vesicle formation. EMBO J 11: 4795–4804, 1992
14. Hermounet S, de Mazancourt P, Spiegel AM, Farquhar MG, Wilson BS: High level expression of transfected G protein α_{i3} subunit is required for plasma membrane targeting and adenylyl cyclase inhibition in NIH 3T3 fibroblasts. FEBS Lett 312: 223–228, 1992
15. Schulze W, Buchwalow IB, Wolf WP, Will-Shahab L: Comparative immunocytochemical demonstration of G proteins in rat heart tissue. Acta Histochem 96: 87–95, 1994
16. Will-Shahab L, Rosenthal W, Schulze W, Küttner I: G protein function in the ischaemic myocardium. Eur Heart J 12: 135–138, 1991
17. Eschenhagen T: G proteins and the heart. Cell Biol Internat 7: 723–749, 1993
18. Asano T, Kamiya N, Semba R, Kato K: Ontogeny of the GTP-binding protein G_o in rat brain and heart. J Neurochem 51: 1711–1716, 1988
19. Asano T, Semba R, Kimiya N, Ogasawara N, Kato K: G_o, a GTP-binding protein: immunochemical and immunohistochemical localization in the rat. J Neurochem 50: 1164–1169, 1988
20. Scherer NM, Toro MJ, Entman ML, Birnbaumer L: G-protein distribution in canine cardiac sarcoplasmic reticulum and sarcolemma. Comparison to rabbit skeletal muscle membranes and to brain and erythrocyte G-proteins. Arch Biochem Biophys 259: 431–440, 1987
21. Fleming JW, Wisler PL, Watanabe AM: Signal transduction by G proteins in cardiac tissue. Circulation 85: 420–433, 1992
22. Eschenhagen T, Mende U, Nose M, Schmitz W, Scholz H, Haverich A, Hirt S, Döring V, Kalmar P, Höppner W, Seitz HJ: Increase messenger RNA level of the inhibitory G-protein α-subunit $G_{i\alpha\text{-}2}$ in human end-stage heart failure. Circ Res 70: 688–696, 1992
23. Eschenhagen T, Mende U, Nose M, Schmitz W, Scholz H, Schulte am Esch J, Warnholtz A: Long term β-adrenoceptor-mediated upregulation of $G_{i\alpha}$- and $G_{o\alpha}$-mRNA levels and pertussis toxin sensitive G-proteins in rat heart. Mol Pharmacol 42: 773–783, 1992
24. Böhm M, Eschenhagen T, Gierschik P, Larisch K, Lensche H, Mende U, Schmitz W, Schnabel P, Scholz H, Steinfath M, Erdmann E: Radioimmunochemical quantification of $G_{i\alpha}$ in right and left ventricles from patients with ischaemic and dilated cardiomyopathy and predominant left ventricular failure. J Mol Cell Cardiol 26: 133–149, 1994
25. Gilman AG: G proteins and dual control of adenylate cyclase. Cell 36: 577–579, 1984
26. Neer EJ: G proteins: Critical control points of transmembrane signals. Prot Science 3: 3–14, 1994
27. Krupinski J, Coussen F, Bakalyar HA, Tang WJ, Feinstein PG, Orth K, Slaughter C, Reed RR, Gilman AG: Adenylyl cyclase amino acid sequence: possible channel- or transporter-like structure. Science 244: 1558–1564, 1989
28. Katsushika S, Chen L, Kawabe J-I, Nilakantan R, Halnon NJ, Homcy CJ, Ishikawa Y: Cloning and characterization of a sixth adenylyl cyclase isoform: Types V and VI constitute a subgroup within the mammalian adenylyl cyclase family. Proc Natl Acad Sci USA 89: 8774–8778, 1992
29. Slezak J, Geller SA: Cytochemical demonstration of adenylyl cyclase in cardiac muscle: effect of dimethyl sulfoxide. J Histochem Cytochem 27: 774, 1979
30. Slezak J, Geller SA: Cytochemical studies of myocardial adenylate cyclase after its activation and inhibition. J Histochem Cytochem 32: 105–113, 1984
31. Schulze W: Cytochemistry of adenylate cyclase. Quantitative analysis of the effect of cytochemical procedures on adenylate cyclase in heart tissue homogenates. Histochemistry 75: 133–143, 1982
32. Schulze W: Methods for histochemical localization of adenylate cyclase and guanylate cyclase in heart membranes. In: N.S. Dhalla (ed.). Methods in Studying Cardiac Membranes. CRC Press Inc, Boca Raton, Florida Vol II: 83–109, 1984
33. Schulze W, Will-Shahab L: Zum Verhalten der Adenylatzyklase im isch ämischen und nekrotischen Herzgewebe. Acta histochem (Suppl) XXX: 157–165, 1984

Molecular and Cellular Biochemistry **147**: 169–172, 1995.

Renaissance of cytochemical localization of membrane ATPases in the myocardium

Ján Slezák[1], Ludmila Okruhlicová[1], Narcisa Tribulová[1], Wolfgang Schulze[2] and Naranjan S. Dhalla[3]

[1]*Institute for Heart Research, Slovak Academy of Sciences, Bratislava, Slovak Republic;* [2]*Max Delbruck Centrum for Molekulare Medizin, Berlin-Buch, Germany;* [3]*Division of Cardiovascular Sciences, St. Boniface General Hospital Research Centre, Faculty of Medicine, University of Manitoba, Winnipeg, Canada*

Abstract

ATPases of cardiac cells are known to be among the most important enzymes to maintain the fluxes of vital cations by hydrolysis of the terminal high-energy phosphate of ATP. Biochemically the activities of Ca^{2+}-pump ATPase, Ca^{2+}/Mg^{2+}-ecto ATPase, Na^+,K^+-ATPase and Mg^{2+}-ATPase are determined in homogenates and isolated membranes as well as in myofibrillar and mitochondrial fractions of various purities. Such techniques permit estimation of enzyme activities *in vitro* under optimal conditions without precise enzyme topography. On the other hand, cytochemical methods demonstrate enzyme activity *in situ*, but not under optimal conditions. Until recently several cytochemical methods have been employed for each enzyme in order to protect its specific activity and precise localization but the results are difficult to interpret. To obtain more consistent data from biochemical and cytochemical point of view, we modified cytochemical methods in which unified conditions for each ATPase were used. The fixative solution (1% paraformaldehyde – 0.2% glutaraldehyde in 0.1 M Tris Base buffer, pH 7.4), the same cationic concentrations of basic components in the incubation medium (0.1 M Tris Base, 2 mM $Pb(NO_2)_3$, 5 mM $MgSO_4$, 5 mM ATP) and selective stimulators or inhibitors were employed. The results reveal improved localization of Ca^{2+}-pump ATPase, Na^+-K^+ ATPase and Ca^{2+}/Mg^{2+}-ecto ATPase in the cardiac membrane. (Mol Cell Biochem **147**: 169–172, 1995)

Key words: membrane ATPases, cytochemistry, myocardium

Introduction

Several membrane ATPases are of special interest because of their role in maintenance of both the electrical and physical properties of myocardial cells. Biochemically, the activities of Na^+,K^+-ATPase (Na^+ pump), Ca^{2+}, Mg^{2+}-ATPase (Ca^{2+} pump-activated by micromolar concentration of Ca^{2+} in the presence of Mg^{2+}) and Ca^{2+}/Mg^{2+}-ATPase ('basic' ATPase – stimulated by millimolar concentrations of Ca^{2+} or Mg^{2+}) have been assayed in the isolated heart sarcolemma [1–6]. Cytochemical methods represent powerful tools for the precise mapping of the distribution of these enzymes and are able to localize their activity at ultrastructural level *in situ* [7–15]. When studying biochemical and cytochemical distribution of enzyme activities, differences between purified membrane fractions and *in situ* activities in tissues can be encountered. This can be caused by different procedures and techniques of both methods. To obtain more consistent results from biochemical and cytochemical points of view, we have modified the methods and prepared unified basic method which is common for all ATPases tested.

Material and methods

The hearts from Wistar rats (weighing 250–300 g) were quickly excised and fixed by perfusion solution containing 1% paraformaldehyde (PFA), 0.2% glutaraldehyde (GA),

Address for offprints: J. Slezák, Institute for Heart Research, Slovak Academy of Sciences, Dúbravska cesta 9, 842 33 Bratislava, Slovak Republic

5 mmol/l tetramisole, 10 mmol/l ouabain, 7% dimethylsulfoxide (DMSO), 0.44 mol/l sucrose in 0.1 mol/l Tris-HCl buffer, pH 7.4 for Ca^{2+}, Mg^{2+}-ATPase as well as Ca^{2+}/Mg^{2+}-ecto ATPase for 3 min at 4°C; ouabain-free fixation was used for Na^+,K^+-ATPase. Small blocks of left ventricle (0.5 mm^3) were incubated in the basic incubation solution of the same composition (0.1 mol/l Tris-Base pH 7.4, 2 mmol/l $Pb(NO_3)_2$, 5 mmol/l $MgCl_2$, 5 mmol/l ATP-Tris) for all of the ATPases during 30 min at 37°C. Specific cationic concentrations for different enzymes were used: 5 mmol/l $CaCl_2$ for Ca^{2+}/Mg^{2+}-ecto ATPase, 5 µmol/l $CaCl_2$ for Ca^{2+},Mg^{2+}-ATPase and 50 mmol/l KCl + 100 mmol/l NaCl for Na^+,K^+-ATPase. To discriminate between specific enzymes localized at different cellular structures, the selective inhibitors were used: 1 mmol/l EGTA (Ca^{2+}/Mg^{2+}-ecto ATPase inhibitor), 7 nmol/l thapsigargin (Ca^{2+},Mg^{2+}-ATPase inhibitor [16]) and 10 mmol/l ouabain for Na^+,K^+-ATPase. Incubation with deletion of substrates and Ca^{2+} or Mg^{2+} served as controls. After completing the incubation, samples were washed in 0.1 mol/l sucrose (pH adjusted to 7.4 with 0.1 mol/l cacodylate buffer), post-fixed in basic fixation solution, washed, then post-fixed for 30 min in 40 mmol/l OsO_4 at 4°C, dehydrated with alcohol and embedded in Epon 812.

Results

In ventricular muscle fixed and incubated at pH 7.4 by a new unified method, the reaction product of Ca^{2+},Mg^{2+}-ATPase was seen on the sarcolemma (Sl) including the intercalated disc (ID) as well as on the junctional sarcoplasmic reticulum (JSR) (Fig. 1). In the presence of thapsigargin the precipitate was localized only on Sl (Fig. 2). In a medium that contained 1 mmol/l EGTA, no precipitate was observed on Sl or JSR. Sometimes, the ultrastructural changes were observed especially dehiscence of ID (Fig. 3) as a result of the absence of Ca^{2+}.

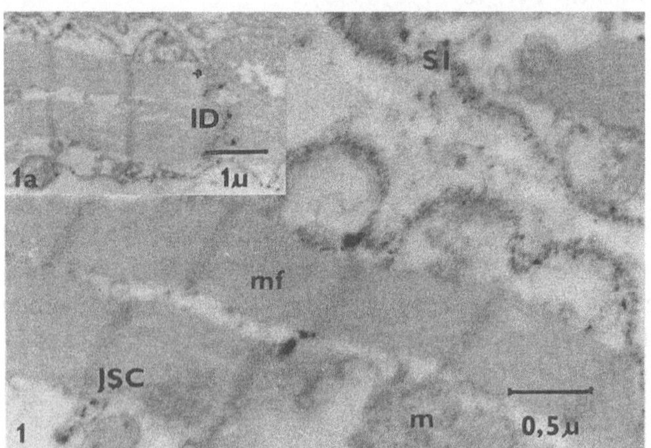

Fig. 1. Ca^{2+},Mg^{2+}-ATPase activity in myocardial cell. The deposits of the reaction product were localized on the sarcolemma (Sl) and the junctional sarcoplasmic reticulum (JSR). Figure 1a Specific precipitate was found on the intercalated disc, however no precipitate was localized on the nexus (arrows) (Fig. 1a). ID – intercalated disc, m – mitochondria, mf – myofillaments.

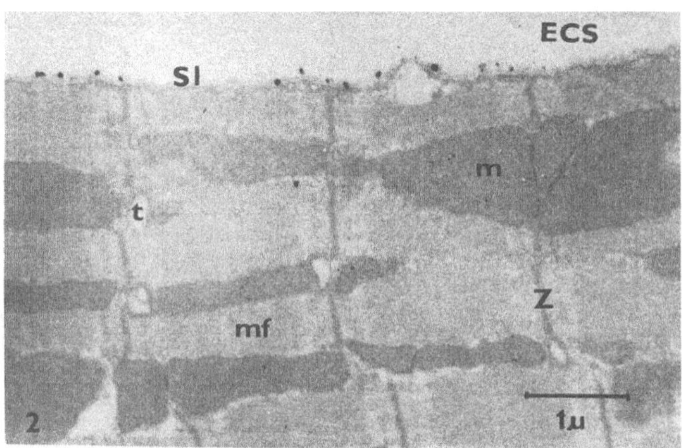

Fig. 2. Inhibition of Ca^{2+},Mg^{2+}-ATPase on the sarcoplasmic reticulum in the presence of 7 mmol/l thapsigargin. ECS – extracellular space, m – mitochondria, mf – myofillaments, t – T system, Sl – sarcolemma, Z – Z band.

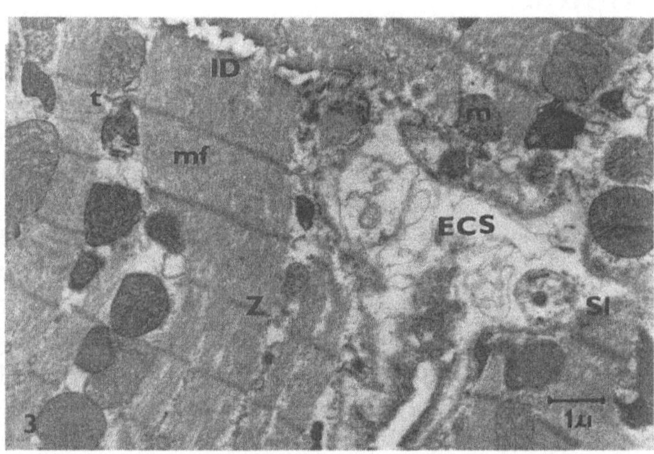

Fig. 3. Effect of 1 mmol/l EGTA on Ca^{2+},Mg^{2+}-ATPase activity. No specific precipitate was present in cellular components. Ultrastructural changes on intercalated disc (ID) were observed – dehiscence of ID. ECS – extracellular space, m – mitochondria, mf – myofillaments, t – T system, Sl – sarcolemma, Z – Z band.

Incubation of the tissue in the medium for Ca^{2+}/Mg^{2+}-ecto ATPase resulted in the precipitate localized only on Sl (Fig. 4). The same result was obtained in the presence of 7 nmol/l thapsigargin. Na^+,K^+-ATPase activity was demonstrated on Sl as well as on JSR (Fig. 5), however ouabain resistant reaction was found on JSR.

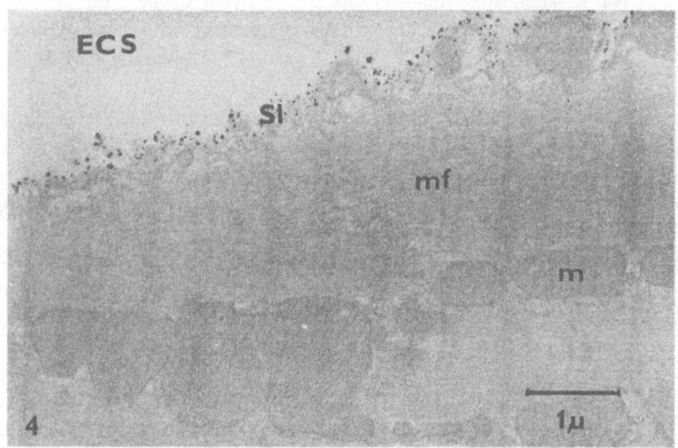

Fig. 4. Cytochemical localization of Ca^{2+}/Mg^{2+}-ecto ATPase activity on the sarcolemma (Sl). ECS – extracellular space, m – mitochondria, mf – myofillaments.

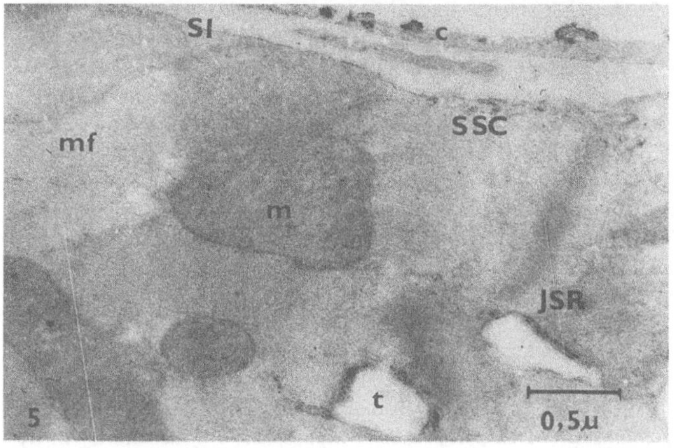

Fig. 5. Na^{+},K^{+}-ATPase activity was localized on the sarcolemma (Sl), t – tubules (t), subsarcolemmal cisternae (SSC) and the junctional sarcoplasmic reticulum (JSR). m – mitochondria, mf – myofillaments.

In control experiments, the exclusion of Ca^{2+}, Mg^{2+} or substrate from the incubation medium resulted in complete loss of cytochemical reaction products (Fig. 6). The results are in agreement with biochemical and most cytochemical studies.

Discussion

There are numerous studies demonstrating the ultrastructural localization of membrane-bound ATPase activities in the myocardium [7, 9–12, 14, 15, 17, 18]. These studies have used different procedures and techniques [17, 19–21] but the basic effort was to provide the optimal reaction conditions for the activity of respective enzymes and their protection during fixation and visualization. In most studies, the reaction product consisted of precipitates of heavy metals. Originating from biochemical assays we modified the cytochemical methods for membrane ATPases and developed a unified procedure common for all above mentioned ATPases. By using this technique, as well as by applying selective inhibitors, we were able to discriminate between localizations of specific enzymes in cardiomyocytes at a subcellular level. We decided to minimize the inhibition effect of GA and PFA as well as lead on the enzyme activities and to provide a suitable composition of the basic incubation medium which could be common for all enzyme reactions. The high sensitivity of ATPase systems (especially Na^{+},K^{+}-ATPase) to aldehydes is well known [22–24] and in fact higher concentrations of aldehydes are detrimental to the enzymes.

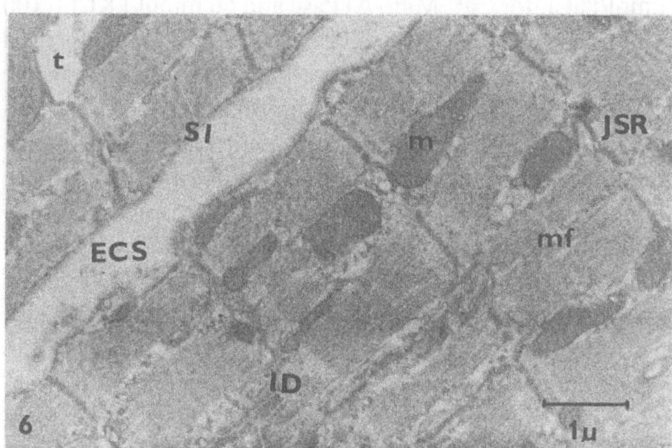

Fig. 6. No specific precipitate was seen in control samples from ventricular heart muscle incubated in unified medium. Sl – sarcolemma, t – t-tubules, JSR – junctional sarcoplasmic reticulum, ID – intercalated disc, ECS – extracellular space, m – mitochondria, mf – myofillaments.

We have used short fixation-stabilization by perfusing the heart with a cold fixture at low concentration (1% PFA + 0.2% GA) with DMSO added and we did not observe any greater alteration in the distribution or activity of the enzyme. In fact the residual activity after fixation was sufficient for the demonstration of ATPases. To inhibit unspecific activity of alkaline phosphatase we added levamisole to the fixation solution as well as to the incubation medium. The reduction of enzyme activity by toxic effect of lead used as precipitating agent can be avoided by using low concentrations and by chelation of free lead by Tris buffer. Our results revealed that membrane ATPases are sufficiently resistant to the inhibitory effect of 2 mmol/l Pb^{2+} in our basic medium.

The basal incubation medium at pH 7.4 for histochemical experiments contained the same concentrations of compounds (ATP-Tris as the substrate, Mg^{2+} ions, Pb^{2+} as a capture ion) and cations (5 mmol/l $CaCl_2$ for Ca^{2+}/Mg^{2+}-ecto ATPase, 5 μmol/l aCl_2 for Ca^{2+},Mg^{2+}-ATPase and 50 mmol/l KCl – 100 mmol/l NaCl for Na^+,K^+-ATPase), which are commonly used for biochemical studies. Selective inhibitors – thapsigargin for Ca^{2+},Mg^{2+}-ATPase of SR, ouabain for Na^+,K^+-ATPase and EGTA for Ca^{2+}/Mg^{2+}-ecto ATPase helped us to discriminate the specific activities and to detect distribution of these membrane ATPases. Our results are in agreement with the results of biochemical [1–3, 5] as well as most cytochemical studies [7, 10–12, 17, 18]. The results of this study indicate that the new unified method is suitable for the cytochemical detection of membrane ATPases in the myocardium which can be compared to biochemical results. Further experiments are necessary to verify the results with other techniques like immunochemistry and fluorescence immunohistochemistry.

Acknowledgements

We thank Mrs. A. Brichtová for technical assistance. This work was supported by Grant Agency for Science (Grant No. 2/999237/93).

References

1. Dhalla NS, Zhao D: Cell membrane Ca^{2+}/Mg^{2+} ATPase. Prog Biophys Molec Biol 52: 1–37, 1988
2. Dhalla NS, Zhao D: Possible role of sarcolemmal Ca^{2+}/Mg^{2+} ATPase in heart function. Magnesium Res 2(3): 161–172, 1989
3. Vrbjar N, Soos J, Ziegelhöffer A: Secondary structure of heart sarcolemmal proteins during interactions with metallic cofactors of (Na,K)-ATPase. Gen Physiol Biophys 3: 317–325, 1984
4. Vrbjar N, Ziegelhöffer A, Breier A, Soos J, Ďzurba A, Monosikova R: Quantitative relationship between the protein secondary structure in cardiac sarcolemma and the activity of the membrane-bound Ca-ATPase. Gen Physiol Biophys 4: 411–416, 1985
5. Ziegelhöffer A, Breier A, Džurba A, Vrbjar N: Selective and reversible inhibition of heart sarcolemmal (Na,K)-ATPase by p-Bromphenyl isothiocyanate. Evidence for a sulfhydryl group in the ATP-binding site of the enzyme. Gen Physiol Biophys 2: 447–456, 1983
6. Tuana BS, Džurba A, Panagia V, Dhalla NS: Stimulation of heart sarcolemmal calcium pump by calmodulin. Biochem Biophys Res Commun 100: 1245–1250, 1981
7. Borgers M, Schaper J, Schaper W: Localization of specific phosphatase activities in canine coronary blood vessels and heart muscle. J Histochem Cytochem 19: 526–539, 1971
8. Wollenberger A, Schulze W: Cytochemical studies on sarcolemma: Na^+,K^+-adenosine triphosphatase and adenylate cyclase. In: P.E. Roy and N.S. Dhalla (eds). The sarcolemma, Recent Advances in Studies on Cardiac Structure and Metabolism, Vol 9. University Park Press, Baltimore, 1976, pp 101–115
9. Ashraf M, Jones HM, Livingston LH: Localization of ATPase activity in the sarcoplasmic reticulum of myocardium. In: G.W. Bailey (ed.). Proceedings of the 34th Annual Meeting of Electron Microscopy Society of America. Claitor's Publishing Company, Baton Rouge, 1976, pp 84–85
10. Malouf NN, Meissner G: Cytochemical localization of a 'basic' ATPase to canine myocardial surface membrane. J Histochem Cytochem 28(12): 1286–1294, 1980
11. Ando T, Fujimoro K, Mayahara H, Miyajima H, Ogawa K: A new one-step method for the histochemistry and cytochemistry of Ca^{2+}-ATPase activity. Acta Histochem Cytochem 14(6): 705–726, 1981
12. Asano G, Ashraf M, Schwartz A: Localization of Na-K-ATPase in guinea-pig myocardium. J Mol Cell Cardiol 12: 257–266, 1980
13. Ogawa KS, Fujimoto K, Ogawa K: Cytochemical localization of Ca^{++}-ATPase, H^+,K^+-ATPase and Na^+,K^+-ATPase in acid-secreting parietal cell and non-secreting parietal cell. Acta Histochem Cytochem 20(2): 197–216, 1987
14. Itoh S, Yanagashita T, Mukae S, Konno N, Katagiri T: Study on reperfusion injury on sarcoplasmic reticulum in acute myocardial ischemia. Jpn Circ J 56: 384–391, 1992
15. Meyran JC, Graf F: Untrahistochemical localization of Na^+-K^+ ATPase and alkaline phosphatase activity in calcium-transporting epithelium of a crustacean during moulting. Histochemistry 85: 313–320, 1986
16. Inesi G, Sagara Y: Thapsigargin, a high affinity and global inhibitor of intracellular Ca^{2+} transport ATPases. Arch Biochem Biophys 298(2): 313–317, 1992
17. Schulze W, Wollenberger A: Zytochemische lokalisation und charakterisierung von phosphatabspaltenden fermenten im sarkotubularren system quergestreifter muskeln. Histochemi 10: 140–153, 1967
18. Schulze W, Wollenberger A: Cytochemistry of membrane-bound ATPase systems of cardiac muscle. In: E. Bajusz and G. Jasmin (eds). Meth Achiev Exp Path, Vol 5. Karger, Basel, 1971, pp 347–383
19. Wachstein M, Meisel F: Histochemistry of hepatic phosphatase at a physiologic pH. (With special reference to the demonstration of bile canaliculi.) Amer J Clin Pathol 27: 13–23, 1957
20. Ernst SA: Transport adenosine triphosphatase cytochemistry II: Cytochemical localization of ouabain-sensitive, potassium-dependent phosphatase activity in the secretory epithelium of the avian salt gland. J Histohem Cytochem 20: 23–38, 1972
21. Ernst SA, Hootman SR: Microscopical methods for the localization of Na^+,K^+-ATPase. Histochem J 13: 397–410, 1981
22. Torack RM: Adenosine triphosphatase activity in rat brain following differential fixation with formaldehyde, glutaraldehyde, and hydroxyadipaldehyde. J Histochem Cytochem 13: 191–205, 1965
23. Sabatini DD, Bensch K, Barnnett RJ: Cytochemistry and electron microscopy. The preservation of cellular ultrastructure and enzymatic activity by aldehyde fixation. J Cell Biol 17: 19–58, 1963
24. Sommer JR, Hasselbach W: The effect of glutaraldehyde and formaldehyde on the calcium pump of the sarcoplasmic reticulum. J Cell Biol 34: 902–905, 1967

Molecular and Cellular Biochemistry **147**: 173–180, 1995.

Effects of selective α_{1A}-adrenoceptor antagonists on reperfusion arrhythmias in isolated rat hearts

Masahiro Yasutake and Metin Avkiran
Cardiovascular Research, The Rayne Institute, St Thomas' Hospital, London SE1 7EH, UK

Abstract

Stimulation of α_1-adrenoceptors (AR) during ischaemia in the rat heart by exogenous phenylephrine exacerbates reperfusion arrhythmias, an effect apparently mediated by the α_{1A}-AR subtype. We tested whether α_{1A}-AR stimulation by *endogenous* catecholamines, released during ischaemia, could modulate reperfusion arrhythmias, using as pharmacological tools the selective α_{1A}-AR antagonists abanoquil (UK52046) and WB4101. Isolated rat hearts (n = 12/group) were subjected to dual coronary perfusion. After 15 min of aerobic perfusion of both coronary beds, abanoquil or WB4101 was infused selectively into the left coronary bed (LCB) for 5 min. The LCB was then subjected to 10 min of zero-flow ischaemia and 5 min of reperfusion. Effects on PR interval, width of the ventricular complex ($QRST_{90}$) and reperfusion arrhythmias were assessed. Abanoquil at concentrations of 0.03, 0.1 and 0.3 μM tended to reduce the incidence of reperfusion-induced ventricular fibrillation (VF) in a dose-dependent manner from 75% in controls to 58, 33 and 25%, but this effects did not achieve statistical significance. Similarly, WB4101 at 0.1, 0.3 and 1 μM also tended to reduce VF incidence from 67 % in controls to 67, 42% and 33% (NS). The incidence of ventricular tachycardia (VT) was 100% in all groups and ECG parameters were not altered significantly by either drug. These results suggest that, in this denervated isolated heart preparation, α_{1A}-AR stimulation during ischaemia by endogenous catecholamines does not significantly modulate reperfusion arrhythmias. (Mol Cell Biochem **147**: 173–180, 1995)

Key words: α_{1A}-adrenoceptors, regional ischaemia, reperfusion-induced arrhythmias, dual coronary perfusion, abanoquil, WB4101, rat heart

Introduction

Studies in a variety of species have shown that α_{1A}-adrenoceptor blockade significantly decreases the incidence of malignant ventricular arrhythmias associated with either myocardial ischaemia or subsequent reperfusion [1–7], supporting an α_{1A}-adrenergic contribution to arrhythmogenesis. Recently, mammalian postsynaptic α_1-adrenoceptors have been divided into 2 pharmacologically distinct receptor subtypes, termed α_{1A} and α_{1B} [8]. There is preliminary evidence that, *in vivo*, stimulation of the α_{1A}-adrenoceptor subtype may exacerbate of reperfusion-induced arrhythmias, whereas stimulation of the α_{1B}-adrenoceptor subtype may suppress them [9]. This has been supported by the *in vitro* studies in canine Purkinje fibres subjected to α_1-adrenoceptor stimulation by an exogenous agonist during *simulated* ischaemia and reperfusion [10].

Recent studies from our laboratory have shown that, in the intact heart also, application of an *exogenous* α_1-adrenoceptor agonist (phenylephrine) exacerbates reperfusion induced arrhythmias, through α_{1A}-adrenoceptor mediated mechanisms [11]. Since release of endogenous catecholamines is known to occur during ischaemia, predominantly from local sympathetic nerve endings within the ischaemic region [12], the possibility exists that α_{1A}-adrenoceptor stimulation may modulate reperfusion arrhythmias even in the absence of an exogenous agonist. The primary objective of the present study was therefore to test whether α_{1A}-adrenoceptor stimulation during ischaemia by *endogenous* catecholamines could modulate reperfusion arrhythmias. In order to achieve this objective, we used as pharmacological tools two selective antagonists of α_{1A}-adrenoceptors, abanoquil (UK52046) and WB4101. In addition, we used isolated rat hearts subjected to independent perfusion of the left and right coronary beds

Address for offprints: M. Avkiran, Cardiovascular Research, The Rayne Institute, St Thomas' Hospital, London SE1 7EH, UK

[13], which enabled the use of regional ischaemia and reperfusion and administration of the drugs under study selectively into the involved zone.

Material and methods

This investigation was performed in accordance with the Home Office Guidance on the Operation of the Animals (Scientific Procedures) Act 1986, published by Her Majesty's Stationery Office, London.

A. Dual coronary perfusion of isolated rat hearts

Isolated hearts from male Wistar rats (Bantin and Kingman Ltd., N. Humberside, UK) were subjected to independent perfusion of left and right coronary arteries, as described in detail by Avkiran and Curtis [13]. Each coronary bed was initially perfused at a constant perfusion pressure equivalent to 75 mmHg with oxygenated perfusion solution from a temperature-regulated reservoir (37°C). The perfusion solution was of the following composition (in mM): NaCl, 118.5; $NaHCO_3$, 25.0; KC1, 3.2; $MgSO_4$ 1.2; KH_2PO_4, 1.2; $CaCl_2$, 1.4; glucose, 11.0. The solution was filtered (pore size, 5 μm) before use and bubbled continuously with 95% O_2 + 5% CO_2(pH 7.4 at 37°C). Flow to each coronary bed was monitored using in-line flow detectors (Transonic T206 Animal Research Flowmeter with 1N probes, Transonic Systems Inc, New York) with a linear detection range of 0.05–30 ml.min^{-1}. After 15 min of perfusion of both coronary beds, basal flow rate in the left coronary bed was recorded and perfusion of this bed was switched from constant pressure to constant flow at the basal flow rate (supplied by a Gilson Minipuls 3 roller pump). This technique, which has been previously described [14], enabled the infusion of drug solutions or vehicle into the left coronary bed, at a known percentage of the total flow supplied to that bed (see later). The right coronary bed was perfused at constant pressure throughout the experiment. The heart was housed in a temperature-regulated chamber at 37°C throughout the experiment and the right atrium was continuously superfused with oxygenated perfusion solution (37°C) at 8 ml.min^{-1} to maintain sinus rate [13].

B. Drug administration and study protocol

Abanoquil has limited solubility in physiological buffer solutions. Therefore both drugs were dissolved in deionised water at the appropriate concentrations of (0.429, 1.43 and 4.29 μM for abanoquil; 1.43, 4.29 and 14.3 μM for WB4101). When required, these solutions were infused selectively into the perfusion line supplying the left coronary bed, at 7% of the total flow rate delivered to that bed; this resulted in final perfusate drug concentrations of 0.03, 0.1 and 0.3 μM for abanoquil and 0.1, 0.3 and 1 μM for WB4101. Control hearts received vehicle (deionised water) at the same infusion rate.

Each concentration of abanoquil or WB4101 was infused into the left coronary bed for 5 min immediately before ischaemia All hearts (n = 12/group) were subjected to 10 min of regional ischaemia (this was achieved by terminating flow to the left coronary bed) followed by 5 min of reperfusion. Experiments were carried out in a prospectively randomised manner.

C. Measured variables

i. Arrhythmias

Arrhythmias were diagnosed from a unipolar electrogram (ECG) that was obtained through a silver electrode inserted into the free wall of the left ventricle and a reference electrode connected to the aorta. The ECG was continuously monitored on a digital storage oscilloscope (model 1421, Gould Electronic Ltd., Ilford, UK) and recorded on an inkjet recorder (model 2200S, Gould). Chart speed was set at 50 mm·s^{-1} a few seconds before reperfusion so as to obtain a permanent high speed record of the changes in the ECG during early reperfusion. The ECG record obtained during the reperfusion period was retrospectively analysed, in a blinded manner, for the incidence, time-to-onset and duration of VT and VF. All analyses were carried out in accordance with the Lambeth Conventions [15]. VT was defined as 4 or more consecutive premature beats of ventricular origin and VF as a signal in which both rate and amplitude varied from cycle to cycle.

ii. ECG parameters

When required, chart speed was increased to 250 mm·s^{-1} to obtain a permanent high-speed ECG recording. These recordings were utilised to measure the interval from the beginning of the P wave to the beginning of the ventricular complex (PR interval) and the width of ventricular complex. As a separate T wave is not seen in the rat ECG, the width of the ventricular complex was measured at 90% repolarisation (with the maximum positive deflection of the ventricular complex defined as the point of 0% repolarisation) and defined as $QRST_{90}$, as previously described [16]. These parameters were measured at the end of 15 min of perfusion with normal buffer, at the end of 5 min of perfusion with each drug containing solution and at 1, 5 and 9 min after the induction of regional ischaemia. They would not be measured during reperfusion due to the frequent occurrence of severe arrhythmias.

iii. Coronary flow, heart rate and coronary vascular resistance

Throughout the experimental protocol, coronary flow was monitored using the in-line flow detectors. Heart rate was determined at selected intervals from the ECG trace. Left coronary vascular resistance was determined after 10 and 15 min of perfusion with normal buffer at constant pressure and after 1 and 4 min of perfusion with drug-containing buffer at constant flow, from the left coronary flow and the perfusion pressure (which was monitored through a side-arm of the perfusion line supplying the left coronary bed).

iv. Size of ischaemic zone

At the end of each experiment, the left coronary bed was perfused for 30 sec with a solution containing 0.016% disulphine blue dye, at 75 mmHg perfusion pressure. The heart was then removed from the perfusion apparatus, the atria and mediastinal tissue were removed and dye-stained tissue, representing ventricular myocardium subjected to ischaemia and reperfusion, was carefully dissected away from the remainder. The stained and unstained tissues were lightly blotted and weighed. The size of the ischaemic zone, expressed as a percentage of total ventricular weight, was calculated from the equation: (weight of stained tissue/total ventricular weight) × 100. The absolute weights obtained also enabled the calculation of flows in left and right coronary beds on the basis of tissue weights supplied by each bed ($ml \cdot min^{-1} \cdot g^{-1}$).

D. Drug sources

WB4101 was purchased from Sigma (Sigma Chemical Company Ltd., Poole, UK) and abanoquil was a gift from Pfizer (Pfizer Ltd., Sandwich, UK).

E. Exclusion criteria

These criteria, selected to minimise variations in heart rate and size of ischaemic zone (due to atypical coronary anatomy) among the hearts, were as previously described [13, 17, 18]. Hearts were also excluded if there was cross-flow between right and left coronary ostia [13, 17, 18]. In addition, hearts that exhibited ventricular arrhythmias during the final 3 sec of ischaemia before reperfusion were not included in the analysis of reperfusion-induced arrhythmias, because in those hearts it would have been impossible to differentiate arrhythmias induced by reperfusion from those induced by ischaemia. Of a total of 126 hearts entered into this study, 1 was excluded on the basis of heart rate, 2 on the basis of size of ischaemic zone, 1 on account of cross flow and 2 on account of arrhythmias during the final 3 sec of ischaemia. Thus, the overall exclusion rate was 5%.

F. Statistical analysis

All experiments were carried out in a prospectively randomised manner. Gaussian-distributed variables were expressed as mean ± SEM and were subjected to one-way analysis of variance. If a difference among mean values was established, comparison with controls was performed using Dunnett's test. Temporal changes in PQ interval, $QRST_{90}$ and coronary vascular resistance were analysed using analysis of variance for repeated measurements. Binominaly distributed variables, such as the incidence of VT or VF, were compared using the chi-squared test for a 2 × n table. If a significant difference was revealed, each drug-treated group was then compared with the control group using the Fisher exact test, with the Bonferroni correction for multiple comparisons. A value of $p < 0.05$ was considered significant.

Results

A. Reperfusion-induced arrhythmias

Consistent with our earlier studies [14, 17, 18], reperfusion after 10 min of regional ischaemia frequently resulted in the rapid (within a few beats) induction of VT. Reperfusion-induced VT was generally polymorphic in nature, and episodes of VT were usually uninterrupted until either spontaneous reversion to normal sinus rhythm or degeneration into VF.

Figure 1 shows the overall incidence of reperfusion-induced VF in control hearts and hearts that received abanoquil or WB4101 before ischaemia. Both drugs tended to suppress the incidence of VF in a concentration-dependent manner, but the reduction did not reach a level of statistical significance. The incidence of reperfusion-induced VT was 100% in all control and drug-treated groups. Abanoquil tended also to increase the number of hearts that were in normal sinus rhythm at the end of the reperfusion period, from 25% in controls to 75, 67 and 75% at concentrations of 0.03, 0.1 and 0.3 μM (NS). WB4101 had a similar effect, with 50% hearts in the control group in sinus rhythm at the end of reperfusion, relative to 50, 58 and 75% of hearts in the groups that received 0.1, 0.3 and 1 μM of the drug (NS).

B. ECG parameters

In the control group and the groups which received abanoquil at 0.03, 0.1 and 0.3 μM, basal values for PR interval were 39 ± 1, 39 ± 1, 38 ± 1 and 40 ± 1 ms, respectively (NS) and the

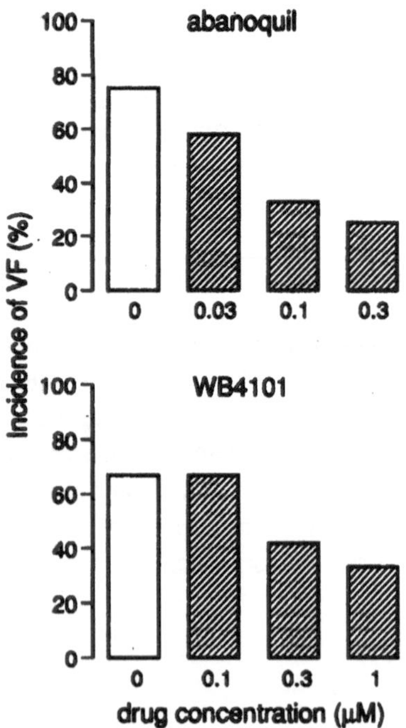

Fig. 1. Effects of abanoquil (0.03, 0.1 and 0.3 μM) and WB4101 (0.1, 0.3 and 1 μM) on reperfusion-induced VF. n = 12 per group.

corresponding values for $QRST_{90}$ were 49 ± 2, 52 ± 2, 48 ± 1 and 50 ± 1 ms, respecdvely (NS). PR interval did not vary significantly throughout the protocol in any group (Fig. 2). In all groups, $QRST_{90}$ prolonged significandy at 1 min after the induction of ischaemia, but tended to return towards the basal values at 9 min after the onset of ischaemia. There was no significant difference between the groups at any time point (Fig. 2).

In the control group and the groups which received WB4101 at 0.1, 0.3 and 1 μM, basal values for PR interval were 38 ± 1, 39 ± 1, 38 ± 2 and 38 ± 1 ms, respectively (NS) and the corresponding values for $QRST_{90}$ were 53 ± 1, 52 ± 1, 52 ± 1 and 51 ± 1 ms, respecdvely (NS). PR interval did not change significantly during drug treatment or the following period of ischaemia (Fig. 2). Administration of WB4101 did not affect $QRST_{90}$. Once again, $QRST_{90}$ prolonged significandy at 1 min after the induction of ischaemia and returned towards the basal values by 9 min; there was no significant difference between the groups at any time point.

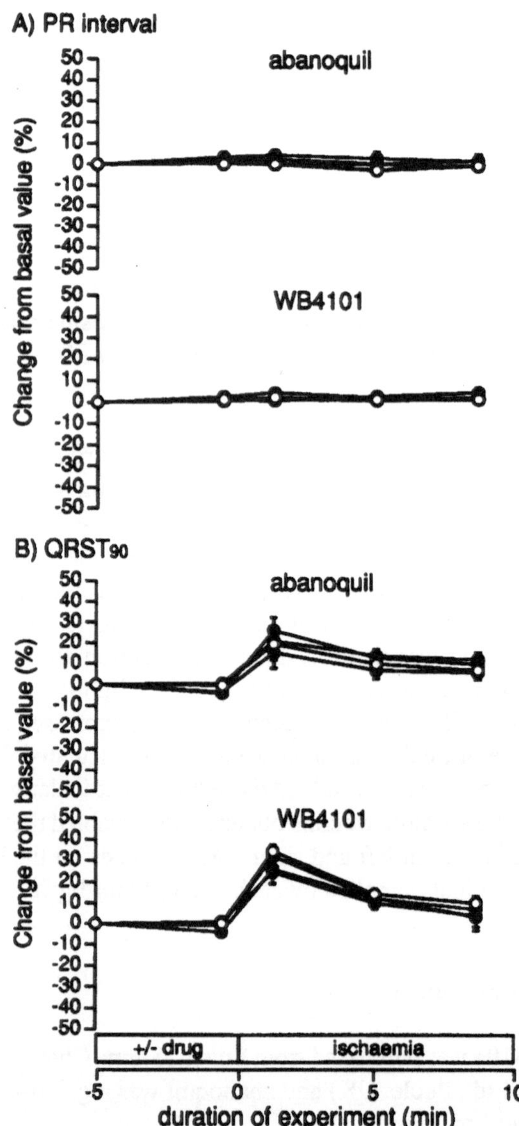

Fig. 2. Effects of abanoquil (0.03, 0.1 and 0.3 μM) and WB4101 (0.1, 0.3 and 1 μM) on A) PR interval, and B) $QRST_{90}$ during pre-ischaemic drug infusion and subsequent regional ischaemia. Open circles indicate control group in each case. n = 12 per group.

increased significandy after switching the mode of perfusion from constant pressure to constant flow; however, there was no significant difference between the study groups in each experimental protocol at any time point (Fig. 3).

C. Coronary vascular resistance

Basal values for left coronary vascular resistance were 20 ± 1, 21 ± 1, 20 ± 2 and 21± 1 mmHg·ml⁻¹·min (NS) in the groups which received 0, 0.03, 0.1 or 0.3 μM abanoquil. Corresponding values for the groups which received 0, 0.1, 0.3 or 1 μM WB4101 were 17 ± 1, 16 ± 1, 18 ± 1 and 19 ± 2 mmHg·ml⁻¹·min (NS). Left coronary vascular resistance was

D. Coronary flow, heart rate and ischaemic zone size

As shown in Table 1, there was no significant difference between control and drug-treated groups in basal left and right coronary flow rates, when measured at the end of the initial 15 min period of perfusion at constant pressure. Thereafter, left coronary flow rate was held constant at the basal value; therefore, there were no significant inter-group differences

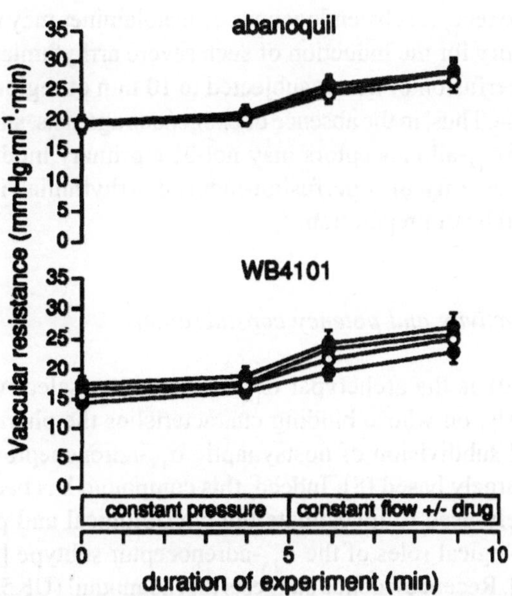

Fig. 3. Effects of abanoquil (0.03, 0.1 and 0.3 μM) and WB4101 (0.1, 0.3 and 1 μM) on vascular resistance in the left coronary bed. Drugs were infused during 5 min of constant flow perfusion prior to the induction of regional ischaemia. Open circles indicate control group in each case. n = 12 per group.

in left coronary flow rate for the rest of the experimental protocol. Flow rate in the right coronary bed (which was perfused at constant pressure throughout) did not change significandy during the period of zero-flow ischaemia in the left coronary bed. Right coronary flow rate increased during reperfusion commensurate with the severity of reperfusion arrhythmias, probably due to reduced extravascular compression as previously described [17, 18].

Basal heart rate also did not differ significantly between control and drug-treated groups (Table 1). The infusion of abanoquil or WB4101 into the left coronary bed for 5 min immediately prior to the onset of ischaemia had no effect on heart rate, regardless of drug concentration. Heart rate did not change significantly in any of the study groups during the period of regional ischaemia and could not be measured during early reperfusion due to the rapid onset of ventricular arrhythmias in the majority of hearts. There was no difference between control and drug-treated groups in the size of the ischaemic zone (Table 1).

Discussion

The present study has demonstrated that, in isolated rat hearts, application of two selective α_{1A}-adrenoceptor antagonists (abanoquil and WB4101) into the zone subjected to ischaemia and reperfusion does not afford significant protection against reperfusion-induced VF. This finding suggests that stimulation of α_{1A}-adrenoceptors by endogenous catecholamines may not be involved in modulation of the severity of reperfusion-induced arrhythmias in this particular model.

A. Myocardial a_{1A}-adrenoceptors

i. Receptor subtypes

Mammalian postsynaptic α_{1A}-adrenoceptors have been divided into 2 pharmacologically distinct receptor subtypes, termed α_{1A} and α_{1B} [8]. The α_{1A}-subtype exhibits high affinity for WB4101, a competitive antagonist, and relative insensitivity to chloroethylclonidine (CEC), an alkylating agent. In contrast, the α_{1B} subtype exhibits low affinity for WB4101 and is inactivated irreversibly by CEC. Radioligand binding studies [19] have shown both adrenoceptor subtypes to be present in rat ventricular myocardium, with an α_{1A}:α_{1B} ratio of 30:70. Although molecular cloning studies [20] suggest the existence of additional α_1-adrenoceptor subtypes, their expression and functional role in ventricular myocardium have not been fully characterised.

ii. Role in modulating reperfusion arrhythmias

There is substantial evidence that catecholamines can modulate the severity of ventricular arrhythmias induced by both ischaemia and reperfusion, through α_1-adrenoceptor-mediated mechanisms (for review, see Corr *et al.* [21]). Since

Table 1. Basal coronary flow, heart rate and ischaemic zone size in the 8 study groups (n = 12/group)

Drug	Concentration (μM)	Coronary flow rate (ml·min⁻¹·g⁻¹) LCB	RCB	Heart rate (beats·min⁻¹)	Ischaemic Zone Size (%)
Abanoquil	0 (control)	11.6 ± 0.8	12.2 ± 0.5	338 ± 8	55 ± 3
	0.03	12.0 ± 0.7	12.6 ± 0.5	324 ± 7	54 ± 3
	0.1	11.5 ± 0.8	12.1 ± 0.7	336 ± 12	56 ± 4
	0.3	11.7 ± 0.9	12.3 ± 0.4	340 ± 11	55 ± 3
WB4101	0 (control)	13.5 ± 0.8	13.9 ± 1.2	343 ± 10	55 ± 3
	0.1	13.4 ± 0.9	14.4 ± 0.9	362 ± 8	61 ± 3
	0.3	12.1 ± 0.7	13.6 ± 0.9	340 ± 8	60 ± 3
	1	12.6 ± 0.9	12.5 ± 0.4	329 ± 9	56 ± 3

LCB: left coronary bed, RCB: right coronary bed. Values are mean ± s.e.m.

release of endogenous catecholamines is known to occur during ischaemia (for review, see Schömig *et al.* [22]), this would be expected to affect significantly susceptibility to severe arrhythmias not only during ischaemia but also during subsequent reperfusion. Indeed, a number of *in vivo* studies have shown that blockade of α_1-adrenoceptors by non-selective antagonists (such as phentolamine and prazosin) can inhibit reperfusion-induced arrhythmias [1, 2, 5–7]. Although the cellular mechanisms of this protective effect are unclear, inhibition of Ca^{2+} overload [23] may play a causal role since Ca^{2+} overload has been proposed as a key mediator of reperfusion-induced ventricular arrhythmias [24].

There is recent preliminary evidence that, in cat hearts subjected to regional myocardial ischaemia *in vivo*, stimulation of the α_{1A}-adrenoceptor subtype may be responsible for the α_1-adrenoceptor-mediated exacerbation of reperfusion-induced arrhythmias, whereas stimulation of the α_{1B}-adrenoceptor subtype may have the opposite effect [9]. This hypothesis has been supported by a number of *in vitro* studies, utilising exogenous α_1-adrenoceptor agonists. Thus, in both isolated cardiac myocytes [25] and Purkinje fibres [10], exposure to an α_1-adrenoceptor agonist during *simulated* ischaemia has been shown to result in the induction of delayed after depolarisations and triggered activity during subsequent 'reperfusion'. At least in the Purkinje fibre, this proarrhythmic effect of α_1-adrenoceptor stimulation appears to be mediated by the WB4101-sensitive α_{1A}-adrenoceptor subtype [10]. Similarly, our recent studies [11] in isolated rat hearts subjected to regional ischaemia have shown that exposure to exogenous α_1-adrenoceptor agonists before and during ischaemia increases the incidence of reperfusion-induced VF in a dose-dependent manner. This effect could be reversed by WB4101 but not by CEC or atenolol (β_1-adrenoceptor antagonist), supporting a causal role for stimulation of the α_{1A}-adrenoceptor subtype in the proarrhythmic effects of exogenous α_1-adrenoceptor agonists. However, in that study, a 7 min period of ischaemia was employed which, by design, resulted in a low incidence of reperfusion-induced VF in controls [11] and was probably too short to result in a significant release of endogenous catecholamines [26]. Thus it was not possible to assess the role α_{1A}-adrenoceptor stimulation by endogenous catecholamines could play in modulating the severity of reperfusion-induced arrhythmias.

In the present study, a 10 min period of ischaemia was employed, which resulted in a high incidence of reperfusion-induced arrhythmias in control hearts. Nevertheless, pre-ischaemic infusion of the α_{1A}-adrenoceptor-selective antagonists abanoquil and WB4101 into the ischaemic/reperfused zone failed to protect significantly against reperfusion-induced VF. This would suggest that, in the present model, while stimulation of α_{1A}-adrenoceptors by exogenous agonists may be *sufficient* to induce VF during reperfusion of hearts subjected to 7 min of regional ischaemia, stimulation of these receptors by endogenous catecholamines may not be *necessary* for the induction of such severe arrhythmias during reperfusion of hearts subjected to 10 min of regional ischaemia. Thus, in the absence of exogenous agonists, stimulation of α_{1A}-adrenoceptors may not be a primary modulator of the severity of reperfusion-induced arrhythmias in this isolated heart preparation.

B. Selectivity and potency considerations

WB4101 is the archetypal α_{1A}-adrenoceptor-selective antagonists, on whose binding characteristics the pharmacological subdivision of postsynaptic α_{1A}-adrenoceptors has been largely based [8]. Indeed, this compound has been extensively used to investigate the physiological and pathophysiological roles of the α_{1A}-adrenoceptor subtype [9, 10, 27, 28]. Recent evidence suggests that abanoquil (UK52046) also exhibits significant selectivity for the α_{1A}-adrenoceptor subtype, exhibiting an approximately 100 fold greater affinity for these receptors relative to α_{1B}-adrenoceptors (cf. approximately 200 fold with WB4101) [29]. Indeed, the affinity of abanoquil for α_{1A}-adrenoceptors may be greater than that of WB4101 [29], a factor that was taken into account in determining the concentration ranges used in the present study.

Since our previous studies [11] and preliminary experiments prior to the present study (results not shown) had indicated that WB4101 could have non-specific effects at concentrations ≥ 3 μM, the maximum concentration used in the present study was limited to 1 μM. Nevertheless, since 1 μM WB4101 has been shown to reverse significantly the proarrhythmic effect of 10 μM exogenous phenylephrine [11], it should have been sufficient to antagonise significantly any proarrhythmic effect of α_{1A}-adrenoceptor stimulation by endogenous catecholamines. Thus, it is unlikely that the lack of effect of WB4101 on reperfusion-induced arrhythmias in the present study was due to insufficient drug concentration. Similarly, since the affinity of abanoquil for α_{1A}-adrenoceptors appears to be 50 fold greater than that of WB4101 [29], the concentrations of abanoquil used should have been sufficient for significant α_{1A}-adrenoceptor blockade.

C. Potential limitations of study

The present study was carried out in a denervated isolated heart preparation. Thus, it may be argued that this may result in attenuated catecholamine release during ischaemia and, consequently, underestimation of the role of α_{1A}-adrenoceptor stimulation by endogenous catecholamines in modulation of reperfusion arrhythmias. While acknowledging this, it should be pointed out that significant catecholamine release has been shown to occur during ischaemia in similar isolated rat heart preparations [26].

It is also possible that our observations may be species-specific. Steinfath and colleagues [30] have shown that rat ventricular myocardium possesses a 5–8 fold greater α_1-adrenoceptor density, relative to guinea pig, mouse, pig, calf and man. Thus, extrapolation of our findings to other species should be undertaken with considerable caution. In this regard, there is some evidence, albeit in a different model (global low-flow ischaemia) that abanoquil may protect the isolated guinea pig heart against reperfusion-induced arrhythmias [3].

D. Concluding comments

In the isolated rat heart, in which exposure to exogenous α_1-adrenoceptor agonists before and during regional ischaemia has been shown to exacerbate reperfusion-induced arrhythmias via α_{1A}-adrenoceptor-mediated mechanisms [11], two distinct antagonists that exhibit selectivity for α_{1A}-adrenoceptors have been shown to be ineffective in reducing the severity of reperfusion-induced arrhythmias. This would suggest that, at least in the present model, stimulation of α_{1A}-adrenoceptors by endogenous catecholamines released during ischaemia is not a primary modulator of the severity of reperfusion-induced arrhythmias.

Acknowledgements

This project was supported in part by the St Thomas' Hospital Heart Research Trust (STRUTH) and The David and Frederick Barclay Foundation. Masahiro Yasutake is an International Research Fellow from the Nippon Medical School, Tokyo, Japan. Metin Avkiran is a British Heart Foundation (Basic Science) Lecturer.

References

1. Sheridan DJ, Penkoske PA, Sobel BE, Corr PB: Alpha adrenergic contributions to dysrhythmia during myocardial ischemia and reperfusion in cats. J Clin Invest 65: 161–171, 1980
2. Culling W, Penny WJ, Cunliffe G, Flores NA, Sheridan DJ: Arrhythmogenic and electrophysiological effects of alpha adrenoceptor simulation during myocardial ischaemia and reperfusion. J Mol Cell Cardiol 19: 251–258, 1987
3. Flores NA, Sheridan DJ: Electrophysiological and antiarrhythmic effects of UK 52,046-27 during ischaemia and reperfusion in the guinea-pig heart. Br J Pharmacol 96: 670–674, 1989
4. Uprichard AGC, Harron DWG, Wilson R, Shanks RG: Effects of the myocardial-selective α_{1A}-adrenoceptor antagonist UK-52046 and atenolol, alone and in combination, on experimental cardiac arrhythmias in dogs. Br J Pharmacol 95: 1241–1254, 1988
5. Schwartz PJ, Vanoli E, Zaza A, Zuanetti G: The effect of antiarrhythmic drugs on life-threatening arrhythmias induced by the interaction between acute myocardial ischemia and sympathetic hyperactivity. Am Heart J 109: 937–948, 1985
6. Thandroyen FT, Worthington MG, Higginson LM, Opie LH: The effect of alpha- and beta-adrenoceptor antagonist agents on reperfusion ventricular fibrillation and metabolic status in the isolated perfused rat heart. J Am Coll Cardiol 1: 1056–1066, 1983
7. Wilber DJ, Lynch JJ, Montgomery DG, Lucchesi BR: α-Adrenergic influences in canine ischemic sudden death: effects of α_1-adrenoceptor blockade with prazosin. J Cardiovasc Pharmacol 10: 96–106, 1987
8. Minneman KP: α_1-Adrenergic receptor subtypes, inositol phosphates, and sources of cell Ca^{2+}. Pharmacol Rev 40: 87–119, 1988
9. Brittain-Valenti K, Danilo P, Rosen MR: Moduladon of reperfusion-induced arrhythmias by α_1-adrenergic receptor subtypes in the cat. Circulation 84: II-494 (abstract), 1991
10. Molina-Viamonte V, Anyukhovsky EP, Rosen MR: An α_1-adrenergic receptor subtype is responsible for delayed after depolarizadons and triggered activity during simulated ischemia and reperfusion of isolated canine Purkinje fibers. Circulation 84: 1732–1740, 1991
11. Yasutake M, Avkiran M: Exacerbation of reperfusion arrhythmias by α_1 adrenergic stimulation: a potential role for receptor-mediated activation of sarcolemmal sodium-hydrogen exchange. Cardiovasc Res 29: 222–230, 1995
12. Schömig A: Catecholamines in myocardial ischemia: systemic and cardiovascular release. Circulation 82 (Suppl II): II-13-II-22, 1990
13. Avkiran M, Curtis MJ: Independent dual perfusion of left and right coronary arteries in isolated rat hearts. Am J Physiol 261: H2082-H2090, 1991
14. Yasutake M, Ibuki C, Hearse DJ, Avkiran M: Na^+/H^+ exchange and reperfusion arrhythmias: protection by intracoronary infusion of a novel inhibitor. Am J Physiol 267: H2430–H2440, 1994
15. Walker MJA, Curtis MJ, Hearse DJ, Campbell RWF, Janse MJ, Yellon DM, Cobbe SM, Coker SJ, Harness JB, Harron DWG, Higgins AJ, Julian DG, Lab JJ, Manning AS, Northover BJ, Parratt JR, Riemersma RA, Riva E, Russell DC, Sheridan DJ, Winslow E, Woodward B: The Lambeth conventions: guidelines for the study of arrhythmias in ischaemia, infarction, and reperfusion. Cardiovasc Res 22: 447–455, 1988
16. Ridley PD, Curtis MJ: Anion manipulation: a new antiarrhythmic approach; action of substitution of chloride with nitrate on ischemia- and reperfusion-induced ventricular fibrillation and contractile function. Circ Res 70: 617–632, 1992
17. Avkiran M, Ibuki C: Reperfusion-induced arrhythmias: a role for washout of extracellular protons? Circ Res 71: 1429–1440, 1992
18. Ibuki C, Hearse DJ, Avkiran M: Mechanisms of the antifibrillatory effect of acidic reperfusion: role of perfusate bicarbonate concentration. Am J Physiol 264: H783–H790, 1993
19. Knowlton KU, Michel MC, Itani M, Shubeita HE, Ishihara K, Brown JH, Chien KR: The α_{1A}-adrenergic receptor subtype mediates biochemical, molecular, and morphologic features of cultured myocardial cell hypertrophy. J Biol Chem 268: 15374–15380, 1993
20. Schwinn DA, Lomasney JW: Pharmacologic characterization of cloned α_1-adrenoceptor subtypes: selective antagonists suggest the existence of a fourth subtype. Eur J Pharmacol 227: 433-436, 1992
21. Corr PB, Yamada KA, DaTorre SD: Modulation of α-adrenergic receptors and their intracellular coupling in the ischemic heart. Basic Res Cardiol 85: 31-45, 1990
22. Schömig A, Haass M, Richardt G: Catecholamine release and arrhythmias in acute myocardial ischaemia. Eur Heart J 12 (Suppl. F): 38–47, 1991
23. Sharma AD, Saffitz JE, Lee BI, Sobel BE: Alpha adrenergic-mediated

accumulation of calcium in reperfused myocardium. J Clin Invest 72: 802–818, 1983

24. Opie LH, Coetzee WA: Role of calcium ions in reperfusion arrhythmias: relevance to pharmacologic intervention. Cardiovasc Drugs Ther 2: 623–636, 1988
25. Priori SG, Yamada KA, Corr PB: Influence of hypoxia on adrenergic moduladon of triggered activity in isolated adult canine myocytes. Circulation 83: 248–259, 1987
26. Schömig A, Fischer S, Kurz T, Richardt G, Schömig E: Nonexocytotic release of endogenous noradrenaline in the ischemic and anoxic rat heart: mechanism and metabolic requirements. Circ Res 60: 194–205, 1987
27. Endoh M, Takanashi M, Norota I: Role of α_{1A}-adrenoceptor subtype in production of the positive inotropic effect mediated via myocardial α_1-adrenoceptors in the rabbit papillary muscle: influence of selective α_{1A}-subtype antagonists WB4101 and 5-methylurapidil. Naunyn Schmiedebergs Arch Pharmacol 345: 578–585, 1992
28. Lee JH, Steinberg SF, Rosen MR: A WB4101-sensitive α_1-adrenergic receptor subtype modulates repolarization in canine Purkinje fibers. J Pharmacol Exp Ther 258: 681–687, 1991
29. Greengrass PM, Russell MJ, Wyllie MG: A novel α_1-adrenoceptor antagonist. Br J Pharmacol 104: 321p. (abstract), 1991
30. Steinfath M, Chen YY, Lavicky J, Magnussen O, Nose M, Rosswag S, Schmitz W, Scholz H: Cardiac α_1-adrenoceptor densities in different mammalian species. Br J Pharmacol 107: 185–188, 1992

Molecular and Cellular Biochemistry **147**: 181–185, 1995.

The role of catecholamines on intercellular coupling, myocardial cell synchronization and self ventricular defibrillation

Mordechai Manoach, Dalia Varon and Mordechai Erez
Sackler School of Medicine, Department of Physiology and Pharmacology, Tel-Avic University, Tel-Aviv, Israel

Abstract

Ventricular fibrillation (VF) is one of the most life threatening events. Although in humans VF is generally sustained (SVF) requiring artificial defibrillation, in various mammals and in some cases in humans VF terminates by itself, reverting spontaneously into sinus rhythm. Since VF is one of the main causes of sudden death, one of the important clinical problems today is if and how we can transform the fatal SVF into a self limited transient one (TVF).

From electrophysiological studies carried out on anaesthetized open chest animals, we have found that TVF requires a high degree of intercellular coupling and synchronization.

Cardiac myocytes are electrically coupled with adjacent cells. The intercellular coupling is a focus of low electrical resistance which allows rapid transmission of electrical impulses between cells. Any decrease in intercellular coupling decreases the ability of the heart for self defibrillation. The cell-to-cell coupling decreases with age, ischemia, VF and variations in physiological conditions probably due to an increase in intercellular resistance (Ri), widening in the internexal gaps, decrease in electrotonic space constant (λ) etc. All of these factors are known to be affected by intracellular concentration of free Ca^{++} ($[Ca^{++}]$).

On the basis of studies carried out on various mammals at different ages, we hypothesized that the ability of the heart to defibrillate depends on the cardiac catecholamine level [CA], during VF. This hypothesis is supported by the facts, known from the literature, that increase in [CA] decreases intracellular free Ca^{++} concentration, decreases Ri and increases λ. By these effects, increase in [CA] enhances intercellular coupling and intercellular synchronization, and thereby, according to our hypothesis, leads to spontaneous ventricular defibrillation – TVF.

During VF the sympathetic activity is enhanced but in some cases the [CA] does not reach the level needed for TVF. In order to help the heart in its effort to elevate the [CA] during VF, we proposed to treat these cases with drugs which inhibit the reuptake of [CA]. The facts that administration of [CA] reuptake inhibitors, before the induction of VF, and/or intracoronary infusion of adrenaline, during VF, transforms SVF into TVF, emphasized the validity of our hypothesis. (Mol Cell Biochem **147**: 181–185, 1995)

Key words: catecholamine, cardiac intercellular synchronization, intercellular coupling, antiarrhythmic compounds, catecholamine reuptake inhibitors, self ventricular defibrillation

Ventricular fibrillation (VF) is one of the most life threatening events in clinic. Once it appears death is imminent unless defibrillation is applied within minutes. Ventricular fibrillation has also been observed in experimental animals. It has been shown that in certain species (rats and mice) VF terminates by itself and reverts spontaneously into sinus rhythm [1] while in dogs self defibrillation has not been observed [2, 3]. The ability of the heart to defibrillate spontaneously has been related to its ventricular muscle mass [4–6] i.e. small hearts can defibrillate spontaneously since they do not contain the minimal number of cells necessary to maintain fibrillation [7]. Following this concept it was suggested that human heart, with a relatively large muscle mass, cannot defibrillate spontaneously and that electrical defibrillation is the only method that

Address for offprints: M. Manoach, Sackler School of Medicine, Department of Physiology and Pharmacology, Tel-Aviv University, Tel-Aviv, Israel can terminate VF in humans. Since it was suggested that no drug treatment can save human life, once VF starts, many antiarrhythmic drugs have been developed in order to prevent initiation of VF by decreasing the incidence of ventricular premature beats and/or ventricular arrhythmias which can lead to VF.

Recent publications questioned the relevance of the size of cardiac mass to the ability of the heart to defibrillate by itself [8]. They indicate that self terminating VF appears also in relatively large muscle mass [9] and even in humans [10–14], while certain small hearts, like those of avians [15] and old rabbits and guinea pigs [16] exhibited SVF. In 1949 Shwartz [10] described self ventricular defibrillation in humans. He named this 'unexpected phenomenon' –Transient VF (TVF). During the last decade, many clinical publications have reported the existence of spontaneous termination of VF like arrhythmias, which had been called by several names, according to their typical ECG patterns [17–20]. In 1966, Bacaner [21] showed that administration of bretylium can transfer SVF into TVF in anaesthetized dogs. This 'chemical ventricular defibrillation' has also been described in humans [22].

The fact that the goal of elimination of VF by antiarrhythmic drugs, has not been achieved, that TVF exists in humans and that drugs can transform SVF into TVF, led us to search for the mechanisms involved in TVF looking for a new approach of antifibrillatory treatment namely 'drug induced spontaneous ventricular defibrillation'.

In order to achieve this goal we performemd a series of experiments carried out on various species at different ages [8, 9, 23, 24]. The results obtained in these experiments indicate that:

1. There is no significant difference between the cardiac muscle mass of animals that exhibited TVF and those that respond with SVF.
2. The ability to defibrillate spontaneously is the normal feature of young mammals.
3. This ability decreases with age.
4. The age span in which TVF appears varies among species i.e. it is very short in dogs and prolonged in rats.

Searching for a common factor that can explain these intra and inter species differences, we hypothesized that TVF can be related to the type of cardiac autoregulation [8, 9, 25] i.e. TVF appears in animals with predominantly sympathetic autoregulation while SVF appears in animals with predominant vagal autoregulation. The intraspecies (age dependent) variations between TVF and SVF could be explained by the fact that the cardiac autoregulation in young mammals is dominantly sympathetic and turns to a vagal predominance with age. The inter species variations (dogs vs rats) could be explained by the facts that the autoregulation of the dog heart (even in very young dogs) is predominantly vagal while this of the rat heart (even very old one) is sympathetic predominant. According to these facts we suggested that TVF requires high cardiac catecholamine level and low cholinergic one [8, 9, 23, 24].

The validity of this hypothesis is emphasized by our findings that administration of either beta adrenergic blockers (eg. propranolol or pindolol) or a parasympathomimetic agonist (eg. acetylcholine or metacholine) in mammals exhibiting TVF, prolongs the duration of TVF and even transforms it into SVF [8, 23, 24, 26].

On the basis of these age and drug effects on the type of VF, we studied the mechanisms involved in SVF and TVF, in the same animal and the same physiological conditions [8, 9, 24]. The comparative studies were carried out in anaesthetized open chest cats of both sexes. The results showed that: In TVF, the cells of both ventricles exhibit a relatively slow rate and quite well-synchronized electrical fibrillating activity, while in SVF the ventricles fibrillate at a higher rate and in a less synchronized manner [27, 28]. Spontaneous defibrillation takes place when almost all myocardial cells are simultaneously in the refractory period [29]. TVF is manifested when the fibrillating heart exhibits 'coarse' mechanical fibrillating movements, when large parts of the ventricular mass act together [29, 30].

'Synchronized fibrillation' means that electrical signals are spread quite simultaneously throughout the cells of both ventricles in a manner that culminates in bringing the myocardial cells to act in phase [18, 28]. This can occur only in hearts with good intercellular communication. The ventricular cells synchronization, therefore, might be related mainly to the propagation velocity of the electrical synchronizing signal between the ventricular cells. This synchronization can be altered by changing the membrane properties of the cells, the intercellular coupling resistance between cells and the space constant of electrotonic decay (λ) [31–34].

The intercellular conductivity seems to depend mainly on the intercellular axial resistance in the direction of propagation (R_i), and non-junctional sarcoplasmatic membrane resistance (R_m) [35–37]. The influence of R_i and R_m on the cellular coupling are in opposite directions [35]: a decrease in R_i increases the intercellular conductivity and thereby intercellular coupling [36], while a drastic fall in the resistivity of R_m can produce cell decoupling [37].

It has been shown that in normally coupled cells, the action potential transfer occurs with no measurable delay [38], while a progressive increase in R_i can lead to a conduction delay, and finally, to a sudden failure of propagation [39]. Since R_i is a sum of cytoplasmatic (R_{cyt}) and junctional (R_j) resistances [38, 40, 41], changes in R_i reflect variations in one or both parameters. Today it is generally accepted that R_j is the major regulating factor of the intercellular coupling [42], and action potential propagation velocity depends on the cell-to-cell connections [34].

VF like any high-rate myocardial activity, increases intercellular Ca^{++} concentration ($[Ca^{++}]_i$) [34, 41, 43], decreases intercellular conductivity and alters cellular membrane properties [32, 37, 39, 44], that in turn increases intercellular separation and uncoupling [42, 45, 46]. On the basis of the studies carried out on mammals of various species and ages [8, 9, 23, 24], we hypothetized that self ventricular defibrillation requires, during VF, a high cardiac adrenergic level and a low cholinergic one. Could high level of catecholamine increase intercellular coupling and synchronization, and could it prevent the deteriorating effect of VF? According to the literature, catecholamine administration shortens ventricular conduction time [35], enhances the electrical coupling between cardiac cells [35], decreases intercellular resistance and increases the space constant of electrotonic decay (λ) even in hypoxia [32], and increases gap junction conductance [47], all of which increase intercellular synchronization. These effects of catecholamines can be related to their effects on $[cAMP]_i$, Na^+/K^+ pump and Na^+/Ca^{++} exchange [37, 43, 46]. Administration of achetylcholine decreases R_m and λ, and thereby quickly abolishes the intercellular electrical coupling [32, 37].

Since VF increases the free $[Ca^{++}]_i$ and thereby decreases contractile force and intercellular coupling, increase in [CA] during VF decreases the VF induced intercellular uncoupling and decline in contractile force at least in part by decreasing the free $[Ca^{++}]_i$. Through these effects CA prevents also the intercellular space widening and increases the intercellular coupling. De Mello [37] suggests that NE that increases [cAMP]i can induce a quick (within second) increase in junctional conductance probably by acting directly on gap junctional molecules.

If increase in sympathetic activity during the onset of VF is a common phenomena and elevation of [CA] has a potent defibrillating effect, why is spontaneous defibrillation in humans so rare?

The answer lies in the following explanation: The [CA] is the result of the amount released from sympathetic nerve endings, plus the amount that arrives with the blood circulation, minus the catecholamine reuptake and [CA] enzymatic catabolism. If the level of catecholamines is enhanced (either by endogeneous secretion or due to i.v. administration) the elevated local adrenergic level decreases due to its reuptake and/or enzymatic catabolism.

The equation that determines the extraneuronal net cardiac catecholamine level under any physiological conditions is:

[CA] = [released] + [circulating] – [reuptake] – [catabolised] [51]

It should be emphasized that the reuptake of catecholmines into the sympathetic nerve endings is fast and efficient. According to literature [52, 53] the reuptake of noradrenaline is about 70% and for adrenaline about 50%. How do these factors that determine [CA] change during VF and how do they affect the type of VF?

During VF, the sympathetic nerve activity is enhanced. Since during VF there is no blood circulation, and catecholamines (whether secreted from the adrenal medulla or introduced by i.v. administration) can not reach the cardiac muscles, the local cardiac level of catecholamines can be increased only by the noradrenaline released from the adrenergic nerve endings.

According to our hypothesis [51], if [CA] is above a 'threshold' level required for spontaneous defibrillation, VF terminates by itself, but if [CA] is below this level – VF is sustained.

When the adrenergic autoregulating activity is high (as in rats or young mammals) or the reuptake is prevented by certain drugs, the elevated level of cathecholamines during VF remains sufficiently high and the ability of the heart to defibrillate is enhanced. However, in elderly, since their sympathetic control and catecholamine efflux have been decreased while their extraneuronal reuptake has been increased, and beta adrenergic receptors' reactivity was decreased [54–57] – the net elevation of catecholamine is not sufficient to transform VF into TVF. We hypothetised, therefore, that in elderly [CA] during VF is not high enough and therefore, for transforming in elderly SVF into TVF, the net [CA] during VF should be increased [25, 58, 59]. Increase of the catecholamine level, during VF, could be obtained either by a drug-induced inhibition of catecholamine reuptake and/or chemical catabolism, or by direct intra-coronary injection of CA during VF. Administration of catecholamines, *per se*, before the initiation of VF, can not change the type of VF unless the reuptake is inhibited and the local level of catecholamines in heart muscle is kept high.

In order to examine this hypothesis, we treated mammals, which exhibited SVF, with various compounds known to inhibit catecholamine reuptake i.e. several tricyclic antidepressants, phenothiazines, cocaine and amphetamine [25, 59]. The results indicated that VF induced following the treatment, terminated by itself (TVF). The defibrillating activity of the compounds has been directly related to their potency to inhibit catecholamine reuptake (IC_{50}) [59].

Looking for a more direct examination of the role of [CA] in the transformation of SVF into TVF, high dose of adrenaline has been injected intracoronarly, during SVF [59, 60]. The results showed that intracoronary injection of 0.5 mg adrenaline in cats, during VF, transforms SVF into TVF. Similar preliminary result has been obtained clinically during post-operative cardiac resusitation [60].

All these results clearly support our hypothesis, and indicate that increase in cardiac catecholamine level, during VF, can serve as a method for enhancement of the ability of the heart to defibrillate by itself.

Is it possible, that catecholamines, or high sympathetic activity, considered to be highly arrhythmogenic [61], might have such an important cardio-protective effect and if so can they, *per se* change the type of VF? The beneficial effect of catecholamines in the defibrillating process has been observed in clinics for a long time. Adrenaline is currently used (intracardially) when VF exhibits a 'fine' type of VF and the electrical defibrillation is not effective [62, 63]. The administration of adrenaline changes the 'fine' type of VF into a 'course' one and thereby increases the efficiency of the electrical defibrillation. Moreover, isoprenaline, is the prescribed drug of choice in Torsade de Pointes [17]. Anyhow, in our proposed antifibrillating treatment we do not suggest to elevate [CA] but to support the heart to increase [CA] during VF, when it tries to do so.

Conclusion

It was found that the predominant factor which enhances the ability of the heart to defibrillate spontaneously is sufficient high cathecholamine level during VF. It could be achieved either by enhanced local sympathetic activity and/or by inhibition of cathecholamine reuptake or its catabolism, on the background of low parasympathomimetic level. This elevation of cardiac catecholamine level during VF, increases the intercellular electrical coupling and thereby increases the intercellular synchronization.

The establishment of the cardio-protective effects of catecholamines and the related mechanisms involved in self ventricular defibrillation, may unveil a new therapeutic approach leading to the development of a new class of antiarrhythmic defibrillating drugs.

References

1. Wigers CJ: Studies of ventricular fibrillation caused by electrical shock. Am Heart J 5: 351–365, 1929
2. Wigers CJ: The mechanism and nature of ventricular fibrillation. Am Heart J 20: 399–412, 1940
3. Bacaner M: Experimental and clinical effects of bretylium tosylate on ventricular fibrillation, arrhythmias and heart block. Geriatrics 26: 132–148, 1971
4. Zipes DP, Fisher J, King RM, Nicoll AB, Jolly WW: Termination of ventricular fibrillation in dogs by depolarizating a critical amount of myocardium. Am J Cardiol 36: 37–44, 1975
5. Damiano RJ, Asano T, Smith PK, Cox JL: Effect of the right ventricular isolation procedure on ventricular vulnerability to fibrillation. J Am Coll Cardiol 15: 730–736, 1990
6. Watanabe Y: Antifibrillatory action of several antiarrhythmic agents. In: S. Hayase and S. Murao (eds). Cardiology. Proceedings of 8th World Congress of Cardiology. Excerpta Medica, Amsterdam, 1979, pp 924–928
7. West TC, Landa JF: Minimal mass required for induction of a sustained arrhythmias in isolated atrial segments. Am J Physiol 202: 232–236, 1962
8. Manoach M, Varon D, Neuman M, Netz H: Spontaneous termination and initiation of ventricular fibrillation as a function of heart size, age, autonomic autoregulation and drugs: A comparative study on different species of different age. Heart and Vessels 2 (Suppl): 56–68, 1987
9. Manoach M, Netz H, Erez M, Weinstock M: Ventricular self defibrillation in mammals: age and drug dependence. Age and Ageing 9: 112–116, 1980
10. Schwartz SP, Orloff J, Fox C: Transient ventricular fibrillation. Am Heart J 37: 21–35, 1949
11. Goble AJ: Paroxysmal ventricular fibrillation with spontaneous reversion to sinus rhythm. Brt Heart J 27: 62–68, 1965
12. Spielman SR, Farshidi A, Horowitz LN, Josephson ME: Ventricular fibrillation during programmed ventricular stimulation incidence and clinical implications. Am J Cardiol 42: 913–918, 1978
13. Patt MV, Podrid PJ, Friedman PL, Lown B: Spontaneous reversion of ventricular fibrillation. Am Heart J 115: 919–923, 1988
14. Clayton R, Higham D, Murray A, Campell R: Self terminating ventricular fibrillation. (Abs) J Am Coll Cardiol 19: 265A, 1992
15. Kobrin VI, Kudinova ED: Electrical activity of intact myocardial cells in different animals during normal activity and ventricular fibrillation. In: Proceedings of the international symposium on problems in comparative electrophysiology. Syktivkar, USSR, 1979, p 36 (abst)
16. Dessertenne F: La tachycardie ventriculaire a deux foyers opposes variables. Arch Mal Coeur 59: 263–272, 1966
17. Krikler DM, Curry PLV: Torsade de pointes, an atypical ventricular tachycardia. Br Heart J 38: 117–120, 1976
18. Ranquin R, Parizel G: Ventricular fibrillo-flutter. Torsade de pointes; an established electrocardiographic and clinical entity. Angiology 28: 115–118, 1977
19. Sclarovsky S, Strasberg B, Levin R, Agmon J: Polymorphous ventricular tachycardia: clinical features and treatment. Am J Cardiol 44: 339–344, 1979
20. Selzer A, Wray HW: Quinidine syncope. Paroxysmal ventricular fibrillation occuring during treatment of chronic atrial arrhythmias. Cir Res 25: 17–26, 1964
21. Bacaner M: Bretylium tosylate for suppression of induced ventricular fibrillation. Am J Cardiol 17: 528–534, 1966
22. Sanna G, Archidiacono R: Chemical ventricular defibrillation of the human heart with bretylium tosylate. Am J Cardiol 32: 982–986, 1973
23. Manoach M, Netz H, Varon D, Amitzur G, Weinstock M, Kauli N, Assael M: Factors influencing spontaneous initiation and termination of ventricular fibrillation. Jap Heart J 27: 365–375, 1986
24. Manoach M, Beker B, Erez M, Varon D, Netz H: Spontaneous termination of electrically induced ventricular fibrillation. In: P.W. Mecfarlane (ed.). Progress in electrocardiology. Pitman Med Pub, England, 1979, pp 361–365
25. Manoach M, Erez M, Varon D: Editorial review. Properties required for self ventricular fibrillation: Influence of age and drugs. Cardiol in the Elderly 1: 337–344, 1993
26. Amitzur G, Manoach M, Weinstock M: The influence of cardiac cholinergic activation on the induction and maintenance of ventricular fibrillation. Basic Res Cardiol 79: 690–697, 1984
27. Manoach M, Netz H, Beker B, Kauli N: Vector-cardiographycal discrimination between sustained and transient ventricular fibrillation. In: F. dePadua and P.W. Macfarlane (eds). New frontiers of electrocardiology. Research studies press, J Wiley and Sons, England, 1981, pp 139–143
28. Chen-Menaker A, Einav S, Manoach M: Signal processing of the ECG during ventricular fibrillation. In: Proceedings IEEE Conv, Haifa, Israel, 1985, 2.4,6: 1–3
29. Manoach M, Wyatt RF: Intracellular myocardial recordings *in vivo* during sustained and transient ventricular fibrillation. In: P. d'Alche (ed.). Advances in electrocardiology. University of Caen, France, 1985, pp 401–403

30. Manoach M, Kauli N, Pinchasov A, Beker B, Netz H, Varon D: Electrical and mechanical differences between sustained and transient ventricular fibrillation (abst). Israel J Med Sc 16: 225, 1980
31. Spach MS, Miller WT, Geselowitz DB, Barr RC, Kootsey JM, Johnson EA: The discontinuous nature of propagation in normal canine cardiac muscle. Cir Res 48: 39–54, 1981
32. Bredikis J, Bukaskas F, Veteikis R: Decreased intercellular coupling after prolonged rapid stimulation in rabbit atrial muscle. Cir Res 49: 815–820, 1981
33. Arnsdorf ME: Membrane factors in arrhythmogenesis: Concepts and definitions. Prog Cardiovas Dis 19: 413–429, 1977
34. Spach MS, Kootsey JM, Sloan JD: Active modulation of electrical coupling between cardiac cells of the dog. Cir Res 51: 347–362, 1982
35. De Mello WC: Increased spread of electrotonic potentials during diastolic depolarization in cardiac muscle. J Mol Cell Cardiol 18: 23–29, 1986
36. Kukushkin NI, Bukauskas FF, Sakson ME, Nasokoya VV: Anisotropy of stationary velocity and delaye of extrasystole waves in the dog heart. Biofizika 20: 687–692, 1975
37. De Mello WC: Cell-to-cell coupling assayed by means of electrical measurements. Experientia 43: 1075–1079, 1987
38. Weingart R, Maurer P: Cell-to-cell coupling studied in isolated ventricular cell pairs. Experientia 43: 1091–10994, 1987
39. Liberman M, Kootsey JM, Johnson EA, Sawanobori T: Slow conduction in cardiac muscle. Biophys J 13: 37–55, 1973
40. Delese J: Cell to cell communication in the heart: structure function correlations. Experientia 43: 1068–1075, 1987
41. Wojtezak J: Contractures and increase in internal longitudinal resistance of cow ventricular muscle induced by hypoxia. Cir Res 44: 88–95, 1979
42. Weingart R: The action of ouabain on intercellular coupling and conduction velocity in mammalian ventricular muscle. J Phisiol (Lond) 264: 341–365, 1977
43. Chen CM, Gettes LS: Combined effect of rate, membrane potential and drug on maximal rate of rise (Vmax) of action-potential upstroke of guinea pig papillary muscle. Cir Res 38: 464–469, 1976
44. Sano T: Mechanism of cardiac fibrillation. Pharmacol Ther 2: 407–513, 1976
45. De Mello WC: Effect of intracellular injection of calcium and strontium on cell communication in heart. J Physiol (Lond) 250: 231–245, 1975
46. De Mello WC: Influence of the sodium pump on intercellular communication in heart fibers: effect of intracellular injection of sodium ion on electrical coupling. J Physiol (Lond) 263: 171–197, 1976
47. Burt JM, Spray DC: Adrenergic control of gap junction conductance in cardiac myocytes (abst). Circulation 78: Suppl II, 258, 1988
48. Lamont SV, Barritt GJ: Effects of Adrenaline on a compartment of slowly-exchangeable Ca in the perfused rat heart. Cardiovas Research 17: 88–95, 1982
49. Entman ML, Levey GS, Epstein SE: Mechanism of action of epinephrine and glucagon on the canine heart. Evidence for increase in sarcotubular calcium stores mediated by cyclic 3′,5′ AMP. Cir Res 25: 429–438, 1969
50. Li T, Vassalle M: The negative inotropic effect of calcium overload in cardiac Purkije fibers. J Mol Cell Cardiol 16: 65–77, 1984
51. Manoach M, Varon D, Erez M: A self-protecting servo-mechanism involved in spontaneous ventricular defibrillation. J Basic and Clin Physiol and Pharmacol 4: 273–280, 1993
52. Eisenhofer G, Esler MD, Cox HS, Meredith IT, Jennings GL, Brush JE Jr, Goldstein DS: Differences in the neronal removal of circulating epinephrine and norepinephrine. J Clin Endocrinol Metab 70: 1710–1720, 1990
53. Goldstein DS, Brush JE Jr, Eisenhofer G, Stull R, Elser M: *In vivo* measurement of neuronal uptake of norepinephrine in the human heart. Circulation 78: 41–48, 1988
54. Spadary RC, De-Moraes S: Aging and rat pacemaker sensitivity to beta adrenoceptor agonists. Braz J Med Biol Res 20: 591–594, 1987
55. Daly RN, Goldberg PB, Roberts J: The effect of age on presynaptic alpha$_2$ adrenoceptor autoregulation of norepinephrine release. J Gerontol 44: 859–866, 1989
56. Fleisch JH: Age related decrease in both adrenoceptor activity of the cardiovascular system. Treads Pharmacol Sci 2: 337–339, 1981
57. Abrass IB, Davis JL, Searpace PJ: Isoproterenol responsiveness and myocardial beta adrenergic receptors in young and old rats. J Gerentol 37: 156–160, 1982
58. Manoach M, Varon D, Neuman M, Erez M: Minireview. The cardioprotective features of tricyclic antidepressants. Gen Pharmac 20: 269–275, 1988
59. Manoach M, Erez M, Wozner D, Varon D: Ventricular defibrillating properties of catecholamine uptake inhibitors. Life Sc 51: PL 159–164, 1992
60. Manoach M, Tager S, Erez M, Varon D, Vaugham Williams M: The defibrillating effect of high cardiac catecholamine level. (Abst) J Moll Cell Cardiol 24: Supp V, S69, 1992
61. Lombardy F, Verrier RL, Lown B: Relationship between sympathetic neural activity, coronary dynamics and vulnerability to ventricular fibrillation during myocardial ischemia and reperfision. Am Heart J 105: 958–965, 1983
62. Jones DL, Klein GL: Ventricular fibrillation: The importance to be course. J Electrocardiol 17: 393–399, 1984
63. Wagner GS, McIntosh HD: The use of drugs in achieving successful DC cardioversion. Prog Cardiovasc Dis 11: 431–442, 1969

Molecular and Cellular Biochemistry **147**: 187–192, 1995.

Inhibition of (Na/K)-ATPase by electrophilic substances: Functional implications

Albert Breier[1], Attila Ziegelhöffer[2], Tania Stankovičová[2a], Peter Dočolomanský[1], Peter Gemeiner[3] and AlenaVrbanová[1]

[1]Institute of Molecular Physiology and Genetics, Slovak Academy of Sciences; [2]Institute for Heart Research, Slovak Academy of Sciences; [3]Institute of Chemistry, Slovak Academy of Sciences, 842 38 Bratislava, Slovak Republic

Abstract

The effect of electrophilic substances: p-bromophenylisothiocyanate (PBITC); fluoresceinisothiocyanate (FITC); [4-isothiocyanatophenyl-(6-thioureidohexyl)-carbamoylmethyl]-ATP (ATPITC); 2,4,6-trinitrobezenesulfonic acid (TNBS); 1-(5-nitro-2-furyl)-2-phenylsulfonyl-2-furylcarbonyl ethylene (FE1); 1-(5-phenylsulfonyl-2-furyl)-2-phenylsulfonyl-2-furylcarbonyl ethylene (FE2) and 1-(5-phenylsulfonyl-2-furyl)-2-phenylsulfonyl-2-tienocarbonyl ethylene (FE3) on the sarcolemmal (Na/K)-ATPase isolated from guinea-pig hearts was studied. FITC and PBITC were found to inhibit competitively the activation of (Na/K)-ATPase by ATP. Being for the enzyme inhibitor and substrate at the same time ATPITC does not offered clear kinetic behavior. However, the activation of (Na/K)-ATPase by sodium and potassium ions was inhibited non-competitively by all three isothiocyanates. These data indicated that isothiocyanates may interact predominantly in the ATP-binding site of the enzyme molecule. In contrary to isothiocyanates TNBS and FE1 (FE2 and FE3 were ineffective) inhibited the activation of (Na/K)-ATPase by ATP non-competitively i.e., their interaction in the ATP-binding site seemed to be improbable. Nevertheless, TNBS and FE1 both manifested affinities to that moiety of (Na/K)-ATPase molecule which is binding potassium. More specific was the effect of FE1 that showed clearly competitive inhibition of potassium-stimulation of the enzyme activity. FE1 exerted also an ouabain-like effect on the mechanical activity of isolated perfused guinea-pig heart. This result indicates that FE1 seems to exert a selective inhibition of the (Na/K)-ATPase not only in vitro but also in integrated cardiac tissue. (Mol Cell Biochem **147**: 187–192, 1995)

Key words: (Na/K)-ATPase, electrophilic reagents, cation binding site, isolated perfused heart, 5-nitrofurylethylene

Abbreviations and symbols: PBITC – p-bromophenylisothiocyanate; FITC – fluoresceinisothiocyanate; ATPITC – [4-isothiocyanatophenyl-(6-thioureidohexyl)-carbamoylmethyl]-ATP; TNBS – 2,4,6-trinitrobenzenesulfonic acid; FE1 – 1-(5-nitro-2-furyl)-2-phenylsulfonyl-2-furylcarbonyl ethylene; FE2 – 1-(5-phenylsulfonyl-2-furyl)-2-phenylsulfonyl-2-furylcarbonyl ethylene; FE3 – 1-(5-phenylsulfonyl-2-furyl)-2-phenylsulfonyl-2-tienocarbonyl ethylene; DMSO – dimethylsufoxide

Introduction

Owing to its important role in membrane transport and excitation-contraction coupling Mg^{2+}-dependent, Na^{+}-, K^{+}- activated adenosinetriphosphate phosphohydrolase (E.C. 3.6.1.3.): (Na/K)-ATPase became the center of numerous studies. They have been dealing with (Na/K)-ATPase from the aspects of its biological function in health and disease, the role as receptor for cardiac glycosides, genetic control and expression as well as chemical constitution and topology of the enzyme molecule (for review see [1-4]. A combination of data originating from resolution of primary structure [5]

[a]*Present adress:* Comenius University, Faculty of Pharmacy, Department of Pharmacology and Toxicology, Bratislava, Slovak Republic
Address for offprints: A. Breier, Institute for Molecular Physiology and Genetics, Slovak Academy of Sciences, Dúbravská cesta 9, 842 33 Bratislava, Slovak Republic

and affinity labeling of the ATP-binding sites by means of isothiocyanates [6–9] enabled to reveal the localization of the isothiocyanate-sensitive ATP-binding sites on the molecule of (Na/K)-ATPase (near the Lys 501). However, in spite of their key importance for reactivity of (Na/K)-ATPase, less frequently were studied the features and chemical reactivity of the cation-binding sites of the enzyme.

In an earlier paper [10] we have predicted the presence of some nucleophilic aminoacid residues (particularly cysteine and lysine but also serine) in cation-binding sites of the (Na/K)-ATPase. The negative interference of 2,4,6-trinitrobenzenesulfonic acid (TNBS, a primary amino group-selective reagent [11]) with stimulation of the enzyme activity by potassium ions [12–14] is supporting the latter prediction. Accordingly, it may also be assumed that other electrophilic reagents such as isothiocyanates [15] or furylethylenes [16, 17] could also interact with the predicted nucleophilic aminoacid residues in the cation binding site of (Na/K)-ATPase. However, unfortunately, isothiocyanate-sensitive amino [69] or sulfhydryl groups [18, 19] have been demonstrated also in the ATP binding site of the enzyme.

The goal of the present study was to check whether the electrophilic reagents described above are capable to inhibit (Na/K)-ATPase by selective depression of its activation by sodium or potassium. To test the latter possibilities we have compared the inhibitory effect of these substances on the reaction kinetics of the heart sarcolemmal (Na/K)-ATPase.

Materials and methods

Materials

1-(5-nitro-2-furyl)-2-phenylsulfonyl-2-furylcarbonyl ethylene (FE1), 1-(5-phenyl-sulfonyl-2-furyl)-2-phenylsulfonyl-2-furylcarbonyl ethylene (FE2) and 1-(5-phenylsulfonyl-2-furyl)-2-phenylsulfonyl-2-tienocarbonyl ethylene (FE3), were synthesized in the Department of Organic Chemistry, Slovak Technical University in Bratislava by Lehnert condensation method [21]. [4-Isothiocyanatophenyl-(6-thioureidohexyl)-carbamoylmethyl]-ATP (ATPITC) was prepared in the laboratory of Dr. P. Gemeiner [22]. p-Bromophenylisothiocyanate (PBITC) was prepared from p-bromoaniline by thiophosgene procedure according to the method of Gemeiner and Drobnica [23]. All other chemicals were of analytical grade and were purchased from Sigma (USA) and Lachema (Czech Republic).

Preparation of the (Na/K)-ATPase

Sarcolemmal membrane fraction from guinea pig heart was isolated using the method of hypotonic shock combined with NaI treatment 13]. As determined by the activities of 5′-nucleotidase (marker for plasma membrane), succinic dehydrogenase, oligomycine-sensitive Mg^{2+}-ATPase (markers for mitochondrial membrane) and Ca^{2+}-ATPase (marker for sarcoplasmic reticulum membrane) present in the final sarcolemmal fraction, the latter has been contained to less than 3% by other sub-cellular membrane particles (not shown). The activity of (Na/K)-ATPase of final membrane fraction amounted 25 μmol P_iP/h.mg approximately.

Estimation of the (Na/K)-ATPase activity

(Na/K)-ATPase activity was determined as the difference between the amount of orthophosphate liberated by splitting of ATP (0.125–4.000 mmol/l) in the presence of NaCI (1.25–100.0 mmol/l), KCI (0.125–10.000 mmol/l) and MgC_2 (4 mmol/l), and in the presence of 4 mmol/l of $MgCl_2$ only. Orthophosphate was estimated according to Tausky and Shorr [24]. Protein content in the membrane fractions was determined according to Markwell *et al* [25]. Enzyme reaction was running for 10 min at 37°C, in 0.5 ml of 50 mmol/l imidazole-HCI buffer, pH = 7.0 in the presence of 30–40 mg of sarcolemmal membrane protein. It was started by addition of ATP (in the case of kinetic measurements substrate was added simultaneously with the inhibitor) and stopped by ice-cold trichloroacetic acid (12%). In some experiments inhibitors were pre-incubated with the membrane fraction in absence or presence of ATP and magnesium ions for 30 min at room temperature. The pre-incubation medium containing excess of the free inhibitor was then removed by centrifugation (10 min at 3000 G). After re-suspending the pellet in 50 mmol/l imidazole-HCI buffer (pH = 7.0) centrifugation was repeated. The final pellet was again re-suspended in the same buffer and used for estimation of the (Na/K)-ATPase activity. Furylethylenes and PBITC were dissolved in DMSO. The final concentration of solvent in the reaction medium did not exceed 0.5 % v/v. All other inhibitors were used dissolved in water. Parameters of enzyme kinetics were obtained by nonlinear regression of the Dixon and Michaelis-Menten relationships. The latter was equipped with the Hill cooperativity constant (n) in the case of stimulation of ATPase activity by sodium or potassium. The type of inhibition was verified using classical Dixon linearization.

Isolated perfused guinea-pig heart

Heparinized (200 IU i.v.) male guinea-pigs weighing 270–330 g were anaesthetized with diethyl ether. After cannulation of the aorta, the heart was excised and perfused at constant flow rate (10 ml/min) and 37°C using the Langendorff technique. Krebs-Henseleit (KH) solution containing in mmol/l: 118 NaCl, 15 $NaHCO_3$, 2.68 KCl, 1.66 $MgSO_4$,

2.0 $CaCl_2$, 1.18 $KH2PO_4$ and 11 glucose, pH = 7.40 ± 0.05, filtered through a 5 mm millipore filter and gassed with a mixture of 95% O_2 + 5% CO_2 served as perfusion medium. Left ventricular pressure, measured by means of an intraventricular balloon, as well as perfusion pressures were linked to pressure transducers (Gould P23, USA), monitored by LDP 186 (Tesla, Czech Republic) and registered by Chiracard 600 T (Chirana, Slovakia). Epicardial electrogram was recorded with the aid of a pair of platinum electrodes. After 15 min of stabilization perfusion with KH-medium in non-recirculating regime, hearts were switched over to 20 min. perfusion with KH solutions containing in addition either dimethyl sulfoxide (DMSO, 0.1 mmol/l) or FE1 (0.1 pmol/l) plus DMSO. Myocardial function was evaluated by monitoring the following variables: Left ventricular systolic pressure (LVP_S), left ventricular diastolic pressure (LVP_D), the difference between LVP_S and LVP_D (LVP_{S-D}) as a measure of mechanical activity, and the electrical activity of the heart (beats/min.). Coronary flow resistance was calculated as LVP_D/coronary flow (provided by the pump). Six consecutive measurements of the above variables were made during each perfusion, the total time interval being of 20 min. The effect of FE1 was evaluated as a difference between data obtained in the presence of FE1 plus DMSO and those in the presence of DMSO alone. Results were expressed as means ± S.E.M. Statistical significance between consecutive measurements in each type of experiment was checked by means of the F-test and between different experimental groups by Student's *t*-test.

Results and discussion

Among the substances tested the effect of isothiocyanates (PBITC, FITC and ATPITC) was confined to the ATP-binding site of the (Na/K)-ATPase. This was indicated also by the protective effects of Mg-ATP complex or magnesium ions itself against the inhibition of the enzyme by isothiocyanates (Table 1). Concerning the protective effect of magnesium it was shown [26] that the ions stabilize the ATP-binding sites of the (Na/K)-ATPase. Earlier studies localized the interaction of PBITC [18] and FITC [6–8] in the ATP-binding site of the (Na/K)-ATPase. In terms of enzyme kinetics, both inhibitors showed competitive type of inhibition (Table 2), characterized by considerable increase in the apparent Michaelis constants for ATP (K_{mATP}) without any changes in apparent maximal reaction velocity (V_{maxATP}). In contrary to PBITC and FITC, ATPITC was capable to serve not only as an inhibitor but also as a substrate for the (Na/K)-ATPase, particularly in the absence of ATP (Table 3). Therefore, the combined type of inhibition observed in this case (Table 2) is not in contradiction with the expected action of ATPITC in the ATP-binding locus of (Na/K)-ATPase. In addition, all three isothiocyanates were found to inhibit also the activation of (Na/K)-ATPase with sodium and potassium ions, but by noncompetitive manner only (Table 2). Hence, the expressed electrophilic characters of isothiocyanates [15] as well as their capability to compete with ATP, predestine them to react

Table 1. Effect of Mg^{2+} and ATP on inhibition of (Na/K)-ATPase activity by isothiocyanates.

Preincubation	(Na/K)-ATPase activity [µmol Pi/h.mg]	[% from the control]
control	25.6 ± 1. 9	
effect of PBITC		
10 ^{6} M PBITC	5.7 ± 2.3	22.3
10 ^{5} M PBITC	0.5 ± 0.1	1.9
10 ^{5} M PBITC + 10 ^{3} Mg^{2+}	3.7 ± 0.3	14.4
10 ^{5} M PBITC + 10 ^{3} Mg^{2+} + 10 ^{3} ATP	16.8 ± 1.6	65.6
effect of FITC		
10 ^{5} M FITC	4.6 ± 1.2	18.0
10 ^{4} M FITC	0.3 ± 0.1	1.2
10 ^{4} M FITC + 10 ^{3} Mg^{2+}	6.7 ± 1.0	26.2
10 ^{4} M FITC + 10 ^{3} Mg^{2+} + 10^{3} ATP	19.3 ± 1.4	75.4
effect of ATPITC		
10 ^{5} M ATPITC	3.2 ± 1.1	12.5
10 ^{4} M ATPITC	0.1 ± 0.1	0.4
10 ^{4} M ATPITC + 10 ^{3} Mg^{2+}	1.2 ± 0.6	4.6
10 ^{4} M ATPITC + 10 ^{3} Mg^{2+}+ 10 ^{3} ATP	6.5 ± 0.6	25.3

Mebrane fraction was preincubated with respective inhibitors in the presence or absence of Mg^{2+} and ATP for 30 min at room temperature. After pre-incubation the excess of inhibitors was removed as described in Material and methods.

Table 2. Effect of electrofilic substances on kinetic parameters of (Na/K)-ATPase

Inhibitor	V_{maxATP} [mmol P_i/h.mg]	K_{mATP} [mM]	V_{maxK} [mmol P_i/h.mg]	K_{mK} [mM]	n_K	V_{maxNa} [mmol P_i/h.mg]	K_{mNa} [mM]	n_{Na}
Control	27.2 ± 1.8	0.68 ± 0.03	26.7 ± 1.6	1.51 ± 0.09	1.51 ± 0.12	26.3 ± 1.3	16.4 ± 1.2	1.47 ± 0
10 ^{5} M PBITC	26.9 ± 1.4	2.31 ± 0.19	15.7 ± 1.2	1.60 ± 0.08	1.43 ± 0.11	14.9 ± 1.7	15.8 ± 1.4	1.53 ± 0
10 ^{4} M FITC	27.5 ± 1.9	2.81 ± 0.22	13.7 ± 1.4	1.47 ± 0.12	1.48 ± 0.13	13.1 ± 1.1	17.1 ± 1.1	1.41 ± 0
10 ^{4} M ATPITC	6.8 ± 0.5	0.12 ± 0.01	9.7 ± 0.5	1.52 ± 0.15	1.27 ± 0.12	10.2 ± 1.6	16.7 ± 1.8	1.51 ± 0
10 ^{4} M TNBS	12.9 ± 1.4	0.72 ± 0.06	18.5 ± 1.7	7.81 ± 0.64	1.21 ± 0.11	12.9 ± 0.9	17.8 ± 1.9	1.43 ± 0
10 ^{5} M FE1	18.5 ± 1.3	0.66 ± 0.02	26.5 ± 1.3	6.03 ± 0.57	1.03 ± 0.12	13.1 ± 1.2	15.1 ± 1.6	1.45 ± 0

Table 3. Influence of ATPITC on the activity of the (Na/K)-ATPase

Substrate	(Na/K)-ATPase activity
0.2 mM ATP	8.24 ± 0.32
0.1 mM ATPITC	0.92 ± 0.08
0.2 mM ATP + 0.1 mM ATPITC	5.36 ± 0.40
1 mM ATP	20.42 ± 1.90
1 mM ATP + 0.1 mM ATPITC	6.92 ± 0.57

selectively only with nucleophilic aminoacid residues located preferentially in ATP-binding site of the (Na/K)-ATPase. This is practically eliminating them as potential selective modulators of activation of the enzyme by sodium and potassium ions.

In contrast to isothiocyanates TNBS was inhibiting the activation of (Na/K)ATPase by potassium ions with combined kinetics (Table 2). This was characterized by expressed increase in the apparent Michaelis constant for potassium (K_{mK}) and by approximately 38% diminution in the apparent maximal reaction velocity for potassium (V_{maxK}). Activation of the (Na/K)-ATPase by increasing concentrations of ATP as well as sodium ions were modulated by TNBS non-competitively i.e., indirectly only. The above kinetic characteristics seem to localize the site of action of TNBS in or close to the potassium binding site of the (Na/K)-ATPase molecule. The selectivity of the latter process was already discussed in our earlier papers [10, 13, 14]. Nevertheless, the combined type of inhibition of (Na/K)-ATPase activation by potassium points to presence of some side effects of TNBS occurring in loci distant from the dominant target structure of inhibitor i.e., the potassium binding site of the enzyme.

Among furylethylenes tested only FE1 was able to inhibit the (Na/K)-ATPase in guinea pig heart sarcolemma to a considerable extent (not shown). The inhibitor proved to act competitively on activation of the enzyme by potassium ions and noncompetitively on its activation by sodium ions and by ATP (Table 2). The selective effect of FE1 on potassium-activation of (Na/K)-ATPase was subsequently studied by means of inhibition kinetics using both, direct and linearized Dixon plots (Fig. 1). The latter plots revealed unambiguously that FE1 inhibits: i) competitively the activation of (Na/K)-ATPase by potassium and ii) non-competitively the activation of the enzyme by both sodium ions and ATP. Similar results were obtained when we have studied the interaction of FE1 with a highly purified (Na/K)-ATPase preparation from dog kidney (not shown).

It is well recognized that ouabain is inhibiting (Na/K)-ATPase by altering its activation with potassium [27] and this represents the molecular principle of its positive inotropic effect (for review see [2]). Therefore, it was of interest to check whether FE1 that was acting *in vitro* in a similar mode as ouabain, may induce an ouabain-like effect on isolated perfused guinea-pig heart. Because of its hydrophobic char-

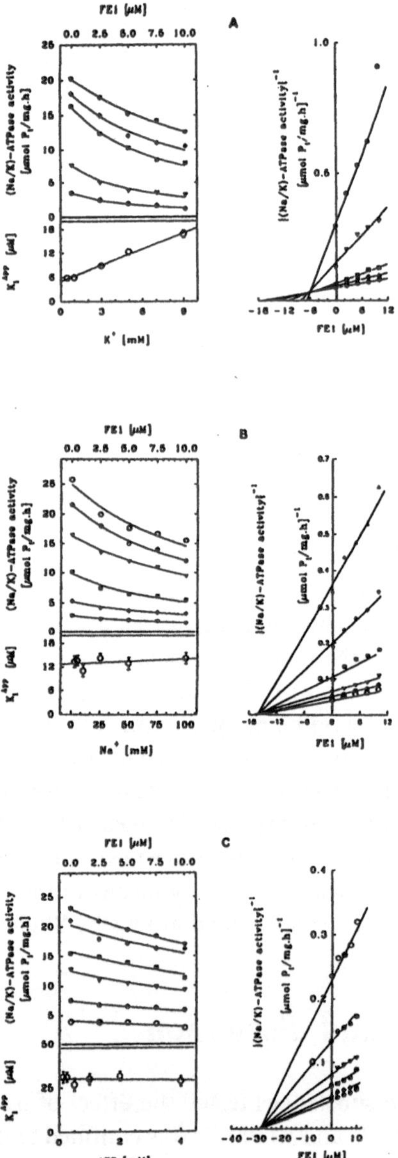

Fig. 1. Kinetics of inhibition of (Na/K)-ATPase by FE1. Panels A: Effects of inhibitor on activation of (Na/K)-ATPase by potassium ions; Panels B: Effects of inhibitor on activation of (Na/K)-ATPase by sodium ions; Panels C: Effects of inhibitor on activation of (Na/K)-ATPase by ATP. In all cases represent: i) The left upper panel – a direct plot of activation of the (Na/K)-ATPase by either (as indicated) the potassium ions (asterisks 9, diamonds 5, squares 3, triangles 1 and circles 0.5 mM) or the sodium ions (asterisks 2.5, diamonds 5, squares 10, triangles 25, full circles 50 and empty circles 100 mM) or by ATP (asterisks 4, diamonds 2, squares 1, triangles 0.5, full circles 0.25 and empty circles 0.125 mM) in the presence of increasing concentrations of FE1; ii) The left lower panel – dependence of the $K_{0.5}$ values on concentrations of either potassium ions or sodium ions or ATP. Values of $K_{0.5}$ were obtained by non-linear regression of experimental data given on the respective left upper panels using the equation $V = V_0/(1+i/K_{0.5})$. V and V_0 represent the reaction velocities of (Na/K)-ATPase in the presence of FE1 in concentrations *i* and *i* = O. Note please – when in this plot the $K_{0.5}$ values show concentration dependence (like in panel A), the inhibition may be considered as competitive and the K_i equal to the intercept on the abscissa. Oppositely, when the $K_{0.5}$, values do not show concentration dependence (like in panels B and C), the inhibition may be considered as non-competitive and the K_i is equal to $K_{0.5}$; iii). Right hand panels – Classical Dixon plots.

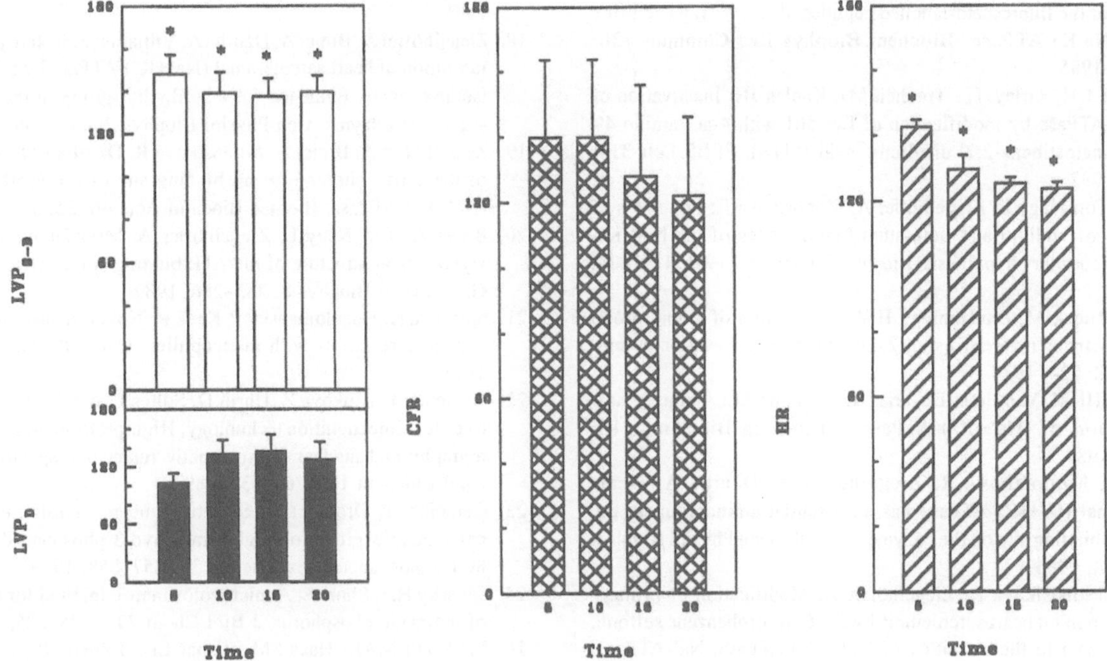

Fig. 2. Time-dependence of FE1-effect on haemodynamic parameters of isolated perfused guinea pig heart. Concentration of FE1 was 0.1 μmol/l, for perfusion protocol see Materials and methods. Panel A: Effect of FE1 on the mechanical activity of the heart expressed as the difference between systolic and diastolic pressures in the left ventricle ($LVP_{S\,D}$ empty bars) and the left ventricular diastolic pressure (LVP_D full bars). Panel B: Effect of FE1 on coronary flow resistance expressed as the ratio between LVP_D and the coronary flow (provided by the pump). Panel C: Effect of FE1 on heart rate. Data are expressed in per cent and are given as means ± S.E.M. from 8 independent measurements. Control values: $LVP_{S\,D}$ = 3.8 ± 0.5 kPa, LVP_D = 1.2 ± 0.1 kPa, CFR = 0.574 ± 0.026 kPa min/ml, HR = 210 ± 10 beats/min. Significance: *($p < 0.01$) against control (at time 0).

acter FE1 had to be dissolved in DMSO prior to be added to the perfusate (final concentration of DMSO was 0.1 mM). However, same concentration of DMSO present in the perfusate did not affect the functional variables of the isolated perfused heart significantly (not shown). Presence of 0.1 μmol/l FE1 and 0.1 mmol/l DMSO in perfusate led to significant ($p < 0.01$) increase in mechanical activity of the heart and in heart rate (Fig. 2). The accompanying slight elevations in coronary flow resistance and end-diastolic pressure were not found statistically significant ($p > 0.1$). The observed considerable increase in mechanical activity of the heart induced by FE1 resembles to certain extent the effect observed after application of ouabain (Goldberg, 1966). These results obtained with FE1 on isolated perfused guinea-pig heart clearly indicated that FE1 is capable to hit the molecule of (Na/K)-ATPase not only in isolated membrane preparation but also in functioning isolated perfused heart. This would mean that FE1 may selectively recognize and attack the molecule of (Na/K)-ATPase also in integrated cardiac tissue and may represent an useful tool in studying the place and mechanism of activation of the enzyme by potassium ions.

Acknowledgement

This work was supported in part by Slovak Grant Agency for Science (grants No. 2/57/193, 240/93 and 241/93).

References

1. Dhala NS, Zhao D: Cell membrane Ca^{2+}/Mg^{2+} ATPase. Prog Biophys Mol Biol 52: 1–37, 1988
2. Skou JC, Esmann M: The Na,K-ATPase, J Bioenerg Biomembr 24: 249–261, 1992
3. Ziegelhöffer A, Vrbjar N, Breier A: How do the ATPases work? Biomed Biochim Acta 45: S211–S214, 1986
4. Isnesi G, Kirtley MR: Structural features of cation transport ATPases. J Bioenerg Biomembr 24: 271–283, 1992
5. Shull GE, Shwartz A, Lingrel JB: Amino-acid sequence of the catalytic subunit of the (Na^{+}+K^{+})-ATPase deduced from a complementary DNA. Nature (London) 316: 691–695, 1985
6. Farley RA, Tran CM, Carilli CT, Hawke D, Shively JE: The aminoacid sequence of a fluorescein labelled peptide from the active site of (Na,K)-ATPase. J Biol Chem 259: 9532–9535, 1984
7. Kirley KL, Wallick ET, Lane LK: The aminoacid sequence of the fluorescein isothiocyanate reactive site of lamb and rat kidney Na^{+} and K^{+} dependent ATPase. Biochem Biophys Res Commun 125: 767–773, 1984

8. Ohta T, Morohashi M, Kawamura M, Yoshida M: The aminoacid sequence of the fluorescein-labelled peptides of electric ray and brine shrimp (Na,K)-ATPase. Biochem Biophys Res Commun 130: 221–228, 1985
9. Pedemonte CH, Kirley TL, Treuheit MJ, Kaplan JH: Inactivation of the Na,K-ATPase by modification of Lys 501 with 4-acetamido-4′-isothiocyanatostibene-2,2′-disulfonic acid (SITS). FEBS Lett 314: 97–100, 1992
10. Breier A, Turi Nagy L, Ziegelhöffer A, Monošíková R: Principles of selectivity of sodium and potassium binding sites of the NA^+/K^+-ATPase. A corollary hypothesis. Biochim Biophys Acta 946: 129–134, 1988
11. Wand H, Rudek M, Dautzenberg H: Determination of amino group insoluble carrier material using 2,4,6-trinitrobenzenesulfonic acid. Z Chem 18: 224, 1978
12. De Pont JJHHM, Van Emst-De Vries SE, Bonting SL: Amino groups modification of (Na^++K^+)-ATPase. J Bioenerg Biomembr 16: 263–281, 1984
13. Breier A, Monošíková R, Ziegelhöffer A, Džurba A: Heart sarcolemmal (N^++K^+)-ATPase has an essential aminogroup in the potassium binding site on the enzyme molecule. Gen Physiol Biophys 5: 537–544, 1986
14. Breier A, Monošíková R, Ziegelhöffer A: Modification of primary aminogroup in rat heart sarcolemma by 2,4,6-trinitrobenzene sulfonic acid in respect to the activities of (Na^++K^+)-ATPase, Na^+-ATPase, K^+-pNPPase. Function of potassium binding sites. Gen Physiol Biophys 6: 103–108, 1987
15. Drobnica L, Kristián P, Augustin J: The chemistry of the -NCS group. In: S. Patai (ed.). The Chemistry of Cyanates and their Thio Derivatives. John Wiley and Sons, NewYork, 1977, pp 1003–1221
16. Baláž Š, Végh D, Šturdik E, Augustin J, Liptaj T, Kováč J: Substitution reactions of some (5-nitro-2-furyl)ethylene derivatives with thiols. Collect Czech Chem Commun 52: 431–436, 1987
17. Rosenberg M, Šturdik E, Liptaj T, Bella J, Végh D, Považanec F, Sitkey V: Reactions of 1-(5-nitro-2-furyl)-2-nitroethylene with amino and hydroxyl groups. Collect Czech Chem Commun 50: 470–481, 1985
18. Ziegelhöffer A, Breier A, Džurba A, Vrbjar N: Selective and reversible inhibition of heart sarcolemmal (Na^++K^+)-ATPase by p-bromophenyl isothiocyanate. Evidence for a sulfhydryl group in the ATP-binding site of the enzyme. Gen Physiol Biophys 2: 447–456, 1983
19. Ziegelhöffer A, Breier A, Monošíková R, Džurba A: Some properties of the active site and cation binding site of the heart sarcolemmal (Na^++K^+)-ATPase. Biomed Biochim Acta 46: S553–S556, 1987
20. Breier A, Turi Nagy L, Ziegelhöffer A, Monošíkova R, Džurba A: Hypothetical structure of the ATP-binding site of (Na^++K^+)-ATPase. Gen Physiol Biophys 8: 283–286, 1989
21. Špirková K, Dočolomanský P, Kada R: Novel trisubstituted ethylenes and their reactions with nucleophiles. Chem Papers 46: 329–332, 1991
22. Gemeiner P, Bíliková Z, Uhrin D, Šoltes L, Mosbach K: ATP derivatives for biorecognition technology: High-performance liquid chromatography and nuclear paramagnetic resonance spectra. Biotechnol Appl Biochem 11: 176–183, 1989
23. Gemeiner P, Drobnica L: Selective and reversible modification of essential thiol groups of D-glyceraldehyd-3-phosphate dehydrogenase by isothiocyanates. Experientia 35: 857–858, 1979
24. Taussky HH, Shorr E: A microcolorimetric method for determination of inorganic phosphorus. J Biol Chem 202: 675–685, 1953
25. Markwell MAK, Haas SM, Bieber LL, Tolbert NE: A modification of the Lowry procedure to simplify protein determination in membrane and lipoprotein samples. Analytical Biochem 87: 206–210, 1978
26. Mardh S: Stabilizing effects by Mg^{2+} on Na,K-ATPase. Acta chem scand B36: 269–271, 1982
27. Lüllmann H, Peters T, Reuner G, Ruther T: Influence of Ouabain and dihydroouabain on the circular dichroism of cardiac plasmalemmal microcosms. Naunyn Schmiedelberg's Arch Pharmacol 290: 1–19, 1975

Molecular and Cellular Biochemistry **147**: 193–196, 1995.

Index to Volume 147

Arras M, *see* Weihrauch D *et al.*

Ausma J, Cleutjens J, Thoné F, Flameng W, Ramaekers F, Borgers M: Chronic hibernating myocardium: Interstitial changes 35

Avkiran M, *see* Yasutake M

Ballantyne, *see* Kukielka GL *et al.*

Bate A, *see* Hudlická O *et al.*

Bernátová I, *see* Gerová MP *et al.*

Bhatt SK, *see* Shao Q *et al.*

Böhm M: Alterations of β-adrenocepter-G-protein-regulated adenylyl cyclase in heart failure 147

Borgers M, *see* Ausma J *et al.*

Breier A, *see* Ziegelhöffer A *et al.*

Breier A, Ziegelhöffer A, Stankovičová T, Dočolomanský T, Gemeiner P, Vrbanová A: Inhibition of (Na/K)-ATPase by electrophilic substances: Functional implications 187

Brown MD, *see* Hudlická O *et al.*

Buchwalow IB, *see* Schulze W *et al.*

Cleutjens J, *see* Ausma J *et al.*

Cook JM, *see* Goldstein S *et al.*

Dhalla AK, *see* Singh N *et al.*

Dhalla NS, *see* Shao Q *et al.*

Dhalla NS, *see* Slezák J *et al.*

Dočolomanský T, *see* Breier A *et al.*

Doležel S, *see* Gerová MP *et al.*

Džurba A, *see* Vrbjar N *et al.*

Džurba A, *see* Ziegelhöffer A *et al.*

Entman ML, *see* Kukielka GL *et al.*

Erez M, *see* Manoach M *et al.*

Flameng W, *see* Ausma J *et al.*

Fu MLX, *see* Schulze W *et al.*

Gemeiner P, *see* Breier A *et al.*

Gerová MP, Pecháňová O, Stoev V, Kittová M, Bernátová I, Juráni M and Doležel S: Biomechanical signals in the coronary artery triggering the metabolic processes during cardiac overload 69

Gohlke P, *see* Linz W *et al.*

Goldstein S, *see* Sabbah HN *et al.*

Goldstein S, Sharov VG, Cook JM and Sabbah HN: Ventricular remodeling: insights from pharmacologic interventions with angiotensin-converting enzyme inhibitors 51

Haverich A, *see* Steinhoff G
Hučin B, *see* Pelouch V *et al.*
Hudlická O, Brown MD, Walter† H, Weiss JB and Bate A: Factors involved in capillary growth in the heart 57

Irlbeck M, *see* Zimmer H-G *et al.*

Juráni M, *see* Gerová MP *et al.*

Kittová M, *see* Gerová MP *et al.*
Kolbeck-Rühmkorff C, *see* Zimmer H-G *et al.*
Krause EG and Szekeres L: On the mechanism and possible therapeutic application of delayed adaptation of the heart to stress situations 115
Kukielka GL, Youker KA, Michael LA, Kumar AG, Ballantyne CM, Smit CW and Entman ML: Role of early reperfusion in the induction of adhesion molecules and cytokines in previously ischemic myocarium 5
Kumar AG, *see* Kukielka GL *et al.*

Lesch M, *see* Sabbah HN *et al.*
Linz W, Wiemer G, Schaper J, Zimmermann R, Nagasawa K, Gohlke P, Unger T and Schölkens BA: Angiotensin-converting enzyme inhibitors, left ventricular hypertrophy and fibrosis 89

Manoach M, Varon D and Erez M: The role of catecholamines on intercellular coupling, myocardial cell synchronization and self ventricular defibrillation 181
Matsubara T, *see* Shao Q *et al.*
Michael LA, *see* Kukielka GL *et al.*
Milerová M, *see* Pelouch V *et al.*
Morwinski R, *see* Schulze W *et al.*

Nagasawa K, *see* Linz W *et al.*
Nováková O, *see* Ošťádal B *et al.*

Okruhlicová L, *see* Slezák J *et al.*
Ošťádal B, Pelouch V, Ošťádalová I and Nováková O: Structural and biochemical remodelling in catecholamine-induced cardiomyopathy: Comparative and ontogenetic aspects 83
Ošťádal B *see* Pelouch V *et al.*
Ošťádalová I, *see* Ošťádal B *et al.*

Parrat JR, *see* Ravingerová T *et al.*
Pecháňová O, *see* Gerová MP *et al.*
Pelouch V, Milerová M, Ošťádal B, Hučin B and Šamánek M: Differences between atrial and ventricular protein profiling in children with congenital heart disease 43
Pelouch V, *see* Ošťádal B, *et al.*
Pyne NJ, *see* Ravingerová T *et al.*

Ramaekers F, *see* Ausma J *et al.*
Ravingerová T, Pyne NJ and Parratt JR: Ischaemic precondition in the rat heart: The role of G-proteins and adrenergic stimulation 123
Ravingerová T, *see* Ziegelhöffer A *et al.*

Šamánek M, *see* Pelouch V *et al.*

Sabbah HN, *see* Goldstein S *et al.*
Sabbah HN, Sharov VG, Lesch M and Goldstein S: Progression of heart failure: A role for interstitial fibrosis 29
Schaper J, *see* Linz W *et al.*
Schaper J, *see* Weihrauch D *et al.*
Schölkens BA, *see* Linz W *et al.*
Schulze W, *see* Slezák J *et al.*
Schulze W, Wolf W-P, Fu MLX, Morwinski R, Buchwalow IB and Will-Shahab L: Immunocytochemical studies of the G_i Protein mediated muscarinic receptor-adenylyl cyclase system 161
Seneviratne C, *see* Singh *et al.*
Shao Q, Matsubara T, Bhatt SK and Dhalla NS: Inhibition of cardiac sarcolemma Na^+-K^+ ATPase by oxyradical generating systems 139
Sharov VG, *see* Goldstein S *et al.*
Sharov VG, *see* Sabbah HN *et al.*
Singal PK, *see* Singh *et al.*
Singh N, Dhalla AK, Seneviratne C and Singal PK: Oxidative stress and heart failure 77
Slezák J, Okruhlicová L, Tribulová N, Schulze W and Dhalla NS: Renaissance of cytochemical localization of membrane ATPases in the myocardium 169
Smit CW, *see* Kukielka GL *et al.*
Stankovičová T, *see* Breier A *et al.*
Steinhoff G and Haverich A: Cell-cell and cell-matrix adhesion molecules in human heart and lung transplants 21
Stoev V, *see* Gerová MP *et al.*
Styk J, *see* Ziegelhöffer A *et al.*
Szekeres L, *see* Krause EG

Thoné F, *see* Ausma J *et al.*
Tribulová N, *see* Slezák J *et al.*

Unger T, *see* Linz W *et al.*

Varon D, *see* Manoach M *et al.*
Vrbanová A, *see* Breier A *et al.*
Vrbjar N, Džurba A and Ziegelhöffer A: Influence of global ischemia on the sarcolemmal ATPases in the rat heart 99
Vrbjar N, *see* Ziegelhöffer A *et al.*

Walter† H, *see* Hudlická O *et al.*
Weihrauch D, Arras M, Zimmermann R and Schaper J: Importance of monocytes/macrophages and fibroblasts for healing of micronecroses in porcine myocardium 13
Weiss JB, *see* Hudlická O *et al.*
Wiemer G, *see* Linz W *et al.*
Will-Shahab L, *see* Schulze W *et al.*
Wolf W-P, *see* Schulze W *et al.*

Yasutake M and Avkiran M: Effects of selective α_{1A}-adrenoceptor antagonists on reperfusion arrhythmias in isolated rat hearts 173
Youker KA, *see* Kukielka GL *et al.*

Ziegelhöffer A, *see* Breier A *et al.*
Ziegelhöffer A, *see* Vrbjar N *et al.*

Ziegelhöffer A, Vrbjar N, Styk J, Breier A, Džurba A, Ravingerová T: Adaptation of the heart to ischemia by preconditioning: Effects on energy equilibrium, properties of sarcolemmal ATPases and release of cardioprotective proteins 129
Zimmer H-G, Irlbeck M, Kolbeck-Rühmkorff C: Response of the rat heart to catecholamines and thyroid hormones 105
Zimmerman R, *see* Weihrauch D *et al.*
Zimmermann R, *see* Linz W *et al.*

ADP-Ribosylation: Metabolic Effects and Regulatory Functions

edited by **Joel Moss,** *Laboratory of Cellular Metabolism, National Heart, Lung and Blood Institute, Bethesda, MD, USA*; **Peter Zahradka,** *Div. of Cardiovascular Sciences, St. Boniface General Hospital Research Centre, Winnipeg, MB, Canada*

Considering the current interest in cellular regulation and intracellular signalling systems, it is surprising that the contribution of ADP-ribosylation reactions to the modulation of a variety of specific cell processes, in parallel with other post-translational modifications such as phosphorylation, has not been generally recognized. While it is not feasible to cover all aspects of ADP-ribosylation, the thirty-one articles contained in this volume provide a valuable overview of recent progress in the field within the context of cell control mechanisms. For the convenience of the reader, the various topics have been grouped into several sections: (a) poly(ADP-ribosyl)ation; (b) mono-ADP-ribosylation; (c) toxin mono-ADP-ribosylation; (d) inhibitors and activators; (e) protein modification with ADP-ribose and its analogues; and (f) non-modification forms of ADP-ribose. The contents of the individual chapters reflect the ideas of the contributors, many of whom have spent their careers attempting to resolve the biological functions of ADP-ribosylation. We hope that this publication will serve as a useful reference for those investigators that are new to the area as well as those who are actively studying ADP-ribosylation.

Contents

Preface. **I:** Historical Perspective. **II:** Poly(ADP-ribosyl)ation. **A.** Structure and Enzymology of Poly(ADP-ribose) Polymerase. **B.** Polymer Regulation. **C.** Cellular Functions. **D.** Poly(ADP-ribose) Polymerase Gene Regulation. **III:** Mono(ADP-ribosylation). **A.** ADP-ribosylation Cycle. **B.** Cellular Mono-ADP-ribosylation. **IV:** Toxin Mono-ADP-Ribosylation. **V:** Inhibitors and Activators of ADP-Ribosylation. **VI:** Derivitization of Proteins with ADP-ribose, NAD and their Analogues. **VII:** Cyclic ADP-ribose, NAD Hydrolysis and ADP-ribose Synthesis.

1994, 256 pp. ISBN 0–7923–2951–1
Hardbound USD 190.00/NLG 365.00/GBP 134.95

DEVELOPMENTS IN MOLECULAR AND CELLULAR BIOCHEMISTRY 12

Reprinted from MOLECULAR AND CELLULAR BIOCHEMISTRY, 138:1-2

KLUWER ACADEMIC PUBLISHERS

P.O. Box 322, 3300 AH Dordrecht, The Netherlands
P.O. Box 358, Accord Station, Hingham, MA 02018-0358, U.S.A.